Palgrave Historical Studies in the Criminal Corpse and its Afterlife

Series Editors

Owen Davies
University of Hertfordshire
School of Humanities
Hatfield, United Kingdom

Elizabeth T. Hurren
University of Leicester
School of Historical Studies
Leicester, United Kingdom

Sarah Tarlow
University of Leicester
History and Archaeology
Leicester, United Kingdom

Aim of the Series:
This limited, finite series is based on the substantive outputs from a major, multi-disciplinary research project funded by the Wellcome Trust, investigating the meanings, treatment, and uses of the criminal corpse in Britain. It is a vehicle for methodological and substantive advances in approaches to the wider history of the body. Focussing on the period between the late seventeenth and the mid-nineteenth centuries as a crucial period in the formation and transformation of beliefs about the body, the series explores how the criminal body had a prominent presence in popular culture as well as science, civic life and medico-legal activity. It is historically significant as the site of overlapping and sometimes contradictory understandings between scientific anatomy, criminal justice, popular medicine, and social geography.

More information about this series at
http://www.springer.com/series/14694

Elizabeth T. Hurren

Dissecting the Criminal Corpse

Staging Post-Execution Punishment in Early Modern England

palgrave
macmillan

Elizabeth T. Hurren
School of Historical Studies
University of Leicester
Leicester, United Kingdom

ISBN 978-1-137-58248-5   ISBN 978-1-137-58249-2   (eBook)
DOI 10.1057/978-1-137-58249-2

Library of Congress Control Number: 2016943515

Printed on acid-free paper

This Palgrave Macmillan imprint is published by Springer Nature
The registered company is Macmillan Publishers Ltd. London.

CONTENTS

LIST OF TABLES

LIST OF FIGURES

LIST OF ILLUSTRATIONS

LIST OF MAPS

List of Maps

Abbreviations

AA	Anatomy Act (1832)
ANS	Autonomic Nervous System
BPP	British Parliamentary Papers
De.RO	Devon Record Office
Do.RO	Dorset Record Office
ESRO	East Sussex Record Office
HRO	Huntingdon Record Office
Le.RO	Leicestershire Record Office
Li.RO	Lincoln Record Office
LL	London Lives
LPL	Lambeth Palace Library
LUL	Leeds University Library
MA	Murder Act (1752)
Nor.RO	Norfolk Record Office
Not.RO	Nottingham Record Office
PNS	Parasympathetic Nervous System
RCS	Royal College of Surgeons
SNS	Sympathetic Nervous System
TNA	The National Archives
WT	Wellcome Trust library
WYAS	West Yorkshire Archive Service

ACKNOWLEDGEMENTS

This book would not have been written without the generous support of the Wellcome Trust in London. I am very grateful for the period of research granted to me under the Programme Grant WT095904AIA *'Harnessing the Power of the Criminal Corpse'* and the input of my fellow scholars on the project team based at the University of Leicester. In particular, I have enjoyed working with Professors Sarah Tarlow, Owen Davies and Peter King alongside Drs Zoe Dyndor, Francesca Matteoni, Shane McCorrestine, Floris Tomasini, Richard Ward, as well as Dr Rachel Bennett, who has recently successfully completed her PhD. Many of the thought-provoking publications produced by everyone involved are listed in the bibliography of this book, and for those that seek a more interactive engagement with the overall aims of the project please do get in touch via our web-link: http://www2.le.ac.uk/departments/archaeology/research/projects/criminal-bodies-1.

All academic books have their critics. There will be those who will seek to learn more about medical education, crime and justice, the curative powers of the body, magic and medicine, anatomical literature, image creation, and the sundry other topics that encompass a history of the body. These are covered in some depth by those authors already listed above, and the *Select Bibliography* therefore attempts to guide readers to material regarding the areas they may wish to learn more about. This book is therefore very much about rediscovering punishment pathways inside a judicial system. Like all who travel historically, the sight-seeing has to be selective, or those on the journey would soon tire of the guide. Hence, what you are about to read, like other books that have preceded it, is an expression of scholarly endeavour; for it seeks to make *a* contribution to *the* contribution of academic life. It is not therefore going to repeat what is already known, but rather is based on a substantial amount of new research in the archives because it is seeking to ask new questions of things that might seem familiar, but are not. Having written about the impact of the Anatomy Act of 1832 on Victorian society, this book steps back to the intervening Georgian period from the Murder Act of 1752. This is a time period that has seldom been studied from the vantage point of the criminal corpse in the way that the next seven chapters recount. There is therefore no overlap with this author's previous publications. Instead the approach has been very much about responding to remarks made to me by Hayden White over dinner at High Table at New College Oxford in 2012: 'If you are a morally sensitive person, then you would not want to see or know about some criminal acts of human history. Disbelief is triggered when knowledge rushes in to smooth over the unbelievable. This would be something you don't see *and* something you don't *want* to see. How do you then write a history of undocumented events—the sort of truth you disavow because it is so unbelievable—something that was deliberately destroyed, edited or refashioned for public consumption—If this really happened, what should I do as an historian?' I soon discovered that the criminal corpse was a central character in a dissection drama that long ago was cleaned up for public tastes. I hope that by inviting you to take part in the punishment options, their realism proves startling, for that will create new conversations, renew historical models, and refine ways of seeing the criminal past.

For the past fifteen years, since becoming an academic, I have always been amazed by the kindness of strangers. Private houses have opened, cups of Earl Grey tea were brewed, and invitations to delicious lunches extended. I am in point of fact also indebted to all those libraries, county record offices, and national repositories that welcomed me over the past two years. In particular, I would like to express my sincere thanks to the Wellcome Trust library, the archivist at the Royal College of Surgeons, as well as numerous County Record Offices in the Midlands, most particularly Leicestershire Record Office. Many County Record Offices now find themselves under exceptional financial constraints with staff maintaining their professionalism in the face of some extraordinary funding cutbacks. It is a testimony to their continued kindness and unique expertise that all scholars still benefit from their accumulated knowledge of the source material. I am all too aware that this may not be for much longer and that the research career that I embarked on in the 1990s might not be feasible in the near future. It is surely one of the saddest outcomes of the global financial crisis in 2008 that the short-sightedness of profiteers in financial markets has in seven years undermined the public access of documents collected on English local history for the past 500 years. The irony of e-globalisation, open access and its digital highway, is that for all its technological innovation it cannot prevent the locking up of knowledge written on paper for centuries.

Often academics rush around from one lecture to another, time being precious, and there never being enough hours in the day to get things done. Those therefore that people our personal lives often have to be that extra bit understanding. I would consequently like to pay a short tribute to friends and family. This book is dedicated with love to my 'baby brother', David John Wright (because a 'big sister' is for life, not just for Xmas!), and my beautiful god-daughter Ellen Rose De Banke. Life for them both promises much and I look forward to seeing their bright futures unfold. The foundation-stone of home and the things that are heartfelt there, could not continually be regenerated without Professor Steven King and HMKB. It will, I know, make them both laugh to read that at last *you know who* has organised *the* party—*phew*!

Pro Sempra Contra

Dr Elizabeth T. Hurren (eh140@le.ac.uk)

Reader in the Medical Humanities
Rutland, July 2015.

PREFACE

There comes a time in every academic career when the amount of reading a scholar has done can be a hindrance to genuine creative thinking. For this reason, the ability to see familiar things with renewed vision has always been valued by historians since Antiquity. Challenging the status quo in the medical humanities has often been about finding simpler solutions to complex human situations. Confounding traditional values is not however an easy task when established views have strong roots in the public imagination. Historically, curious coincidences have sometimes been as informative as years of pain-staking research. As frustrating as this is for those that work diligently in the darkest corners of research laboratories and academic libraries, it is nevertheless irrefutable that some of the most unexpected observations have been the key to solving the mysterious. Without the wild card alerting us to the solutions to troubling scientific conundrums, there might never have been new ways of seeing and believing in the human capacity for medical innovation.

For centuries, serendipity was a scientific secret. It was seldom exposed to public scrutiny. Academe is now much more open about disclosing the fortunate nature of the misfortunate. The poet Robert Graves, in 1929, celebrated this process of intellectual liberation in his famous poem *Broken Images* when he described how confusion was an emancipation. In the course of which, natural curiosity (a major theme in this book) always sustains more and more creative thinking.[1] In all sorts of scientific settings historians are continually reminded of the timely nature of human failure. Basic errors, being mistaken, intellectual setbacks, and the sorts of practical misfortunates that befall all researchers, can ironically open up hidden knowledge. The courageous ability to think foolish thoughts, to have a change of heart, or do a volte-face are as energising to the human spirit of endeavour in modern biomedicine, as they once were to medical men working with limited equipment in the early modern era. If this book encourages its readership to think more about these very necessary historical skills, then it will have achieved its ultimate goal.

Imagine then a criminal corpse lying on a dissection table in eighteenth century England. The condemned has been punished for a murder conviction by being hanged on the gallows and is about to be dissected by anatomists in the Georgian era. The body in question has been washed clean. As a member of the audience you are about to be being given access to it before the post-mortem punitive rites begin. You believe this to be a just outcome because it has been decreed that this is what will happen to homicide perpetrators under new capital legislation known as the Murder Act (1752) that will remain in force until the Anatomy Act (1832). But as you figuratively gaze at the body you find it very hard to see the clean flesh. Your fresh historical eyes are distracted by skin that is pinpricked all over. The effect is a bit like pointillism, the technique of Impressionist painters who applied small dots of paint with a brush to a canvas. Up close, it is very difficult to see what image of punishment is being performed in this criminal tragedy. The chief character, the corpse, though centre-stage is punctured with so many moral ideals, medical musings, and scientific speculations that it is difficult to see the physical circumstances of the punishment choreography itself. It is also hard to separate out a state policy of deliberate medical obfuscation from genuine scientific confusion; or, to know with certainty whether what is about to take place should be taken at face value or not. Nor can you apprehend easily to what extent the physical facts have a basis in a material reality or are about to become an historical cliché. Nowadays, at this historical distance the pinpricks seem to make visible sense, but as a viewer you are wary of this perspective. It seems misleading to decide things from the vantage point of historical hindsight; and you are right to distrust conventional images of criminal dissections distorted in eighteenth century studies.

Traditionally the criminal corpse has been an academic dartboard for over three centuries. Since the 1960s in particular, the history of the body has become so pin-marked with theoretical ideas that few historians can actually see the punished corpse up close anymore. In cultural studies there has often been a lack of appreciation that a theory evolves to fit the known facts. If it has merit, it will survive the discovery of unknown facts too. At no point in its history should the chosen theory stand in for, or displace empiricism. Nor should the theoretical perspective ever be a fixed point of academic reference. It can inform but its real task is to stimulate debate. It may become an accepted theory by testing it against archive material but over time it must remain open to revision. This much seems obvious. But in a history of the criminal corpse accepted theories are often presented as material facts without scholars checking their medical veracity. This shortcoming has a particular Anglo-centric flavour. Too often, crime and cultural studies have effectively abandoned the condemned body left beneath the hanging-tree. The logical reaction is to step aside, read about why this has happened, and thus redress the historical gap. But the real difficulty is that there is no reliable historical account of the entire punishment journey of the criminal corpse because scholars have so seldom focused on it in England. New ways of understanding this criminal past are not necessarily rediscovered in university libraries.

From the outset this book signposts, rather than exhausts its chapters with historiography. This is a deliberate choice, so that the capacity for curiosity and surprise is not distorted. It happens that the criminal corpse sits at the intersection of a number of very large historical disciplines. These include crime and punishment; medicine and science; religion and theology; magic and popular belief; as well as social structure, politics and state power in the 'civilising processes' of early modern European history. To read this backstory requires two character traits: resilience and tolerance. Shelves of books heaving with historical texts have to be assimilated. It is not an exaggeration to say that it would take the average reader about six months just to get to grips with the scale of the task. The real difficulty with this scholarly endeavour is not that it would be futile—the reader would encounter many thought-provoking opinions—but at its conclusion there would be a lot to filter out. And the information left would not necessarily provide new insights into the sorts of unanswered questions that historians have never tackled in the historical archives. In other words, you could expend a vast amount of energy finding out what everyone already knows about the criminal corpse in the course of which your precious capacity for original thinking could be beleaguered by so-called facts that are nothing of the sort. It is not therefore narrow-minded to want to dissect the criminal corpse from the eighteenth century in the twenty-first century. It was after all the foundation of so much medical and scientific speculation in the Enlightenment. Avoiding the historical pitfall of the confirmation bias of accepted theories requires a more enlightened approach.

The late Christopher Hitchens said that the Enlightenment exalted all human beings to question everything, and, having done so, to keep doing so, repeatedly:

> I would define the Enlightenment as the belief in free enquiry, the belief in the scientific experiment, the ability to conduct and test such things, and the belief that this in itself was an emancipation – not just from disease or ignorance or stupidity – though there were such conflicts – but that it was time that human beings took responsibility for themselves, rather than relying on a divine suasion of any sort, and that essentially secular insight was what made possible the American revolution and its French equivalent.[2]

In the long-eighteenth century, the criminal corpse in so many respects became the bedrock of that free enquiry. It is then an historical irony that it has been poked and prodded with all sorts of theoretical discourses (usually about power and disempowerment) that are not based on a concerted study of the physical characteristics of actual post-execution punishments. In discovering the body about to be dissected, some historians forgot that they were not the ones originally charged with carrying out the dissecting! It was the task of penal surgeons, often working in non-descript dissection venues, to enact the penalties of the capital legislation, and yet we still know very little about their working practices. Broadly-speaking, a lot has been written about how these medical men arrived in the punishment room, the sort of educational credentials they obtained to get there, and the best places to learn human anatomy in early modern Europe—with Padua, Montpellier, Paris, Edinburgh and London leading the medical field. That however is where the medical story tends to become fuzzy. Once anatomists stood on the threshold of eighteenth century dissection rooms, historians of crime and medicine tend to lose interest in what actually happened next across England. The fine detail it is argued could distract attention from wider scientific trends in European culture. Historically this is nonsensical. If, as the late Roy Porter observed the criminal body was the basis of so much Enlightenment thinking, then to figuratively hand it over to a medical fraternity and neglect to ask exactly what they did with it, is a serious omission.

There remain major gaps in our understanding of the role that punishment played in the furtherance of human anatomy in early modern England compared to Europe (the latter being well-documented). This book has tried to develop historical antennae that seek to re-engage with those things in a history of dissection that are often contrary to expectations from 1752. Did for example everyone die on the gallows at a time of so much scientific uncertainty is a fundamental question that merits substantive archival answers. The chapters that follow therefore firstly contain a lot of scholarship cited in the bibliography. This has however been carefully sifted down to its essentials. You will not find long summaries of previously published work on the nature of a medical education or the relationship between political upheaval and statistical trends in crime rates. Others (including this author in her previous books) have written at length about these factors elsewhere. Instead major gaps in our historical understanding have been concentrated on. To do otherwise it is argued would be counter-productive to original thinking. This book then secondly is the result of a deliberate research fusion. There was no mapping out in advance of potential archives to concentrate on to the exclusion of others. To maintain the element of surprise, lots of different types of sources have been brought together as never before, and this is again a strategic decision to stimulate creative thinking. A third objective is that the new material builds upon but also continually questions conventional historical opinions. This might seem an obvious thing for an historian to do but it is surprisingly radical for this neglected topic area. Crime histories have tended to ignore the criminal corpse, and have left cultural historians to fumble around in the dark of dissection rooms making careless medical statements. There have been countless shelves of books printed on what early modern historians thought happened next, but they seldom examine what actually did occur from a medico-legal standpoint. Added to which, there is little human sense of the experiential nature of the post-execution choreography, rituals, or its material aftermath.

There are then things that this book is seeking to do very differently, not for the sake of being contrary, but to test the limits of our human understanding. There will be times when it is necessary to step back and look at the condemned body intensely, but not exclusively from the vantage point of what has already been written about it. Humanism—the ability we all share to know something about our body's natural functions which have stood the test of time—will play a large part in what follows. The histories of emotion, pain and punishment also feature predominately. These emphasise the need to engage with the five senses and their historical biological continuities. Some findings will be disturbing. But they are all about asking: have we got that right, can we see it from another perspective, and what if we have taken too much for granted by forgetting that bodies looked and felt different in early modern times? No book is then perfect, especially one based on so much human imperfection and violent behaviour. In eighteenth century England, unconventional medical men were determined to work with the dead-end of the morally deficient. They envisaged that the deviant could delight scientific knowledge. Hence this monograph has been inspired by a Latin exhortation once very familiar to eighteenth century anatomists that dissected the criminal corpse: *pro sempra contra*—surely now is the time '*for the balance of things contrary to expectations*'.

NOTES

1. Robert Graves (1929), 'In Broken Images' published in (1961), *Poems Collected by Himself* (London and New York: Doubleday).
2. Christopher Hitchens (2005) talk on 'Thomas Jefferson: Enlightenment, Nation-Building and Slavery', The Film Archives, You-tube, 1hr 05 minutes, https://www.youtube.com/watch?v=99-72amEijM.

PART I

Introduction

Taceant colloquia –

Effugiat risus –

Hic locus est ubi mors gaudet succurrere vitae.

Let Conversation Cease –

Let Laughter Flee –

This is the place where Death delights to help the Living.

[Saying attributed to Giovanni Morgagni eighteenth century anatomist—said to be copied on the walls of some dissection rooms after the Murder Act (1752) in England—now standard in many pathology rooms around the world—http://www.forensicmed.co.uk/pathology/]

CHAPTER 1

The Condemned Body Leaving the Courtroom

On 23 May 1754 a Scottish soldier named Ewen MacDonald was quartered at Newcastle. He spent his off-duty time drinking at a popular public house. By nightfall he was drunk and disorderly. What happened next was to make medico-legal history. In an inebriated state MacDonald started to brawl with some pub regulars. There was a physical altercation that spilled into the street and then into an alleyway. The soldier drew a knife in what he would later claim was self-defence. Lunging out he stabbed a local cooper called Robert Parker, piercing the jugular vein in the victim's neck. The drunken violence then escalated out of control. MacDonald was surrounded by a crowd of angry men threatening him with revenge. Cornered into two brick walls, he assaulted one of his attackers by breaking his arm and verbally abused the others. In the interim, the earlier injured party was forsaken in the affray and bled to death on the ground. By the early hours of the morning the Newcastle coroner was called to the fatal scene and at first light an 'inquest returned a verdict of wilful murder'.[1] As a serving-solider, MacDonald should have been arrested by the military authorities but the Murder Act (1752, 25 George II, c. 37) had recently come into force in Newcastle. Local magistrates decreed that the military suspect for being so violent must be imprisoned in the Borough gaol. A homicide charge would be held over to be tried at the next Assizes, leaving six weeks to determine the medical evidence and collate its findings with witness statements. By Michaelmas, there was considerable newspaper interest in

© The Author(s) 2016
E.T. Hurren, *Dissecting the Criminal Corpse*, Palgrave
Historical Studies in the Criminal Corpse and its Afterlife,
DOI 10.1057/978-1-137-58249-2_1

the pending murder trial, and whether, if found guilty, a public dissection at Newcastle Surgeon's Hall would be ordered by the presiding judge. If so, it would be one of the first official cases of post-mortem punishment to take place in the Northern counties of Georgian England.

MacDonald was duly found guilty on 28 September 1754. The judge passed the death sentence under the Murder Act: 'You will be hung by the neck *until dead*, and thence to be *dissected and anatomized*'. An extensive report of the execution and its post-mortem rituals appeared in the *Newcastle General Magazine:*

> Ewen MacDonald was executed on the town moor, Newcastle pursuant to his sentence at the Assizes....this most unfortunate young man, who was only nineteen years of age appeared all the time of his confinement deeply effected with a true sense of guilt...but at the gallows his behaviour in endeavouring to throw the executioner from off the ladder was unbecoming to one just on the brink of eternity...His body was taken to Surgeon's Hall and there dissected.[2]

The following day, popular broadsheets also featured an execution-day special. They reported that the dissection of MacDonald was a troublesome affair for the Newcastle officials:

> It was said that after the body was taken to Surgeon's Hall and placed ready for dissection, that the surgeons were called to a case at the Infirmary, who, on their return, found MacDonald so far recovered as to be sitting up; he immediately begged for mercy, but a young surgeon not wishing to be disappointed of the dissection, seized a wooden mallet with which he deprived him of life.[3]

The story of *Half-hanged MacDonald* featured prominently in the national and regional press coverage of the Murder Act. Yet, its medical controversy has been neglected in eighteenth-century crime studies. To the authorities in Newcastle a bungled execution seemed to undermine the deterrence objective of the new legislation. Newspaper editors questioned the medical circumstances of capital death given their obvious shortcomings. There was much speculation in the press about the force of the prisoner's willpower. Maybe he had a dangerous inner strength different to ordinary people that could defy the executioner. Some commentators thought that it was possible that the soldier had not been guilty of homicide but manslaughter; had God intervened to save his life? If so, it seemed immoral for divine justice to be confounded by surgeons determined to obtain a criminal corpse to dissect. Regional broadsheets carried detailed accounts from witnesses at the crime scene. These claimed that not only was MacDonald provoked into a bar fight but that exonerating evidence had been dismissed in court. The local forces of law and order meantime maintained that resentment against the Murder Act was running so high that the resuscitation story had been fabricated to sell more newspapers. Puzzlement was also expressed when the *Local Record of Newcastle* said shortly afterwards: 'It was further reported, as the just vengeance of God that this young man [penal surgeon] was soon after killed in the stable by his own horse. They used to show a mallet at Surgeon's Hall as the identical one used by the surgeon' to hasten death.[4] Here then was a dramatic storyline that had engendered conflicting accounts, unsettling emotions, and unresolved controversy concerning the punishment of the condemned. In every respect this vignette encapsulates the key themes of this book.

'RECULER POUR MIEUX SAUTER'[5]: STEPPING BACK, TO SET FORTH

Across a broad spectrum of early modern histories the criminal corpse has become an iconic cultural symbol and political standard-bearer for customary notions of law and order.[6] The legal nemesis of the condemned has been intertwined with religious beliefs that were shaped by theologies of dying, death, and the afterlife.[7] Hence the executed criminal was an integral feature of state power and its punishment rituals.[8] It had a close association with the broad impulse of a 'scientific revolution' by 1700. Over the next century, extensive newspaper coverage, pamphlet literature and popular street ballads, featuring executions were connected in the popular imagination to medical professionalization and Enlightenment values across Europe.[9] Human anatomy teaching thus became essential for a European medical education, with Paris, Edinburgh and London (in that order of priority) attracting fee-paying students anxious to obtain extra qualifications as physicians and surgeons from dissecting criminal corpses.[10] By the mid-Georgian period the general thrust of these intellectual trends was epitomised by the Murder Act[11] and this is why it has stimulated extensive historical debate about its judicial authority, discretionary powers, and geographical coverage in England.[12] There remains however a significant lacunae in our historical appreciation of the central role that the condemned body played in the public performance of the capital code's punishment schemata from the gallows to the grave. In our opening story, if the soldier's condemned body had not been traced onwards to its dissection, then the medico-legal controversy of *Half-hanged MacDonald* would be half-finished at best or more likely left out altogether of eighteenth-century crime studies. The latter is common in the majority of cases that feature in criminal histories for Georgian England. That tends to limit our historical appreciation of a whole range of complex questions concerned with the moral authority of the capital code from its anti-mortem to post-mortem outcomes.[13] Before beginning therefore to revisit that punishment choreography in its entirety, this introduction sets out the generic themes that the next six chapters will be exploring to provide an overview of this book's novel approach.

A central and unifying feature of the chapters that follow is to revisit how exactly medico-legal officials knew that someone was dead at a time when timing death with limited medical equipment was still a scientific mystery in early modern England. It was a physical fact that basic biology shaped the legal remit of the new capital legislation. Executioners and penal surgeons had to work with the bodily limits of metabolism, physiology, and organ vitality, whether on the gallows or on arrival at dissection venues. Attempts were made to resolve these logistical medical issues by employing experienced executioners. The hangman's technology was altered with the change-over from a short to a long-drop, and tying the rope more securely with a trefoil rather than overhand knot. Even so, particular modes of execution, their medical competency, and mutable nature, remained obscure. It is noteworthy then that the majority of criminal historians have been reluctant to do in-depth studies of spatial execution sites or punishment spaces, and the diversity of medical opinions about the boundaries of life and death that informed/mis-informed medico-legal officials.[14] Steve Poole has recently pointed out that the introduction of a 'long-drop' was not efficient or humane with 'old and 'new' practices often running in parallel in provincial and metropolitan life. The rituals of procession were accordingly maintained up to the 1830s. These reflected ongoing concern about how to put to death the condemned in a humane fashion. It remains then a defining feature of eighteenth century crime histories that the corpse has often been left beneath the hanging-tree *looking* dead, but not *truly dead,* in the way that contemporaries understood those difficult medical definitions. It was well-known that outward appearances could be deceptive, especially in cold weather when the condemned body went into extreme-hypothermia. The historical literature, by abridging post-execution rites, has created a mistaken impression that penal surgeons only handled dead bodies from the gallows and that capital penalties from a medical standpoint were straightforward once a criminal stopped jerking on the hangman's rope. This book upends such medico-legal clichés in crime studies.

Few historians have traced the fascinating transitional language that expressed a great deal of contemporary concern about whether or not the condemned could survive capital death, or not, and what their capacity for pain was and therefore how powerful the criminal justice system could claim to be. In medico-legal circles there was a confusing vocabulary in vogue: the '*half-hanged*', '*dead-alive*', '*in the name of death*', '*truly dead, or pained*', '*death, the uncertain, certainty*' and so on. This imaginative discourse was a reflection of shifting ideas about how biologically-speaking it was possible to be stuck in-between an earthly and spiritual world in early modern times. Mystical beliefs and their mutability gave rise to open-ended debates about how to time physical punishments. It was these logistical issues that had to be stage-managed by penal surgeons as well. They were required to double-check the biological status of the executed on arrival at a dissection location. If the criminal displayed any life-signs, however faint, then the surgeon had an ethical quandary. They could break the Hippocratic Oath or practice the new art of resuscitation in vogue from the 1780s. Some took the decision that it was more merciful to commit human vivisection with the lancet when brain function appeared to have failed and could not be measured with medical instruments. Others questioned whether the judicial authority of the courts gave them a special medical prerogative to experiment with the boundaries of life and death. This could be done to improve the quality of life for the law-abiding by testing the limits of death's dominion on murderers found guilty that were less than human. Yet, everything about this penal set-up exuded moral complexity. It placed a high degree of discretionary justice in the hands of surgeons, to such an extent, that some were uncomfortable with the level of personal culpability and responsibility. Hitherto these basic procedural pains and their ethical worries have seldom been reappraised in the historical literature. Hence rediscovered archival sources, reflecting a rich multiplicity, feature prominently in this book's reconstruction of process and participation on the punishment journeys of the condemned from the gallows to the grave between the Murder Act (1752) and the Anatomy Act (1832) in England.

The widespread observation that contemporaries were often confused by the organic instability of death and dying, leads us to a second generic theme in this book. Many histories of crime and justice take the reader up to the critical point when the executed stopped jerking physically on the eighteenth century gallows. They then cite the admirable research of Vic Gatrell and Peter Linebaugh who some two decades ago traced strong reactions by ordinary people who protested about post-execution rites at the hanging-tree.[15] Stories about how the crowd did confound the medico-legal officials by preventing bodies being handed over to surgeons are now well-known for some symbolic execution sites in central London. In the 1730s at Tyburn there were unedifying scenes when a number of surgeons fought each other, as well as the mob, for exclusive rights to the criminal corpse.[16] These research findings have convinced leading crime and cultural historians that a Northern European sensibility emphasising body-integrity in death held sway in the popular imagination for those living in early modern England, especially once the Murder Act became law.[17] The difficulty with these cultural perspectives is that, as Jonathan Sawday observes, capital punishment by dissection 'enjoyed' but only ever had a 'quasi-legal status' in England.[18] The London Company of Barbers and Surgeons, as well as Oxford and Cambridge universities, obtained bodies according to customary rights laid out in a succession of Royal Charters. These crucially were endorsed by, but did not originate with, Parliament. Riots at Tyburn in the 1730s are then a classic example of how a moral economy seemed to inform the actions of the angry mob. It does not necessarily follow on that the same depth of feeling was held in provincial England, or that the same levels of antagonism were prevalent in a political economy two decades later when the Murder Act reached the statute books. Standard historical views, oft repeated, still require a systematic analysis of archival source material detailing actual post-mortem practices for many metropolitan and provincial areas.

Avoiding then *post-hoc* rationalisation involves carefully examining in context whether in the intervening timespan cultural attitudes to criminal dissections were modified in any respect, and if so, with what outcomes for crime and justice processes from 1752. It is an obvious but important observation to keep in mind that this was a time of considerable intellectual transition as 'scientific' ideas competed with 'Natural Philosophy'. The degree therefore to which the audience at a criminal dissection ever held, or held on to, Northern European sensibilities still needs to be corroborated. Recently, Sarah Tarlow is one of a number of historical archaeologists that have reminded historians of crime and justice that beliefs systems about the body in death could be complex and conflicting.[19] Seemingly incompatible philosophies about the dead body's potency were prolific and determined by local scaffolds of crime and justice. The judicial punishment of the dead, the ceremonial intent of the capital code, as well as the sentience and agency of those present, all ran in parallel.[20] Alongside a prevalent sense of emotional attachment, elements in the crowd could confound the forces of law and order with indifference. This did not cancel out others obstinacy or reluctance to compromise their cultural beliefs about body-integrity. Yet it did problematize the cultural spread of a spectrum of punishment rites in what was a complicated system of law and order. It contained many types of medico-legal performances involving a wide variety of criminal dissections in those provincial parts of early modern England that still remain understudied. To overcome this lack of coverage, this book's second over-arching theme touches on the official reach of penal powers debated in a history of ideas.

The conventional perspective that the crowd objected to post-execution rites wholeheartedly has served a large theoretical corpus covering crime and punishment, synonymous with the writings of Michel Foucault.[21] His influence in crime histories has been very extensive indeed; featuring prominently is the disciplinary nature, deterrence value, and legal penalties that involved the crowd, embodiment, and capital sentencing.[22] The 'historized body' is now seen as a metaphor of power discourses, an historical prism to 'interpret, problematize and destabilize... knowledge/power creation'.[23] The inherent difficulty nevertheless with the wide-ranging scope of these approaches—whether cultural, literary and/or theoretical—is that they tend to rely on a cartload of notional concepts. Often criminal corpses at the gallows have been piled up in carts laden down with circumstantial evidence. As Roy Porter observed the history of the body has become 'the historiographical dish of the day' and regrettably some scholars have lost sight of the actual physical body being punished.[24] This book's chapters contend that from a medical standpoint, there has been a strong confirmation bias in crime studies that have effectively mishandled the criminal corpse. Many scholars for instance never think to question whether the criminal was a corpse or a condemned body in deep physical trauma during its prolonged nemesis. It was the case that many criminals did expire on the gallows, but there is equally a lot of contemporary evidence that many others did not perish at the hanging-tree. They did so when they got to their dissection destination. What is needed then is an English local history of post-execution rites. The dominance and privileging of certain theoretical approaches has not been as helpful as it could have been in an historical sifting of the evidence on *Dissecting the Criminal Corpse.*

Few eighteenth-century historians have reappraised what it really meant to physically cut the criminal corpse by dissection. This is because Foucault in *Discipline and Punish: The Birth of the Prison* (1977 translation) was very critical of 'the purely biological basis of existence' and its empiricism that he believed historical processes had abused in Western society.[25] The overall thrust of such 'anti-essentialist' thinking has been that researchers have tended to ignore those parts of the capital punishment choreography that are more difficult to access in the archives than others. This explains why nobody has made a serious study of those bodies that were distributed for dissection from the English gallows, in the way that this book does. Hence there has been an overwhelming tendency in eighteenth-century studies to privilege the execution spectacle as an instrument of state power, rather than paying enough historical attention to its equally spectacular post-execution encore across early modern England. This book takes therefore a counter-intuitive approach to the theoretical thrust of cultural studies by stepping back into the archives. The aim of this second strand of research is to arrive at an empirical appreciation of the historical picture of post-mortem 'harm' undertaken by penal surgeons post–1752.

The radical uncertainty at the core of the English capital punishment system was that there needed to be an unfinished penal choreography at the hanging-tree. The incomplete punishment journey seemed to give moral authority to the deterrence value of the retribution rites. This medico-legal imperative has stimulated this book's third research strand. Few histories retrace the pace of punishment, its choreography, or medical efficacy. Little is known about what precisely the medical provisos of the legislation entailed, how these were modified over time, and whether they made sense to penal surgeons. What happens again in many standard accounts is that the narrative imperative of the dramatic storyline falls off at the gallows when the body is symbolically dropped. As it figuratively gets released from the rope of the hanging-tree, it falls into a 'cultural compost' of speculative and often unproven sentiments in the historical retelling.[26] In effect, few studies utilise the contemporary source material to rediscover the missing post-mortem scenery. This status quo is ironic because it did not happen this way in many locations. There was in fact a great deal of sustained interest by contemporaries about what was about to occur at criminal dissections. Many thousands of the labouring poor for instance walked with the convicted at arm's length. They viewed the body, watched for life-signs, and stared at the dangerous to make sure they were dead. To enhance its deterrence value the capital code allowed for this level of sustained spectatorship because it was associated with a great deal of physical uncertainty that played on the creative imaginations of those present. Piecemeal steps were created to deliberately confuse the painful processes of punishment and participation on location.[27] Capital punishment was thus never a simple expression of public engagement. By the post-execution stage it had to be act of co-creation too, to be convincing. This meant that the penal choreography needed to stimulate human empathy, inquisitiveness, meddling, hostility, and insecurity about the painful procedures of post-mortem 'harm' on a body that was supposed to be unfeeling. As a result this book's third generic theme contains a fusion of three open-ended and puzzling punishment pathways: firstly, the impact of the struggle and threat that dissection posed in the popular imagination; secondly, the damage it might potentially cause to those present in terms of their physical capacity for pain and painful spectatorship; and, thirdly, the notoriety and infamy it could create for those condemned to die. There is then much more here than a simple linear story of retribution for murder being enacted and reconstructed in six chapters.

There had to be a punishment paradigm that medico-legal officials could work with, and one that was flexible enough to contain a lot of medical ambiguity: something historians still need to reconstruct in its entirety after 1752 and which this book provides. When looking for instance at the complex nature of the changing audience composition at criminal dissections, we need to be mindful that the political contradictions and moral discrepancies being staged in dissection spaces had subtle, as well as overt, power-balances at play: again something that a lot of cultural studies have yet to substantiate in the archives. Just then as the legal narrative of the punishment drama was not necessarily linear, so too the post-execution spectacle did not always have an undeviating medical logic. This reflected the fact that timings shifted as the choreography of punishment was enacted. Procedures at the gallows looked linear but since they also involved a body operating to a cyclical life-cycle closing down, there was a lot of potential for liminal spaces in the processes of punishment. What seemed to matter most to the sorts of social groups present was that their scaffolds of local justice mirrored popular sentiments and these often reflected curative beliefs about the latent power of the criminal corpse: a theme that Owen Davies and Francesco Matteoni have together recently substantiated.[28] It was obvious to most people involved in the maintenance of law and order, that legal rhetoric and medical realities were neither interchangeable, nor predictable. The majority accepted that different types of criminal dissections would be staged in various medico-legal settings. Punishment criteria had to be imperfect and were inconsistent because these factors, seldom studied until now, matched contemporary expectations. Predictably an assorted number of penal surgeons were assembled who interpreted their medico-legal duties according to parish, county, regional and intra-regional, notions of criminal justice. The material afterlives of criminal corpses tended to reflect popular attitudes to class, gender, and society as well. Physical retribution for murder was seldom fixed and often fluid, with discretionary justice being deliberated and renegotiated, refined and reshaped by changing cultural attitudes to the condemned body in terms of religion, philosophy and science. For contemporaries disposing of the criminal corpse involved 'different projects of inquiry – different spectacles – made different bodies visible in anatomy'— and yet crime studies still neglect this medical reality.[29]

England was then covered with deeply symbolic places to dissect the criminal corpse and it is these that all the chapters in some respect will be visiting in considerable archive detail to substantiate a fourth overarching research theme. As well as new source material on the geographical and architectural alignment of chosen dissection spaces in the community, we will also be bringing together familiar sources in novel ways. There will be opportunities to look over the shoulders of attending penal surgeons and to look around the dissection venues they occupied, seeing afresh others present in the room. This included those involved in the secretive side of legal governance which has sometimes hindered the ability of historians to get involved with the sheriffs, their deputies, magistrates, and hangers-on, post-execution. The 'great contrast in their styles of justice compared to the local gentry', and the middling-sort that sat on courtroom juries, meant that once events reached the press the political dance of local retribution became the focus of intense provincial interest.[30] 'Something of these class and geographical differences emerges' when encountering post-execution crowds on location.[31] Their attachment to punishment rites as dramatic post-mortem performances in which they were active participants still merits closer historical scrutiny. For crime histories by tending to précis the punishment choreography at the gallows, have not served histories of the crowd well for the early modern period. Instead, a miasma of criminality hangs over the dead-end of convicted murderers when crowds supposedly departed at the hanging-tree. *Dissecting the Criminal Corpse* takes issue with this false finishing-point by relocating different sorts of medico-legal starting-points. This book takes up the end of a storyline that did not stop half-way through and, in research terms, rather than being an end-point should have been a beginning. Early modern society did not abandon or lose interest in the criminal as it was labelled a corpse, quite the reverse. Mingling with post-execution crowds *in situ* requires historians to retrace stories like those of *Half-hanged MacDonald* that made medico-legal history, which were lost or inconvenient in the grand punishment narrative of eighteenth-century studies.

Historians of the crowd have tended to be inward-looking for the early modern period.[32] Recently, Matthew White has taken issue with those in crime studies that 'aggregate' the 'mob' into a 'faceless crowd', prompting a fifth research strand in this book.[33] He has rediscovered in early nineteenth-century coroners and Old Bailey records, both an ongoing enthusiasm for execution spectacles and strong emotional attachments to punishment rites that were continually expressed across the social spectrum. His observations that 'depictions of...avid execution-going' have been 'consistently two-dimensional and frequently impressionistic, often paying scant attention to social complexities' have been an intellectual stimulus for the evidence-gathering presented on those that got involved with post-execution rites from 1752.[34] The crowds at criminal dissections, have, either, been forgotten altogether, or misplaced, and/or constructed as essentially monolithic. Regrettably this has made them a finishing-point too. Once more, there needs to be a more sinuous starting-point for research on crowds, and one that matches the fluid nature of those that travelled onwards with the condemned, rediscovering their stratified composition in English counties.[35] Returning to the archives, it becomes feasible to put the personality back into the crowd by accompanying them to dissection venues, thereby building on the approach of Vic Gatrell at the hanging-tree.

In developing this fifth research strand it has been viable to achieve things that have seldom been attempted in standard histories of the post-execution crowd. By concentrating on the historical prism of medical death (our first generic theme) and expanding that to encompass its basic biological continuities which human beings share in all time periods, the chapters are able to reconstruct in some detail the synaesthesia of criminal dissections.[36] This book is not suggesting that this was a monolithic experience, but it is asserting that there were a range of sensory encounters that were unavoidable for everyone in the room at a criminal dissection. It is important to stress (as the most recent research in the medical humanities is doing) that not only did the early modern crowd dress their faces and bodies differently, and follow distinctive fashions to clothe their limbs, but they also 'inhabited bodies that felt different and diverged in shape from ours, dictated by nutritional, health and labour regimes ...From skin and bone, they were nothing like us', as William Pooley points out.[37] Yet, they did share the same five senses familiar to everyone down the centuries. This means that historians of the body need to rethink what happened when the crowd came to see a criminal corpse that they expected to be both human, and also less than human. The corpse displayed both accessible and 'profoundly unfamiliar flesh'. Historians of crime have tended to understate the contradictory and confusing nature of the experience of viewing criminality exposed. Strong buttocks or limbs, by way of example, on the dissection table were admired in agricultural areas that valued the physical strength of the blacksmith, the drover, and the ploughman. In towns and cities, the clothing and weaving trades promoted the masculine qualities of very strong muscular arms and robust upper body strength. The ways in which the crowd adopted the bare bones of these experiences as flesh separated could then be somewhat disruptive to their bodily sense. The people present had the free will to react but the intellectual and emotional capacity of the crowd was always shaped by sight, smell, hearing, touch, taste, and so on. The undeniable fact was that executed bodies stank, and this was inescapable for those in close proximity. In some respect all those that partook witnessed bleedings, shavings, and slippages in a punishment purge seldom considered in terms of its post-execution spectatorship. Gatrell's work has provided clues about what bodies might have been like on arrival at dissection venues.[38] He points out that many executions were bungled. The criminal could be in a 'bloody mess' by the time she/he was cut down from the gallows. It is difficult to understand why then historians of crime and justice have not thought more about the actual condition of the criminal corpse that got handed over to the surgeons. After all, the physical circumstances of execution could have influenced punishment decision-making beyond the gallows too. Encountering bodies in a *'good'*, *'bad'*, *contaminated'*, *'dirty'*, *'diseased'* or *'destroyed'* condition, as well as rediscovering so-called *'extras'*, means that the chapters can re-present the 'fleshy' nature of the legal remit in the surgeons' hands.[39]

As then the dissection intensified, at a certain physical point in the punishment processes the condemned became repugnant to smell and offensive to look at. It was essential that they did so otherwise the deterrence value of the spectacle was dubious. Yet again, few studies trace the timing of the crowd's agency, and whether it too had a choreography (punishment rites happening to the corpse alongside how the crowd performed their role), and at what point they turned away as the convicted murderer was despoiled for public consumption. Taking the example of how long it took for many thousands to walk past the corpse exposed to public view, as often reported in newspapers at the time, makes it feasible to revisit crowd dynamics. In smaller towns the numbers reported were far greater than the local population census. And as historians of the crowd have pointed out that spectrum of people would have contained very different personal motivations for being there and required sensitive handling by the medico-legal officials charged with ensuring that chaos did not ensue.[40] The local crowd were insiders and generally curious but benign. Those from the surrounding area could be disruptive as outsiders even though they were generally manageable. Women were often regarded as a litmus test of the ability of the forces of law and order to keep the peace. If they bustled and jostled to view the spectacle, returning up to five times to see the proverbial bad man laid out for inspection in a single day, then there was a very real danger of the 'crowd' becoming a 'mob' that would riot. An added complication was that overlaying those present from the county area, were genuine outsiders, trouble-makers with radical tendencies that often travelled from afar. Sometimes they were the menacing members of the dangerous convict's family. At other times the angry, dispossessed, or unruly came to exploit notoriety by association. This meant that it was essential to get the timing of the post-execution choreography right when up to ten thousand people travelled onwards with the executed body: timing ones departure is a theme elaborated below. It was likewise crucial to entertain, keep the interior dynamics of the crowd moving along, making sure the audience's attention was drawn to different spectacles over several days, and in general, ensuring that post-execution rites had momentum. There was thus a changing profile of medical actors and a strong element of immersive theatre was introduced with people not sitting in seats but walking past the central criminal character. Often there was a timed ticket-entry system too, to better manage audience-flow. When then cultural and criminal histories take the punishment provisions of the Murder Act as read, and invariable, they miss so much about the history of the post-execution crowd in terms of its internal drama. Above all, they neglect to appreciate that 'to be hung by the neck *until dead*, and thence to be *dissected and anatomized*' was not just a complicated set of medical procedures to orchestrate but they did not tend to work in the way many early modern historians have assumed. As we shall see, the choreography had four stages: *social death* (being condemned in court), *legal death* (being hanged), *medical death* (anatomically checked on arrival at a dissection venue), and only then proceeding to *dissection and its material afterlife* (post-mortem 'harm'). For now, what is important to keep in mind is that post-mortem procedures in their public performance—and it was much more public than private for most of our chronological focus—were compelling but circumspectly staged, and yet, stimulated significant levels of '*natural curiosity*'.

Culminating then in a final generic theme (the book's sixth) all the chapters in some respect explore the emotional capacity of the crowd to appropriate the post-execution rites of the Murder Act. To engage with a history of emotions it is essential to intertwine a definitive sense of eighteenth-century '*curiosity*' in the way that Neil Kenny's recent research has done.[41] Throughout we will be building-up a sense of the '*curious*' because for contemporaries it was a multi-layered and mutable set of experiences that often triggered their emotional capacity for awe and revulsion, as well as attraction, fascination, and temptation, and at times, a striking indifference. On the punishment sight-seeing that we are about to embark on across England, there will be opportunities to explore what the differences were between '*natural*', '*public*', '*morbid*' and the worst '*perils of curiosity*' at criminal dissections. A paradigm of curiosity has thus been redesigned for this book having traced its historical wallpaper to stimulating archive material. Those sources alert crime studies to reconsider what William Reddy terms '*emotives*', which are essentially the emotional expressions that language gives voice to and therefore are very relevant for this study's focus.[42] '*Emotives*' differ from the ability of human beings to describe events because that narrative does not necessarily express how someone feels. The act of speaking about an emotional experience will, according to Reddy, always have the potential to be self-exploring and self-altering. Something intangible can become its opposite when voiced, made literally to be within someone's emotional grasp. A classic example of this in terms of crime and justice was for a member of the crowd to be upset about a vicious murder but to be unable to locate, or express, and even suppress, those feelings until at an execution scene. Once the crowd had gathered, the spectacle could be cathartic releasing deep-seated emotions as those present jeered and shouted abuse at the condemned. Throughout the potential existed for emotional tension to be realised, refashioned, re-presented and thus figuratively reached for. All this, could alter the original nature of the experience itself. Once people talked about homicide they sometimes found within themselves an emotional capacity to redefine what they felt and why, in the course of which they rediscovered how to get hold of an emotion at its nucleus. This meant that in the punishment drama, some were more angry, others less so, and the remainder came to a realisation that their anger-level was appropriate for the human situation. This is what is meant by the *archaeology of emotions* in the crowd that we will be engaging with.[43] In this book it is argued that Kenny's and Reddy's concepts of '*curiosity*' and '*emotives*' are very difficult to separate because they were inextricably embedded in the emotional spectrum of the post-execution crowd of early modern England. It was their psychological wherewithal to hold onto a capacity for curiosity in a history of emotions in the transition from a moral to political economy that has framed the way that each chapter has approached those present at criminal dissections. This will add a new dimension to the personality of the post-execution crowd in the same way that synaesthesia can too.

The overall ambition in this book is then to bridge a whole series of liminal spaces inside the criminal justice system—in terms of corporeality, materialism, emotionalism, legalities, mentalities, physical uncertainties and semi-professional identities—that created a dramatic subtext to eighteenth-century criminal life. This backdrop explains why the subject matter of the criminal corpse preoccupied the audiences of many eighteenth-century playhouses that went on to the dissection theatre too. By 1819 *Blackwood's Magazine* published a series of popular articles written by theatre critics about these public performances of human anatomy that had grown in popularity since 1700.[44] In one well-rehearsed dramatic storyline we encounter *Time's Magic Lantern and the Dissector* (Illustration 1.1):

> *Doctor. This body is a good subject. It is lean, and therefore well calculated to shew the muscular system. Lay open the abdomen by two transverse incisions, but beware you do not injure the viscera. Now draw aside the outward integuments and you will observe the position of the bowels.... I shall demonstrate that in my lecture today. Here, throw up the windows and sprinkle the floor with camphor. Remove the putrid thigh... Cast a sheet over this body and wipe the dissecting instruments with a towel. Now stand behind me and await the entrance...*[45]

Illustration 1.1   ©Wellcome Trust Image Collection, Slide Number L0031335, *'The Dead-Alive'*, illustrating a man supposed to be dead arising from his coffin, coloured aquatint, published 1805, after a 1784 drawing by Henry Wigstead, (London: William Holland, Oxford Street); Creative Commons Attribution-NonCommercial-ShareAlike 4.0 International License (CC BY-NC-SA 4.0)

Newspaper reporters often commented on crowds taking an eager part in punishment dramas at the gallows by day and enjoying their theatrical re-enactments at night. There was a grey area between moral edification and grisly entertainment.[46] Perhaps this explains why historians of the eighteenth-century theatre have rediscovered theatre-goers paying to sit on the stage with the actors performing their criminal characters, akin to those that pushed up the spiral staircase at Surgeon's Hall by the Old Bailey having bought a ticket to see the condemned body awaiting dissection.[47] This desire to get closely involved in deadly dramas meant that contemporaries could accumulate knowledge of post-mortem rituals, becoming sophisticated critics of the criminal justice system being revelled on the stage. The *mise-en-scènes* was a broader cultural reflection of the extent to which some form of post-mortem punishment of the criminal body and its dramatic timing had become the norm in Georgian England. It is time then to return to the start of what once was a compelling post-execution rite of passage under the Murder Act. In so doing, we are going to take literally what French thinkers should perhaps have paid more attention to when writing about the history of English capital punishment, '*reculer pour mieux sauter*'.[48] In a history of the body it is sometimes indispensable to 'step backward in order to leap farther forward'.

COGI QUI POTEST NESCIT MORI[49]: S/HE WHO CAN BE FORCED, HAS NOT LEARNED HOW TO DIE

Once the convicted murderer departed from the courtroom they were in an early modern state of '*social death*', according to the Murder Act. They had arrived at this moment of high drama, '*legal death*' according to what was once called the *Bloody Code* covering murder and up to 200 or so capital offences. These broadly-speaking fell into three distinct categories: firstly, petty crime usually associated with the experience of being poor such as stealing food-stuffs and personal chattels for subsistence or to sell on; secondly, property crimes typically house-burglary, highway robbery, as well as sheep-stealing and cattle-rustling; and thirdly, domestic violence or grievous bodily harm leading to manslaughter, and drunken street-brawls culminating ultimately in murder. There were of course many types of homicide charges, with conviction rates, sentencing policies, perpetrator profiles (in terms of age and gender), and socio-economic causes, being debated extensively by historians of crime and justice for the long eighteenth-century. The focus in this book is those that were convicted of murder, sentenced to death, and then were made available to penal surgeons on the cusp or throes of '*medical death*'. These were not capital cases that the judge decided to pardon after the death sentence was passed in court: crime historians concur that about five per cent of those sentenced to capital death were within days subsequently shown judicial mercy under the Murder Act.[50] Any reduction, or indeed cancellation, of the death penalty was generally done once the Assizes was finished but crucially before the judge left town. Typically a lot of pardoning cases resulted in a lesser sentence of 'transported for life' being decreed. This might be as a result of new evidence that came to light once the trial closed. The judge alternatively on reflection might adjudicate that the death sentence was too harsh for an accessory to a felony or did not reflect an unfortunate set of circumstances when an innocent bystander found themselves in court. The majority of judges were always very careful to test the local political temperature on their court circuit. Few wanted to be responsible for a public order situation getting out of control in reaction to a harsh sentencing policy that local people held to be morally objectionable. The pardoning system was however very seldom applied in proven homicide cases. This was a reflection of a sizeable cultural appetite for seeing justice done on the public gallows in cases of murder. The first Biblical commandment, '*thou shall not kill*' was morally sacrosanct in early modern Europe. It was linked in the popular imagination to *lex talionis*—the English *common law of retaliation*—in which punishment to the criminal body should be seen to match that done to a murder victim. Hence, an *eye-for-an-eye* mentality got attached to a common definition of pre-meditated murder, inviting judges to punish the criminal corpse by dissection in even the remotest parts of Georgian society. There was though one sentencing option reserved for murderers that committed odious crimes seen as a social evil: these included child murder; causing the mob to riot; threatening the revenue collection of the state in terms say of piracy, smuggling or hijacking the Royal Mail; or lacking personal remorse for a wicked deed like incest or rape in which a victim died and the stability of the family structure was undermined. In such cases, the judge could sentence the convicted murderer to be *hung in chains*, often called *being gibbeted*.[51] Still, for controversial cases, there was a pardoning system, with questions of leniency and discretionary justice being the subject of legal argument and local autonomy. In Georgian times, it was murder most foul that remained high-profile in the dissection theatres of early modern England.

In getting ready to depart from the courtroom it is essential to think graphically about how contemporaries may have broadly viewed what was about to happen, otherwise this book will be walking out of step with the early modern crowd. It is not advisable at this juncture to get lost in hypothetical musings, but some of the broad contours of the historical circumstances that have been substantiated do provide a useful guide. Of relevance is that there has been considerable historical dispute about the deterrence value of the punishment rites for homicide and whether the state managed its theatrical role set out in the Murder Act, or not.[52] At issue is whether the punitive measures and their visceral nature were counter-productive in the end. The balance of the contemporary evidence seems to suggest that ordinary people became either immune to, disinterested by, or critical of, harsh methods of retributive justice as humanitarian ideals gained moral ascendancy by 1800. In parallel however (as we have seen) a common law attitude got enshrined into capital legislation that reflected just how much a deep-seated fear of criminality gained political currency in Georgian times. The mob by virtue of their numeric size had to be accommodated within a more fluid social fabric. Political toleration did not lessen—rather enhanced—a sense of dread on the part of the governing elites and middling-sorts.[53] Mentalities constrained by such sensitivities—in terms of both rhetoric and reality—in turn shaped medico-legal debates about how to punish the condemned and what physical reprisals to stage in powerful and sometimes disturbing ways. In Georgian England, it was easy to see danger everywhere when political revolution in countries like France by the 1790s had a firm basis in reality. But to protect and preserve the status quo meant there had to a heightened vigilance about more dangerous forms of law-breaking.[54] This carried with it a real danger that 'any exaggerated publicity could turn out to be corrosive of the sort of security that the ruling parties craved'.[55] To avoid this situation happening, those in power needed to be mindful of a 'circular logic of conspiracy' that might gain a cultural foothold and cause more trouble for officials. This soon proved to be the case in respect of the punishment provisions of the Murder Act, as Part I of this book recounts.

To engage with these complexities it is also necessary to appreciate in broad terms that crime histories have then tended to summarise (usually inaccurately) the medical styles of physical reprisals done to the corpse. This is an incongruous research trend since cutting the murderer symbolically became the subject of so much sustained '*natural curiosity*' at criminal dissections over the long eighteenth-century.[56] Roy Porter and Anita Guerrini have queried the educational purpose of anatomical venues in London: whether they were places of rationale enquiry or staged extra entertainments to attract fee-paying elites.[57] The balance of the evidence presented in this book suggests that there should be more scepticism about what exactly went on behind the dissection room door in crime historiography.[58] The prevailing viewpoint that 'making justice' was 'often remade from the margins by magistrates, judges and others at the local level', still does not pinpoint with precision the range of formal and informal mechanisms of capital justice delegated to provincial penal surgeons that feature throughout this book.[59] It is equally important to appreciate the sorts of varied places in which criminal dissections could take place. There was intense local interest by spectators of all social gradations to see the capital sentence taken to its logical conclusion. If excluded (as we shall see in Part II) the crowd did on noteworthy occasions beat down the door to get inside a dissection venue in provincial England. Few were refused admittance because of the very real threat to public order by the mob in decades when making a basic living was hard.[60] Throughout the chapters thus build on the painstaking research of Andrew Cunningham by setting in context the actual activities of penal surgeons under the Murder Act.[61] This book takes the framework of his research on the history of anatomy in the eighteenth-century and adds archive detail to it for England. The literature on standards of eighteenth and nineteenth-century medical education is well-covered elsewhere by this author and others, the aim here being to scrutinise the reality of dissecting in places given only a cursory glance.[62] It will be shown that none of the post-mortem options ineludibly caused 'harm' in the way the capital legislation intended once the condemned departed out of the courtroom.

To start then this symbolic punishment procession, departing to a medico-legal tempo, time and its recurring themes will feature prominently in all the chapters. Time management was one of the implied cornerstones of the Murder Act and yet it has been neglected from a medical standpoint in crime literature. Aspects of it are embedded throughout this book because we are concerned with the biological logistics of the actual medical scene on exiting the courtroom. We will be encountering public clocks ticking down the crucial hour of death on the gallows in market squares. There will be opportunities to engage with penal surgeons checking on a pocket-watch that their criminal body did not go into rigor mortis four hours after being hanged. If it did, then they had to clock-watch until the corpse relaxed enough to manipulate the limbs, generally after twenty-four hours. In winter when colder anatomy sessions slowed down decay the anatomist had to grapple with lower light levels and daytime running out. In summer the heat made the heartbeat race, the hands sweat, and the arm-pits sticky with adrenalin, compelling the dissector to quicken their pace, regardless of the anatomical education that was scheduled. Acquiring dexterity with the knife was dependent upon the art of *good timing;* something echoed in material culture from the time. Clock-makers for instance often decorated their time-pieces with a *memento mori* to reflect what became known as 'skeleton-time'.[63]

Today it is common for many people to think that the timing of death is something that has always been certain and absolute, a one-way material journey turning from the world of the living. Across European religious traditions a sacred symbolism associated with graveyard customs encouraged individuals to think of death as a 'final moment' that could occur anytime.[64] So it is understandable that even with the gradual professionalization of medicine such views predominated in the long eighteenth-century. This book however sees these sorts of general depictions as historically deceptive in terms of actual post-mortem practices. The early modern medical fraternity, 'thought of death as a *becoming* rather than an *ending*'—an elementary biology that is often missing from criminal history.[65] Timing death, in an alternative understanding, involved the recognition that a person 'could die differently according to different chronologies and medico-legal settings', themes explored in Chapters 2 and 3. It was also the case that 'the body-clock could be, for the condemned, wound back; that *'death time'* involves the living, the revivified, and the socially resuscitated too': facets of Chapters 4 and 5. Even then for those that did expire on the gallows, their material afterlives and the latent power of criminal corpses pointed to enduring notoriety in death: outcomes depicted in Chapter 6. For this reason two key historical positions inform this book. On the one hand 'the corpse was a locus of *all* doubt' (as defined by Jonathan Sawday) and on the other hand 'becoming *really dead*...takes time' (as observed by Thomas Laqueur) from the early modern era onwards.[66]

At each criminal dissection ordinary spectators had just one reliable anatomical perspective, their own body clock. Few carried with them an anatomy textbook or had prior knowledge of what a dissected person really looked like compared to a theatrical re-enactment. Some attended several sessions at Surgeon's Hall in London but tickets were hard to get and the audience were encouraged to leave before extensive dissection (see, Chapter 4). Procedures there were also being modified continuously: a trend that Simon Chaplin first substantiated when he noted how lacklustre anatomical standards were in the capital by the 1790s.[67] Few historians have studied this important insight, as this book does in-depth. It will be shown that the majority of people in provincial life were in many ways better informed than their metropolitan counterparts. Most attended more accessible anatomies: at a surgeon's house, in the open space at the front of Assizes courts, at a room in small medical dispensaries, and inside the dead houses of newly constructed infirmaries (see, Chapter 5). To engage with a criminal afterlife intimately meant therefore taking the time to walk past criminal corpses and visibly comparing them to each spectator's embodied experience. The fact that so many people kept doing so sets in context the compelling nature of getting involved (refer, Chapter 6). A hand might look big, normal or small. The head could be striking, shaved, and misshapen, or non-descript compared to everyone else in the room. Eyes open at death were seen as a sign of guilt and a cultural taboo; so much so that anatomists today recognise that 'lid-watching is a lost medical art' and one that was once very familiar to contemporaries that attended post-execution rites.[68] It is these shared perspectives that are still accessible despite their historical distance. In this book the body is seen as a reliable form of source material because everyone in all time periods has a basic level of anatomical knowledge about their body and how to keep it going. Rudimentary anatomical labels like—heart, lungs, brain, kidneys, gallbladder, foot, leg, arm, and so on—would have been well-known to the post-execution crowd. It was their colours, shapes, and what they felt like to touch as the body shut down metabolically that required an enquiring mind and accumulated anatomical knowledge. Even though ordinary people present tended not to write down their general observations at criminal dissections, its essential features were so commonplace that historians can engage with their post-mortem attributes. The body has always been its own resource material and talking about it has continuously been emotive; participating without feelings was insupportable.[69]

The next chapters together reconsider what was well-known, what became known, and what remained a medical mystery at criminal dissections. We will encounter the range of official actors—judicial, executive, and medical—that had to manage execution crowds. Source material will be introduced to illustrate how the body itself was once a mutual historical reference point. *Knowing how this works* was a cornerstone of the Murder Act too and it was what made criminal dissections such a familiar and yet unnerving spectacle. As the body was punished it was physically carried to different locations. The changing medico-legal scenery depended on the essential character of criminal justice in the vicinity. Part I therefore provides historical insights into what it meant to dissect. Part II looks at the locations and scope of that dissection work. It will be shown that venues for criminal dissections differed a lot around the country. Four typologies have been identified. These are based on a geographical re-evaluation of dissection venues from detailed regional comparisons. Anatomical methods on location fell into seven generic types and these created material afterlives. *Dissecting the Criminal Corpse* is hence all about the central medico-legal dilemma of the Murder Act—*"mors certa, hora incerta"*— death is certain, its hour is uncertain. In the complex performance of crime and justice that recurring physical predicament placed a great deal of dramatic licence in the hands of penal surgeons that staffed the Murder Act in 1752 until the Anatomy Act of 1832.

NOTES

1. Moses Aaron Richardson (1843), *The Local Historian's Table Book of Remarkable Occurrences Volume II* (Newcastle-Upon-Tyne and London: Richardson & Smith publishers) pp. 44–5.
2. *Newcastle General Magazine*, September edition, 1754.
3. A full report of this turn of events was eventually published in the *Newcastle Courant*, 14 October 1754.
4. *Local Record of Newcastle*, December edition, 1754.
5. The French saying means to 'pick your battles' or 'bide your time and wait for the opportune moment' to reappraise past events; see also, endnote 48 below discussion.
6. Three books admired for this approach in socio-criminal history are penned by the same author: John M. Beattie (1986), *Crime and the Courts in England 1660–1800* (Princeton: Princeton University Press); John M. Beattie (2002), *Policing and Punishment in London, 1660–1750* (Oxford: Oxford University Press); John M. Beattie (2014 edition), *Urban Crime and the Limits of Terror, and The First English Detectives: The Bow Street Runners and the Policing of London, 1750–1840* (Oxford: Oxford University Press).
7. See, J. A. Sharpe (1998 edition), *Crime in Early Modern England, 1550–1750* (London: Routledge).
8. Those hung in chains on a gibbet or burned for petty treason feature elsewhere as part of a Wellcome Trust programme grant 'Harnessing the Power of the Criminal Corpse': see http://www2.le.ac.uk/departments/

archaeology/research/projects/criminal-bodies-.1 at Leicester University. New findings also feature in, Owen Davies, Elizabeth Hurren and Sarah Tarlow editors of the new series *Palgrave Historical Studies in the Criminal Corpse and its Afterlife,* in which this new major monograph appears.

9. See, Douglas Hay et al (2011 edition), *Albion's Fatal Tree: Crime and Society in Eighteenth-Century England* (London: Verso).

10. See, Neville Bonner (1995), *Becoming a Physician: Medical Education in Britain, France, Germany and the United States, 1750–1945* (Oxford: Oxford University Press); Ole Peter Grell, Andrew Cunningham and Jon Arrizabalaga eds. (2010), *Centres of Medical Excellence? Medical Travel and Education in Europe, 1500–1789* (Surrey: Ashgate).

11. Refer in manuscript copy, *BPP,* 25 Geo II, c.37, 1752, "An Act for Regulating the Disposal after Execution of the Bodies of Criminals," HL/PO/JO/10/2/61, Parliamentary Archives.

12. There is a vast amount written on eighteenth-century crime and justice. Selectively, see, V. A. C. Gatrell (1996 edition), *The Hanging Tree: Execution and the English People, 1770–1868,* (Oxford: Oxford University Press); P. Linebaugh (1996 edition), *The London Hanged: Crime and Civil Society in the Eighteenth Century* (London: Verso); R. Shoemaker (2004), *The London Mob: Violence and Disorder in Eighteenth Century London,* (London: Hambledon); Hay et al. (2011), *Albion's Fatal Tree*; P. J. R. King (2010), *Crime and Law in England, 1750–1840: Remaking justice from the margins* (Oxford: Oxford University Press).

13. This has been extensive historical debate about whether morality went hand-in-hand with a 'civilising force' in early modern society, causing murder rates to fall and thus creating body-supply problems. See, Norbert Elias (1978), *The Civilising Process: The History of Manners* (Basel: Urizen Books); Peter Spierenburg (2001), 'Violence and the Civilising Process: Does it Work?' *Crime, Histories and Societies,* Vol. 5, II, 87–105; Peter Spierenburg (2008), *The Spectacle of Suffering: Executions and the Evolution of Repression: From a Pre-Industrial Metropolis to the European Experience* (Cambridge: Cambridge University Press).

14. I am very grateful to Steve Poole for sharing with me his latest research on 'Hanging at the Scene of the Crime' given to a Wellcome Trust conference in 2013 at Leicester University & to be published in R. Ward ed., (2015) *A Global History of Execution and the Criminal Corpse* (Basingstoke: Palgrave).

15. See, Gatrell, *Hanging Tree* and Linebaugh, *London Hanged.*

16. Peter Linebaugh (1975 edition), 'The Tyburn riot against the surgeons' in Douglas Hay, Peter Linebaugh, John G. Rule, E. P. Thompson and Cal

Winslow eds., *Albion's fatal tree: crime and society in eighteenth century England* (London: Verso), pp. 65–117.

17. See, notably, Ruth Richardson (2001 edition), *Death, Dissection and the Destitute*, (London: Phoenix Press); Helen MacDonald (2005), *Human Remains: Dissection and its Histories* (New Haven: Yale University Press).

18. Jonathan Sawday (1996), *The Body Emblazoned: Dissection and the Human Body in Renaissance Culture* (New York and London: Routledge), p. 60.

19. Sarah Tarlow (2013), *Ritual, Belief and the Dead in Early Modern Britain and Ireland* (Cambridge: Cambridge University Press).

20. These themes are embedded in Sarah Tarlow's monograph (cited in endnote 19). I am very grateful to her for sharing with me core ideas that also feature on the publisher's summary of her book's main contribution on the Cambridge University Press catalogue and website.

21. For an excellent summary of the theoretical debates, see, Roger Cooter (2010), 'The Turn of the Body: History and the Politics of the Corporeal', *ARBOR Ciencia, Pensamiento y Cultural*, CLXXXVI, 743, May-June issue, 393–405.

22. See, Michel Foucault (French edition, 1979) translated to English by Alan Sheridan (1995), *Discipline and Punish: The Birth of the Prison* (New York: Vintage Books); Colin Jones and Roy Porter eds. (1994), *Reassessing Foucault: Power, Medicine, and the Body* (London: Routledge).

23. Cooter, 'Turn of the Body', p. 394.

24. Roy Porter (2001), 'History of the Body Reconsidered' in Peter Burke ed., *New Perspectives on Historical Writing*, (London: Polity Press), pp. 232–260, quote at p. 236.

25. This he clarified in Michel Foucault (1970 edition, English translation), *The Order of Things* (New York and London: Routledge), preface, p. xvii.

26. I am grateful to Adam Nicolson (2011), *The Gentry: Stories of the English* (London: Harper Press) for this phrasing in the introduction of his excellent book, pp. ix–xx, which I argue throughout this monograph can be applied to the cultural history of the long eighteenth-century too.

27. See historical debates summarised in, Steven Wilf (1993), 'Imagining Justice: Aesthetics and Public Executions in late eighteenth-century England', *Yale Journal of Law and the Humanities*, Vol. 5, I, 51–78.

28. I am very grateful for an advance copy of Owen Davies and Francesco Matteoni (2016), *Executing Magic: The Power of Criminal Bodies* (Basingstoke: Palgrave).

29. Andrew Cunningham (1997), *The Anatomical Renaissance: The Resurrection of the Anatomical Project of the Ancients*, (Aldershot, Hampshire: Scolar Press), p. 8.

30. Refer endnote 26 above, for the intellectual debt owed to Nicolson, *Gentry* here.

31. Ibid., p. 90, the phrasing is Nicolson's but the context in this book is medico-legal.

32. On the conduct of the early modern crowd, see, Nicholas Rogers (1990), 'Crowd and People in the Gordon Riots', in Eckhart Helimuth ed., *The Transformation of Political Culture: England and Germany in the late-Eighteenth Century* (Oxford: Studies of the German Historical Institute for Oxford University Press), pp. 39–55; John Stevenson (1985), 'The "moral" economy of the English crowd: myth and reality', in Antony Fletcher and John Stevenson eds., *Order and Disorder in Early modern England* (Cambridge: Cambridge University Press), pp. 218–38; Robert B. Shoemaker (2004), 'Streets of Shame? The Crowd and Public Punishments in London, 1700–1820' in Simon Devereaux and Paul Griffiths eds., *Penal Practice and Culture, 1500–1900:Punishing the English* (Basingstoke: Palgrave), pp. 232–57; George F. Rudé (2005 edition), *The crowd in history: a study of popular disturbances in France and England, 1730–1848* (London: Serif Books); Ian Munro (2005), *The figure of the crowd in early modern London: the city and its double* (Basingstoke: Palgrave); John Walter (2006), 'Crown and crowd: popular culture and popular protest in early modern England', in John Walter ed., *Crowds and Popular Protest in early modern England* (Manchester: Manchester University Press), pp. 14–26; Frederick Burwick (2011), *Playing to the Crowd: London Popular Theatre, 1780–1830* (Basingstoke: Palgrave).

33. Matthew White (2008), '"Rogues of the Meaner Sort?" Old Bailey Executions and the Crowd in the Early Nineteenth Century', *London Journal*, Vol. 33, July, II, 135–153.

34. Ibid., p. 148.

35. In a noteworthy aside, Benjamin Heller (2010), 'The "Mene Peuple" and the Polite Spectator: The Individual in the Crowd at Eighteenth-Century Fairs', *Past and Present*, vol. 208, I, 131–157, by focusing on crowds at fairgrounds, points out that early modern historians have often made too many assumptions about their 'heterogeneous nature and appearance' in ways that contemporaries did not conceive of themselves.

36. Recently Mark Forsyth (2013), *The Elements of Eloquence: How to Turn the Perfect English Phrase* (London: Icon Books), p. 32 has defined 'synaesthesia' as 'either a mental condition whereby colours are perceived as smells, smells as sounds, sounds as tastes and so on, or it is a rhetorical device whereby one sense is described in terms of another'. As we shall see later in this book, both experiences were made feasible by the post-execution

spectacle. In particular, as Forsyth points out, 'Synaesthesia's of smell are jarring and effective, and are probably an easy shortcut to a memorable line', p. 33. Few forgot a stinking criminal body because it was such a formative experience when first encountered.

37. William Pooley (2014), 'The history of the body in nineteenth-century rural France', *Past and Future, Institute of Historical Research Magazine*, XVI, Autumn/Winter, p. 17.

38. Refer, Gatrell, *Hanging Tree*, pp. 29–45, on 'hangings'.

39. In the conclusion to this book we reflect on the new 'fleshy' research of Adam Bencard (2008), 'History in the Flesh', (unpublished PhD, Københavns Universitet. Institut for Folkesundhedsvidenskab. Medicinsk Museion, University of Copenhagen).

40. I am here indebted to two books by John Bohstedt (2010), *The Politics of Provisions: Food Riots, Moral Economy, and Market Transition in England, c. 1550–1850* (London: Ashgate) and his (1993), *Riots and Community Politics in England and Wales, 1790–1810*, (Harvard: Harvard University Press). His thought-provoking chapter, John Bohstedt (1994), 'The Dynamics of Riots: Escalation and Diffusion/Contagion' in M. Potegal and J.F. Knutson eds., *The Dynamics of Aggression: Biological and Social Processes in Dyads and Groups* (New Jersey: Psychology Press, Lawrence Erlbaum Associate Publishers), pp. 257–306 is equally intellectually-stimulating. His study of food riots and crowd control has in particular inspired a rethinking of criminal dissections and the crowd management issue.

41. Two of his books have been intellectually thought-provoking, Neil Kenny (1998), *Curiosity in Early Modern Europe: Word Histories* (Wiesbaden: Harrassowitz) and his (2004), *The Uses of Curiosity in Early Modern France and Germany* (Oxford: Oxford University Press).

42. See, notably, William Reddy (1997), 'Against Constructionism: The Historical Ethnography of Emotions', *Current Anthropology*, Vol. 38, 327–51. Refer also: William Reddy (1998), 'Emotional Liberty: History and Politics in the Anthropology of Emotions', *Cultural Anthropology*, Vol. 14, 256–88; William Reddy (2000) 'Sentimentalism and Its Erasure: The Role of Emotions in the Era of the French Revolution', *Journal of Modern History*, Vol. 72, 109–152; William Reddy (2001), *The Navigation of Feeling: A Framework for the History of Emotions* (New York: Cambridge University Press); and William Reddy (2009),'Historical Research on the Self and Emotions', *Emotion Review*, I, 302–315.

43. Building on, Sarah Tarlow (2012), 'The Archaeology of Emotion and Effect', *Annual Review of Anthropology*, Vol. 41, 169–85.

44. By way of example, see,, Toni-Lynn O'Shaughnessy (1987–8), "A Single Capacity in The Beggar's Opera", *Eighteenth-Century Studies*, Vol. 21, II, Winter, 212–227.

45. Editorial (1819), 'Hiatus, a play of Time's Magic Lantern: No. IX: *The Dissector, Blasquez & Scholar*', *Blackwood's Magazine*, Vol, XXVI, (May) Issue X, 161–4, quote at p. 162.

46. See, below, Illustration 1.1 ©Wellcome Trust Image Collection, Slide Number L0031335, '*The Dead-Alive*', illustrating a man supposed to be dead arising from his coffin, coloured aquatint, published 1805, after a 1784 drawing by Henry Wigstead, (London: William Holland, Oxford Street); creative commons license authorised for academic purposes.

47. This finding was highlighted at a recent exhibition of eighteenth-century plays at the Victorian and Albert Museum: http://www.vam.ac.uk/content/articles/0-9/18th-century-theatre/.

48. It is used figuratively to mean 'pick your battles' or 'bide your time and wait for the opportune moment': both literal and figurative apply to the Murder Act in England.

49. The legal early modern Latin translation is cited here in common English.

50. I am grateful here to Peter King for clarifying this context in conversations with the author. See, also, Gatrell, *Hanging Tree*, pp. 566–89.

51. I am grateful for an advance copy of Sarah Tarlow (2016), *Hanging in Chains: Gibbeting and the Murder Act* (Basingstoke: Palgrave Pivot). For that reason this feature of post-mortem 'harm' is not covered in this book.

52. Refer, again, Wilf, 'Imagining Justice', pp. 51–78.

53. See, notably, Thomas Laqueur (1989), "Crowds, Carnival, and the State in English Executions, 1604–1868," in A.L. Beier, David Cannadine, and James Rosenheim eds., *The First Modern Society: Essays in English History in Honour of Lawrence Stone* (Cambridge: Cambridge University Press), pp. 305–56. Also: John Brewer (1979–80), 'Theatre and Counter-Theatre in Georgian Politics', *Radical History Review*, Vol. 32, 7–40; and Terry Castle (1986), *Masquerade and Civilization: The Carnivalesque in Eighteenth-Century English Culture and Fiction* (London and Stanford: Stanford University Press), pp. 1–51.

54. Highlighted, by, Nicolson, *The Gentry*—see also, endnotes 26 and 30.

55. Ibid., points out that this siege mentality was often at the heart of the property-owning divisions in England; Nicolson describes gentry magistrates and their jurymen from the sixteenth century as being open to new-comers but essentially acting as a 'government bordering on tyranny, in a country filled with sweet musk roses and eglantine', p. 43.

56. On the main theoretical approaches in medical history, see, Roy Porter (2003), *Flesh in the Age of Reason: How the Enlightenment Transformed the Way We See Our Souls and Bodies* (London: W. W. Norton and Co).

57. A. Guerriini (2004), 'Anatomists and Entrepreneurs in Early Eighteenth Century London', *Journal of the History of Medicine and Allied Sciences*, Vol. 59, April, II, 219–39; Roy Porter (1988), 'Seeing the Past', *Past and Present*, Vol. 118,186–203.

58. See, for example, Randy McGowan (2003)'The Problem of Punishment in Eighteenth-Century England' in Simon Devereux and Paul Griffiths eds., *Penal Practice and Culture, 1500–1900: Punishing the English*, (Basingstoke: Palgrave), pp. 210–31; Randy McGowan (2005), 'Making Examples' and the Crisis of Punishment in Mid-Eighteenth-Century England' in David Lemmings ed., *The British and Their Laws in the Eighteenth Century*, (Woodbridge: Boydell), pp. 182–205 in which dissection is seen as less high-profile than gibbeting, and the privilege of a medical elite: viewpoints this book takes issue with and returns to in Chapter 7.

59. This book therefore builds on Peter King's admirable work in *Crime and Law in England*, endnote 12.

60. There is often a strong historical emphasis on the primacy of a stable political economy at the time; see, by way of example, two useful books by James A. Sharp (1990), *Judicial Punishment in England* (London: Faber and Faber) and his (1998 2nd edition) *Crime in Early Modern England 1550–1750* (London: Longman).

61. Andrew Cunningham (2010), The *Anatomy Anatomis'd: An Experimental Discipline in Enlightenment Europe* (Aldershot, Hampshire; Ashgate).

62. Notably, E. T. Hurren (2011 edition), *Dying for Victorian Medicine: English Anatomy and its Trade in the Dead Poor, 1832 to 1929*, (Basingstoke: Palgrave), chapter 3. Important context can also be found in, Susan Lawrence (2002), *Charitable Knowledge: Hospital Practitioners and Pupils in Eighteenth Century London* (Cambridge: Cambridge University Press); Keir Waddington (2003), *Medical Education at St. Bartholomew's Hospital, 1123–1995* (Woodbridge: Boydell); Jonathan Reinarz (2009), *Healthcare in Birmingham: The Birmingham Teaching Hospitals*, 1779–1939 (Woodbridge: Boydell).

63. See, for example, Wellcome Library Collection, Wellcome Images, V0042183, 'A Clock Dial on which a Skeleton holds an Oil Lamp, drawn by Caleb Elwin', reproduced in Cambridge, 1800.

64. Simon Marsden (2007), *Memento Mori: Churches and Churchyards of England* (London: English Heritage).

65. I am very grateful to Dr. Shane McCorrestine for these stimulating ideas which form the central focus of a conference he convened on '*When is Death*' at Leicester University in April 2015, and whose findings will feature in an edited volume by Palgrave in 2016.

66. Sawday, *Body Emblazoned*, p. 88 and Thomas Laqueur (2011), 'The Deep Time of the Dead', *Social Research*, Vol. 78, Fall, III, 799–820, quote at p. 802.
67. Simon Chaplin (2009), 'John Hunter and the Museum Oeconomy, 1750–1800', (unpublished PhD, King's College London).
68. A. D. Macleod (2009), 'Eyelid Closure at Death', *Indian Journal of Palliative Care*, Vol. 15, July–December, II, 108–110, quote at p. 108; Hippocrates wrote about eyes open versus eyes closed, and its cultural taboos, noting that the lower lids often remain open in brain trauma—an observation still made about fatal neurological conditions and at executions by lethal injection in America today.
69. Reddy, 'Against Constructionism: The Historical Ethnography of Emotions'.

CHAPTER 2

Becoming *Really Dead*: Dying by Degrees

Introduction

In the field of criminal history Vic Gatrell alerted a generation of historians to the agency of the English crowd and visceral nature of the execution act.[1] It was a brutalising, demeaning, painful, spectacle of retributive justice throughout the long eighteenth-century. Peter Linebaugh correspondingly highlighted how mob hostility to the surgeons broke out at bungled executions during the 1730s in London.[2] Frequently the hangman was bribed to mishandle an execution either by the convict or their anxious relatives. Those lacking the financial wherewithal to induce a corrupt executioner, had an anxious wait to see if the condemned was in an insensible rather than deceased condition when cut down from the gallows. The task of being a lifesaver often fell to penal surgeons with the expertise to revive someone on the boundaries of life and death which inexperienced hangmen had mishandled. Most contemporaries recognised that the capital code was a game of medical chance because it involved the gamble of '*resurrection...*not an attitude to death but the last chance to escape it'.[3] Hence this chapter explores the execution of the condemned from a number of neglected medical perspectives that involved how to determine capital death under the Murder Act, from 1752 to 1832.

To be 'hung by the neck *until dead*' was a challenging judicial proviso to execute. It often divided opinion in medico-legal circles about which working guidelines would ensure that a convicted murderer was killed

© The Author(s) 2016

E.T. Hurren, *Dissecting the Criminal Corpse*, Palgrave
Historical Studies in the Criminal Corpse and its Afterlife,
DOI 10.1057/978-1-137-58249-2_2

on the eighteenth and nineteenth-century gallows. It was obvious that the appointed executioner was duty-bound to ensure that the condemned died on the rope. Generally this was either by a broken neck and/or strangulation. Afterwards, the corpse was to be despatched whilst the body was warm, sent to the surgeons to be punished in death. In this way the crowd were alive to the possibility of post-mortem 'harm'. The problem with these legal duties was that compliance was very difficult because officially declaring someone lifeless was not an exact science in early modern England. In histories of the period too often it has been assumed that all criminals, once hanged, were a corpse when in fact archive material reveals that a significant number were found to be living, not deceased, under the hanging-tree. The physical status of everyone executed for murder had therefore to be double-checked on arrival at dissection venues in ways still misunderstood. Hence, more historical attention needs to be paid to the chief medical dilemma that was at the crux of the capital legislation. Continually the medico-legal fraternity faced an ethical quandary that is still challenging in modern biomedicine—for some people 'becoming *really dead*...takes time'.[4] Eighteenth-century standards of *medical death* (no life-signs in the heart, lungs and brain), as opposed to *legal death* (the act of being hanged) were thus intermingled with wider considerations of judicial power, reconstruction of the state, and a medical knowledge of death continually in transition. Neglect of the puzzle of medical death is consequently a significant historical oversight, but one that is explicable in the standard literature.

In 1751–2, when the capital legislation was being drafted medico-legal mentalities that looked fixed were in fact very mutable. They fluctuated a lot, even before the new stipulations reached the official statute books. In many respects, the specific duties described were a précis of a very fluid set of 'scientific' aspirations, and it is these that have been neglected. Strictly speaking the condemned body was to be punished by 'taking life' (an eye-for-an-eye retribution) and despoiling the body 'to its extremities' (the literal meaning of dissection). But, as we shall see, these routine measures were never a coherent expression of unifying anatomical methods and practices. Of relevance is that when the new capital statute became law '*old anatomy*' underwent a paradigm shift by pursuing what became known as '*new anatomy*'.[5] The body was no longer studied wholly as an object of God's creation or indeed exclusively as an educational tool for medical students. The intention was to promote original anatomical research by exploiting the potential for scientific endeavour afforded by the criminal corpses of the capital code. Georgian legal officials and medical men thus recognised, as Joanna Innes explains, that eighteenth-century parliamentary legislation was usually drafted in a cursory manner and 'often did not commend itself to eighteenth-century Britons'.[6] It had to be written with 'a sufficient level of generality to cope with diverse local circumstances'. What then looked like a fixed set of punishment rituals involved a great deal of speculation. Over time, the criminal body became a contested medical commodity, synonymous with oscillating ideas about the mutable boundaries of life and death. By the 1780s these encompassed the complex nature of dying, medical death, and physical vitality. Effectively this research emphasis complicated the practical workings of the capital legislation. If the hangman could not execute his duty, then it fell to the penal surgeon to complete the task of execution. This was what really tarnished the medical fraternity in the presence of the post-execution crowd: an outcome often misconstrued in standard histories of the period.

There have then been a lot of basic misconceptions in crime and cultural studies about the practicalities of executing post-mortem 'harm' under the Murder Act.[7] The legal phrase '*dissection and anatomization*' is still used without medical precision. It tends to be misread as a general statement that on safe delivery of a murderer's body the surgeons had the exclusive authority to cut up the corpse in the dissection theatres of Georgian England. This chapter takes issue with this misleading impression. It shows that there was a lot of medico-legal indecision about the '*half-hanged*', '*dead-alive*', and '*nearly dead*'. As life ebbed from the criminal body medical men debated death's bodily ambiguities, particularly how to monitor metabolism and vitality with such limited medical equipment. Effectively, this meant that '*dissection and anatomization*' were redefined as two separate punishment procedures by surgeons working under the capital legislation. The first duty was called 'anatomization' that became associated with a set of checking-mechanisms to determine whether life-signs had deteriorated enough to bring about medical death. The second duty was termed 'dissection' and involved cutting the body 'on the extremities to the extremities' by dismemberment until the murderer was despoiled as a human being. This state of being less than human did not begin until a lifeless state had been established medically, even though the prisoner deserved their punitive treatment according to common law and there would be very little human material to be buried at the end to complete the capital sentence. Surgeons had hence the legal discretion to pragmatically define their working-duties when a body arrived at a dissection venue.

Most acknowledged that it was very difficult to precisely know whether when a criminal stopped jerking on the executioner's rope those physical signs indicated medical death or not. A condemned body might start to expire, but it also could be shutting-down to protect its vital functions in physical trauma. A liminal state of pain was often disputed in the eighteenth-century medical press too, with surgeons in the capital disagreeing with their provincial counterparts about whether murderers died painfully on the gallows or not. After the condemned departed from the courtroom, the physical showcase that followed as punishment rites unfolded could be rather perturbing for the officials in charge. This was the clandestine side of the Murder Act and it is what this chapter is all about.

The complex ways that medico-legal officials defined medical death in the eighteenth and early nineteenth centuries is the detailed subject of Section 1's discussion. This context is necessary to appreciate the medical condition of the condemned body in its criminal setting. Then Section 2 investigates each punishment step *in situ*, starting with a body on the hangman's rope moving into the realm of physical trauma as life ebbed away. The main focus is the working protocols for establishing medical death in the heart-lungs-brain. Any procedural modifications are identified in terms of their continuities and change over the course of this book's chronological focus to better engage with early modern mentalities of death and dying. Section 3 then re-examines what it meant to come to a medical recognition that executed criminals were in a physical state of 'dying by degrees'. This could occur both hanging from the rope and/or underneath the gallows. Medico-scientific thinking was continuously being tested by working hands-on. New ideas about the potency of human vitality nevertheless complicated how the criminal body was viewed on arrival at dissection venues. It will be shown that some penal surgeons did contravene the Hippocratic Oath with the lancet to hasten death. Others disputed their unsavoury duties in private practice and licensed surgical circles; increasingly these sorts of anatomists monitored the dying processes with caution. Records survive of leading penal surgeons making a detailed record of its metabolic timing. Their influential work could not however prevent the controversial issue of medical death by capital punishment being opened up to public scrutiny in the provincial press. The majority of medico-legal officials henceforth adopted a choreographed set of punishment stages so that they could function ethically and practicably under the Murder Act. Its working framework was: *first* hanging (legal death), *then* being anatomized (medical death), and *finally* dissection (post-mortem punishment).[8] Its widespread adoption (as we shall see throughout this chapter) indicates that it is historically inaccurate to elide '*dissection and anatomization*' into a single description of post-mortem punishment.[9] The last moments of existence were seen as a grey ethical area.[10] Amongst anatomists the general attitude was that it was the province of 'the doctor... to determine when officiousness became torture'.[11] The predicament of managing medical death by penal surgeons was to cast a long shadow over the Murder Act from its inception.

'*DEATH: THE UNCERTAIN CERTAINTY!*'[12]: ACCOMPANYING THE DEPARTED INTO MEDICAL DEATH

Near-death experiences were routine occupational hazards for penal surgeons and ones that were well-known to the medico-legal fraternity of mid-Georgian England. Logistical issues remained however largely in the 'private', rather the 'public' sphere to avoid bad publicity. Their rediscovery in the archives has therefore important implications for a wide range of early modern histories of the body and criminal justice.[13] It is a material fact that in all human beings vital metabolic processes have always been, and remain, capable of shutting down to keep someone experiencing trauma alive. In the past, this meant that the choreography of post-execution rites could be as emotive, engaging, and powerful, as the formal execution spectacle itself. What made routine punishments compelling was that sometimes *both* the hangman's rope and the lancet were needed for the condemned to become a corpse. That being the case, it is essential to accompany the departed on their physical shut-down into the realms of eighteenth and early nineteenth-century medical death. At issue were changing definitions of insensibility, arising out of a baffling array of early modern physic that did not have the scientific capacity to monitor medical death with precision or medical technology to time accurately the complex physical survival-mechanisms of the condemned. Under the Murder Act medico-legal officials anxious to enhance their professional status by handling the criminal corpse were often confounded by death's dominion.

Today scientists and historians recognise that defining the dying process in the past was a lot easier than establishing medical death. As Leslie Whetstine, an American bioethicist, observes: 'One of the more problematical issues in intensive care is not so much what death *"is"* but instead *"when"* death occurs and the operational criteria used to confirm it'.[14] If modern biomedicine can still be confounded, small wonder that an eighteenth-century surgeon might be troubled too. A recent article in the *Lancet* retraces some of the common physical signs of medical death in the past and their medical competency in context:

- Cessation of heart action, respiration and vigour of the body
- Lividness of the back (*refuted as also occurring in illness without death*)
- Rigor mortis (*may be confused with spasm and catatonia*)
- Coldness of the body (*regarded as a poor sign since after death the body is thought to rewarm as heat convection from inner organs; the thermometer in 1860 improved the use of this criteria*)
- Non-pulsatile, pale-yellow arteries
- No blood flow from transacted veins or arteries (*it was refuted that blood flowed from dead patients for hours*)
- Grey, black and cloudy corneas
- No muscle movement after electrical stimulation
- Depression or flatness of the loins, buttocks and jaw
- Relaxation and open state of the anus
- Putrefaction—putrid smell, distention of the abdomen, purple or green spots on the body (*may not be different from disease without death*)[15]

It was then common in the eighteenth-century once the dying process could be observed to keep checking when the patient approached medical death. Unsurprisingly there were was a long list of checks and balances which reflected the fact that surgeons had to take account of a wide range of medical opinions in the shadow of the gallows. To engage with that contemporary scene we need to try to re-imagine the range of medico-scientific options once the condemned was suspended from the hangman's rope.

The surgeon standing to the side of the gallows in the shadows waited to approach the condemned without agitating the crowd. At Tyburn before 1752 and after the Murder Act newspaper reporters noted that the visible presence of the medical men could cause the crowd to riot to obtain the body for burial.[16] In a noteworthy example in 1818 involving a notorious case at Godalming in Surrey local newspapers reported that there was such ill-feeling against the two murderers, Chennel and Chalcraft, on the gallows that unlike most homicide cases 'there was no danger... that an overwhelming crowd would create any confusion' and make off with the bodies.[17] The same newspaper reporter commented on a contemporaneous case at 'Greenwich where a murderer named Hussey' had been 'rescued by the mob' because local people thought the law had been too harsh. There was a risk of a public order situation getting out of control, and this meant that by the time the surgeons had a chance to get their hands on the body it was often cut down, lying in a cart on the way to the dissection venue. And this observation is significant because the executioner and sheriff's officers would have had very rudimentary methods to establish medical death at the gallows compared even to surgeons. Leaning then over the body there were three very basic confirmatory tests under the hanging tree:

- Signs of "sensibility"—blowing a strong stimulant such as a hellebore or mustard into the nose—inserting a sharp instrument under the nail—scalding with hot water or oil—trumpeting or loud noises
- Holding a mirror, soap bubbles, feather or candle to the nose to detect respiration...
- A container of liquid was placed on the abdomen, or in a decubitus, on the 11th rib; no motion of the liquid indicated death[18]

It is debatable that any of these were carried out with precision given the unpredictable mood of the assembled crowd and the fact that the weather on the day of execution was an added medical complication.

All of these logistical problems are observed in the case of 'Abraham Dealtry' tried and found guilty of highway robbery during 'Lent on 6 April 1745'; his subsequent execution was staged during a very cold and inclement springtime in York. The *Morning Advertiser* and *London Evening Post* both reported that: 'Body cut down after 10 minutes and put in a coffin ready for burial; signs of life detected on the way to the graveyard; revived and blooded and returned to York Castle Gaol; thereafter sentenced to *Transportation for Life* at Yorkshire Summer Assizes, Summer 1745'.[19] Another well-known controversial execution is that of William Duell aged sixteen who was hanged for rape and murder 'for the space of twenty-two minutes' on a bitter cold winter's day in the capital on 24 November 1740'.[20] Sir Richard Hoare the Sheriff for the City of London noted in his diary that after execution Duell's body was taken by hackney-coach to Surgeon's Hall where on being anatomically checked in order to make 'the necessary preparations for cutting him up' by dissection, vital signs of life were observed.[21] Duell was bled, revived, returned to Newgate, and later reprieved—since he was socially, legally, but not medically dead, his sentence was commuted to *'Transportation for Life'* too. Given the seriousness of his crime, and his youthful, robust physical condition, the shorter hanging-time in cold weather was the subject of considerable debate. In a noteworthy aside to the reported case, one legal official helpfully explained that:

> It is true the sentence was to hang him by the neck *until he were dead* and this has not been done; but that it is not done is owing to the inattention only of the magistrates, whose business it is to see that the body be lifeless before it is carried away.[22]

When duty-surgeons could not stand openly at the gallows because of crowd antagonism, this source suggests that a local magistrate with very limited scientific training (even less than the sheriff and hangman) had to declare medical death. This fall-back position harked back to the medieval tradition of watching over the executed body in public, a legal rite known as the *exitus*.[23] Official medical death in these circumstances was determined by common agreement by those present that vital signs had ceased. There were then lots of local stories about disputed medical death timings that attracted popular attention and these caused a public outcry by 1752. The balance of the evidence in winter executions suggests that the three basic confirmatory tests at the gallows (outlined above) were sometimes done in a perfunctory manner, hurried, or ignored altogether. Staging a hanging was about a medico-legal performance of the character of medical death. This meant that in the popular imagination the crowd travelled onwards with the criminal body to satisfy themselves that the picture created by the criminal justice system of the condemned in death was based in a medical reality.

There is then a consensus amongst leading early modern historians that 'the eighteenth century brought a growing medical interest in death' and so popular opinion held that 'the doctor must manage the process of ceasing to be'.[24] The enigma of medical death on the gallows likewise captivated the general public's imagination. Consequently, 'a quite extensive medical press was operating by 1800—over thirty titles had come and gone; and if few journals achieved longevity before collapsing, or being renamed, new ones were always plugging the gaps'.[25] Access to this level of new information and the possibility of reading about the *'dead alive'*, *'half hanged'* and *'revived'*, reflected more public awareness about the innovative things happening in anatomical circles concerning the mystery of medical death. This explains why dissections were potentially exhilarating events in the locality of a murder. Around 1752 the medical press thus became the conduit of 'that perennial eighteenth-century phenomenon, the participation of the lay public in medical debates' with grisly overtones.[26] Predictably perhaps by the 1820s sales of popular journals featuring tales of near-death experiences had attracted an avid readership. *Blackwood's Magazine* catered to this market by publishing in 1827 the startling story of '*Le Revenant*' (The Ghost). The narrator, a condemned criminal, was hanged for murder in London, but cheated medical death. He described how –'With the first view of the scaffold, all my recollection ceases' until, 'having awoke, as if from a sleep, and found myself in a bed'.[27] On opening his eyes, he thought he had been reprieved but was told that he had been hanged for an hour, cut down, bled, and resuscitated by a 'gentleman' with medical skills: 'My condition is a strange one! I am a living man; and I possess certificates of both my death and burial.' Strictly speaking his *legal death* happened on the scaffold, his *social death* was marked, but he did not experience *medical death*. In so doing, his tale reflected ongoing speculation about whether those executed for murder died on the gallows or inside designated anatomical venues. Incidents started to preoccupy the general public outside of London too.

Thomas Dunn was executed for the murder of Mary Lakin at Leicester city gallows on a bitterly cold day, 25 January 1796. His body was handed over to a famous local surgeon, Dr Thomas Smith Kirkland of Ashby-de-la-Zouche. He got the body because the judge ordered that it be sent back to where the homicide had been committed to be shamed publicly.[28] In his private diary, Dr Kirkland wrote that because of the greater risk of extreme hypothermia in cold weather, and considerable local feeling against a violent murderer, Thomas Dunn's body was 'exposed to view in the White Horse public house' in the town. This was done over 'three days' under the Murder Act. Dr Kirkland recorded that he worked in full public view of several thousand people. The working procedure was that he first did an anatomical operation that he termed '*anatomization*' to check for '*lifelessness*', before he proceeded to dissection. The body was washed, shaved, and he made two incisions—one down the chest—one across the torso area—known as the 'crucial incision'—a cross-like cut to observe how the heart and major organs expired.[29] Many onlookers travelled up from Leicester, seventeen miles away. They did so because Dunn was a soldier in the 5th regiment of the Irish Dragoon guards. He was a physically very robust young man, and someone who might be capable of defying the capital sentence on the gallows. Dr Kirkland admitted that he used the opportunity to further his new anatomical research and showcase his skills to his paying customers in the wider community. Later he privately wrote about how the problem of managing medical death 'added to the sense of occasion and its sensational elements'. The balance of the evidence in cases like this one, and others we will be encountering later in this chapter, confirms that there was a medico-legal choreography, it was well-known, and its primary purpose was to reassure those present that the dangerous were dead.[30] To Dr Kirkland the townsfolk of Ashby-de-la-Zouche were akin to a choir in Greek tragedy, but one not positioned off-stage. By being present their gossip gave public voice to a dreadful anatomical fate. Each medico-legal step dramatized the social, legal and medical death of Thomas Dunn. Later he was dissected, his skeleton boiled down, and displayed in Kirkland's private medical museum.

In many respects the actions of provincial Georgian crowds espoused a version of Aristotle's position that the object of anatomical study was to discover how the soul operated through flesh (be it animal or human).[31] The moment of medical death seemed to mark the afterlife and the mystery of that life-force had to be stage managed by the criminal justice system (elaborated in Section 2, below). If likewise, as Roy Porter and others claim: 'The eighteenth century brought the development of the medical management of death at the bedside'—and there was a vibrant print culture about the medical limits of mortality and the soul's afterlife—then a medicalization of the capital sentencing process was an intrinsic part of a broader cultural trend that 'heightened speculation as to precisely what death was' and who specifically had power over it.[32] An added complication was that 'medicine seemed to be pre-empting the hand of God', but there also continued to be customary 'beliefs about death and the rituals which expressed them'.[33] It was commonplace to try to philosophically debate the nature of life itself. This meant that medical opinions about when the criminal became a corpse became bound up with a better scientific understanding of execution and by extension the limitations of executioners when confronted with near-death situations. Hence when then a leading medical man, Dr A. P. W. Philip, delivered a paper to the Royal Society in London entitled, '*The Nature of Death*', there was a great deal of contemporary interest in his review of working definitions and medical opinions that had been in vogue between the Murder Act and publication of his treatise in 1834.

Philip in his treatise explained that eighteenth-century medical opinion held that there were 'three distinct classes of bodily function'.[34] These were known as 'the sensorial, the nervous, and the muscular'. The general scientific view was that whilst each 'had no direct dependence on each other'—each could technically function without the other—nonetheless the three systems were inter-connected. It was a physical reality that if one shut down there would be a fatal domino reaction in the other and this would 'more or less immediately destroy the organs of all'. Philip stressed the importance of using precise medical terminology when defining the 'two physiological-stages of death'. The first was popularly known as '*the name of Death*' and the second was dubbed '*absolute Death*'.[35] He asserted that if the sensorial, nervous and muscular systems of the body start to close down, then the heart and lungs would fail. These in turn would have a detrimental impact on the 'sensitive system of the brain and spinal marrow' because blood supply would slow down until it ceased to reach the head. Only however when the sensorial functions had terminated—the organs failed to function—*and* the sensitive system displayed no physical evidence of brain activity—did '*the name of Death*' technically become '*absolute Death.*' There were two exceptions—in a 'violent death' at a homicide, or at an 'execution scene'.

By the 1820s contemporary anatomists thought they had 'proved by experiment that in violent death the heart and lungs do not derive their power from the brain and spinal marrow but may be destroyed by impressions made upon them'.[36] In other words, if a body experienced a major trauma, then that violent assault often speeded up the dying process below the neckline. Meanwhile, anatomists were still exploring at what point medical death actually happened in the brain. Opinion was divided because of the mysterious notion of suspended animation and theological debates about the soul and its afterlife. It was possible, some believed, to become stuck in a liminal metabolic space between life and death. In practical terms, hitting someone over the head could cause the heart and lungs to expire, but not necessarily brain death, since it depended on the level of violent force on the lower part of the body. If however someone stabbed the heart or lungs with a knife, puncturing the major blood vessels, it would cause rapid system failure. In these traumatic circumstances the brain might survive a short-time, but not for long. Brutal violence, in the opinion of Dr Philip and his surgical colleagues, was normally the chief cause of starting to die known as the '*name of Death*' in homicide cases. Once the heart and lungs ceased, brainstem death was inevitable but its timing could be delayed. To reach the second stage called '*absolute Death*' a full physiological shut-down was required.[37] Only then could a surgeon declare medical death. Philip importantly pointed out that there was an exceptional type of violent death that disturbed surgeons. Medical death was different in 'the struggle of the criminal after the drop falls' on the gallows.[38] He claimed that contemporary anatomists and hangmen knew there were different standards of legal versus medical death. That being the case, we need to briefly engage with how anatomical standards had developed, before then considering practicalities at the gallows.

Historians of the body concur that from the Renaissance until the Enlightenment era, the main purpose of the art of anatomy—and it was seen as an art form, not a science, part of Natural Philosophy—was to look inside the human body to '*know thyself*'. At a time when attending an anatomical lecture was about exploring the discipline's 'grammar, rhetoric… metaphors and analogies', dissection and its 'medical utility' actually 'ranked below the theological purpose' of the anatomical theatre. The primary educational purpose of an anatomical demonstration was to view God's Creation inside the criminal—'*Know thyself and others* was the extended message'.[39] This basic premise was transformed in the so-called 'scientific revolution' when older anatomical methods and their educational value unravelled. It was during this time that '*anatomization*' started to strengthen its association with 'living anatomy' and the shut-down of vital forces in the 'dying body': this was Thomas Kirkland's intellectual position in Ashby-de-la-Zouche in 1796 (as we saw earlier). The focus of enquiry was now the puzzle of medical death since it was one of the few original avenues of scientific enquiry left to traditional anatomists once 'anatomy became experimentally matter-of-fact knowledge' after William Harvey discovered the heart was a pump in 1628.[40] If functioning anatomy was governed by a number of mechanistic theories then exploring how these metabolically closed down during the dying process was a logical way to resuscitate an anatomical tradition of '*anatomization*'. This type of reinvention was common down the centuries in the anatomical community and therefore not seen as misleading or intellectually false because the discipline had never been monolithic. It is nevertheless one of the ironies of the criminal justice system that the ability of surgeons to manually revive the condemned, mirrored how anatomists tried to resuscitate their intellectual credibility by deploying their accumulated knowledge about how criminals might be able to defy medical death after 1752. As Andrew Cunningham effectively puts it—'Each new resuscitated body resulted from a new and distinctive research programme in anatomy'.[41]

Dissection by contrast became associated with studying the 'pathology of the corpse' after medical death. Confusion has arisen in criminal history about the relationship between anatomy and dissection under the Murder Act because many leading eighteenth-century medical figures practiced both and they never needed to explain their distinctive nature or disciplinary developments since they were well-known to contemporaries. Anatomists, as Roger French points out, by the eighteenth-century had an 'unrivalled technical knowledge of the body'.[42] They were thus ideally suited to hand over the criminal corpse to in the first instance to examine the mysteries of brain function, cardio-respiratory failure, and to monitor their medical demise. This is why they cut the chest open, examined the major organs, and watched the heart stop beating, checking brain function, and only then proceeding to dissection once all signs of life had ceased. One representative example is worth touching on briefly to set these medico-scientific trends in context. The famous case of Elizabeth Ross is well-known in crime historiography. Yet, it is seldom explored from an anatomical angle and the management of medical death. It affords an opportunity to see in broad terms what the key differences were between the working-practices of '*anatomization*' and '*dissection*' under the Murder Act. As has already been outlined, for centuries both terms had been synonymous with each other. They were used to ascertain a cause of death. This was no longer their function by the time the capital legislation came into effect. Separate procedures were required to declare medical death and then proceed to post-mortem work. One leading penal surgeon left behind remarkably detailed records of how this worked. His case notes on Elizabeth Ross are hence an archetypal historical prism of post-execution punishment being carefully managed by medico-legal officials under the Murder Act.

Elizabeth Ross was executed on 9 January 1832 at Newgate for strangling her murder victim Caroline Walsh. Surviving notes of her detailed punishment rites illustrate that the '*anatomization*' procedure was very different from being '*dissected*'. The hanged body was received from the gallows by the Chairman of the Board of Curators at the Royal College of Surgeons, who noted carefully:

> A longitudinal incision was made from the upper part of the sternum to the pubis and two oblique later ones – The first over the oblique muscle; and second in the direction of the pectoralis major. The incisions were directed to be made through the integuments only. These incisions were made in the presence of the Sheriffs and Undersheriffs who attended the delivery of the body and were afterwards sewed up. The body was ordered to be delivered to Mr Luke or his assistants at the London hospital at six o'clock in the same evening precisely.[43]

Illustration 2.1   ©Royal College of Surgeons Library, William Clift Collection, Box 67b.13, '*Sketch of Elizabeth Ross*' and associated dissection notes on criminal corpses; Creative Commons Attribution-NonCommercial-ShareAlike 4.0 International License (CC BY-NC-SA 4.0)

A private dissection room sketch was later made of Elizabeth Ross (see, Illustration 2.1 above) when she arrived at the London hospital.[44] It appears to confirm that her torso had not been cut extensively above the breast area. Initially she was opened with two incisions so that the penal surgeon could confirm to the sheriffs and undersheriffs present that she had reached medical death. It is noteworthy that this anatomical check was carried out in a separate location. In medico-legal circles it was commonly called '*anatomization*' when it involved a criminal death. The post-execution stipulations of the Murder Act were therefore logistically done in reverse—first 'anatomization' and then 'dissection'. The two initial lancet cuts went down the chest below her breastbone and across the stomach—only cutting into the main muscles and not the deep tissue—which enabled those present to watch her heartbeat and pulse ceasing (rigor mortis having set in after a critical four hours, common to every dying person). She was hence declared medically dead, before being sewn up. Only then was she moved to the dissection venue for the third stage of the death sentence. To shield her human dignity a woollen blanket concealed the initial anatomical marks where she has been stitched up. Record linkage work like this, matching detailed dissection notes to iconic anatomical images of the criminal body, reveals a distinctive medico-legal choreography that requires more careful elucidation in the historiography. Those present would not have followed legal, medical and then post-mortem steps if they did not believe it necessary to check that Elizabeth Ross was deceased: a theme we will be returning to in some detail later. Meantime contemporaries knew that natural revival and manual resuscitation did sometimes lead to a full physical recovery and required careful handling after the Murder Act. Medical death in the brain soon became the focus of extensive speculation. To better understand that ongoing context it is necessary to return to what Dr Phillip called 'dying by degrees' in the '*name of Death*'.

Dying by Degrees: Deliberating, Disguising, and Disputing Medical Death

In early modern England, there was a great deal of contemporary conjecture about what today are called *anoxic insults to the brain* which in the past was termed *vitality*: this essentially meant that the biomedicine of oxygen starvation was little understood. Any theoretical speculation was based on very limited scientific knowledge about how to define physical trauma in the brain as it went into a state of medical death. It is self-evident this would have mattered a lot at an execution scene. Another practical consideration was that rudimentary techniques to test for vital signs of life in the dying brain remained unsophisticated. These continued to be: shaking the patient, cutting their thumb to get a blood flow, or using a swan's feather on the neck to test for a breathing stimulus. Around 1721, smelling salts became a standard compound in the surgeon-apothecary's medical bag. A few general practitioners by 1744 had access to electric shock instruments when they were first introduced, but many more had to wait until 1800 when they became more widely available in leading medical institutions. By then, in 1803, what brought controversial brain-revivals connected to the capital code to the attention of medical consumers was the famous public demonstration by Giovanni Aldini of electro-resuscitation techniques on a criminal corpse executed for murder in London. His case-history was not only the blueprint for Mary Shelley's *Frankenstein* but it created public disquiet about the medical side of the Murder Act too.[45] The experiment raised uncomfortable questions about when to declare '*absolute Death*'. If the boundary of medical death was '*the uncertain certainty*'—to use a contemporary phrase—what, many were asking in print, did that mean for the penal surgeon at the post-execution scene?

Making sure the criminal '*looked dead*' was essentially a theatrical re-enactment of rough retributive justice in the eighteenth-century. Everyone was in a hurry to get the hanging-stage done, especially in the winter cold or summer heat. The difficult complication of a loss of sensibility in the brain did perturb many of the officials involved and it went on doing so. Dr Philip in his essay on '*The Nature of Death*' sets those general debates once more in context:

> It is generally supposed that the struggle of the criminal after the drop falls is the measure of his sufferings. The most vigorous suffer most, because in them the sensibility is with most difficulty extinguished; but it is not uniformly in them that this struggle is the greatest. We have reason to believe that it is little, if at all connected with the feelings of the sufferer.[46]

At first glance this seems to be a reasonable assumption given that early modern medical science was improving its working knowledge of cardio-respiratory failure and brain malfunctions. Yet, much of this is also painful conjecture too. Phillip appears to be arguing in 1834 that just because a body jerks on the gallows before going stiff does not mean that the criminal is either sensible of what is happening as s/he dies, or for that matter is in excruciating pain during the whole execution itself, as the penal punishment intended. He could not know at the time but a lack of physical jerking might in fact have indicated a faint trace of brain-stem function (see below): his observations are therefore prescient for someone with limited medical equipment. Instead he tells us what he can claim to know, namely that human beings vary considerably in how they react under extreme stress. Some people have higher pain thresholds than others. The body is capable in certain conditions of terrible fear of closing down to protect what he calls, '*The Laws of Vital Functions*'.[47] Philip states that in his experience 'all such convulsive motions are the same nature'. He calls this a medical condition of "*subsultus tendinum*"—and notes that it is often observed 'in fever' patients. The person suffering from a high fever is not unlike someone being executed on the gallows. Philip remarks they have 'little sensibility' about what is happening to them once they start to convulse. Physical jerking is triggered to protect the body. The limbs might shake and then stiffen, but these can in fact be medical indications that the body is shutting down to protect itself in some way that early modern science still has yet to understand. They do not guarantee that the person is experiencing extreme pain or in a state of '*absolute Death*'.

Evidence like this, indicates that the physical predicament of being stuck in a dying state in the brain was the ongoing subject of lively debate amongst penal surgeons in the private sectors of eighteenth and early nineteenth-century life. What emerges from local disagreements is just how much medical men deliberated, disguised and disputed medical death, especially in the colder climes of eighteenth-century England. In many respects they were hindered by the same practical dilemma as modern doctors:

> Hypothermia can...mimic brain death. At body temperatures below 32[degrees] C (90[degrees] F), a patient may lose all brainstem reflexes, be tachycardic, and be unable to shiver, depending on the degree of hypothermia. Most healthcare professionals have heard the expression, "*a patient is not dead until warm and dead*." A brain death diagnosis can't be made until the patient's core body temperature rises to at least 32[degrees].[48]

An added problem continues to be that: 'In patients suffering severe facial trauma, pupillary assessments can be difficult or impossible to perform due to severe oedema [accumulation of bodily fluid]'. This was the case for those that died on a rope from strangulation. For centuries doctors looked out for what is often called 'waggling in the eyes' to determine brain malfunction that is terminal. This involved: an inspection of the eyes for fixed dilated pupils, absence of corneal reflexes, cloudiness of the cornea, and loss of eye tension. In capital death however where there was often a delay in getting to the criminal body, penal surgeons had to work with faces swollen by the effects of suffocation and this made it more difficult to examine blood pooling of the eye capillaries. In which case there was the option to do a set of further hands-on procedures that have not changed radically since the early nineteenth-century but can be better monitored with more sophisticated medical equipment today:

> An assessment of ocular movements includes oculocephalic and oculovestibular testing (Cranial nerves III, IV, VI). The *oculocephalic reflex*, also known as the *doll's eye reflex*, is assessed next...With the eyes held open, the

patient's head is turned rapidly horizontally and vertically and eye movement is observed. If the oculocephalic reflex is present, the eyes move in the direction opposite to head movement. If the oculocephalic reflex is absent, which is consistent with brain death, the eyes remain midline with head movement.[49]

Of relevance was also the option of doing a cold and hot water test in the ear, a variant of which was once familiar to eighteenth-century penal surgeons too. We will be examining these methods below when looking at contemporary case notes. Meantime, here it is helpful to explain the medical circumstances in general terms first because later we will be looking at what medical officials thought their task was behind closed doors when we examine their recorded timings of bodies shutting-down from the gallows:

> The head of the bed is positioned at 30 degrees. Then, 30 to 50 mL of ice-cold (33[degrees] C) water is injected slowly (over 30 seconds) into each ear, with a 5-minute wait between injections into each ear. Once the water is injected, both eyes are held open to assess for ocular movement. In patients with an intact brainstem, you'll see a slow movement of the patient's eyes to the side of the ice-water irrigation, followed by a rapid corrective movement of the eyes.

> An alternative test for assessing vestibular function is bilateral irrigation with warm water (44[degrees] C). Normally the eyes move away from the side irrigated. Simultaneous bilateral irrigation with cold water normally causes downward deviation of the eyes; bilateral irrigation with warm water normally triggers upward deviation. In brain death, eye movement is absent for at least 1 minute of observation.[50]

As we have seen throughout this chapter the modern biomedicine of brain-stem death is still a very complex area of emergency medicine.[51] The body in extreme cold, or after a head trauma, or when the circulation of the blood is faulty, can reduce its vital systems to a minimum level to try to keep life functioning. Coma is often diagnosed when the brain is rushed with endorphins to deal with extreme pain thresholds. The comatose person is not capable of moving their limbs, some nerves may twitch, but they certainly look dead to the untrained medical eye. The recent use of therapeutic hypothermia in emergency medicine—letting the brain manage major trauma by reducing the temperature of the body to a minimal level after a sudden cardiac arrest following an accident or injury—is one important indication that our historical knowledge of our biological survival mechanisms has been inadequate.[52] In the long eighteenth-century, coroners, doctors, executioners and sheriffs recognised that brain function was a fascinating medical frontier. And the medical protocols they developed have been enduring because they were sensitive to bodily matters that were mysterious, as well as medically-documented.

This medical backdrop explains why at a special meeting of the governors of the Leicester Infirmary on 17 November 1815 (by way of example) the surgeon responsible for carrying out the criminal dissection of corpses on the premises commented:

> Medicine is now extending her Empire over diseases more immediately connected with the mind, as well as those which the body is primarily and principally affected – NOR [sic] is she left without encouraging hopes that she shall at length be permitted to explore that dark region, the Brain, which has hitherto too generally been with-holden [sic] from her view, and presented to less enlightened and less scientific eyes.[53]

Others wrote in the same vein to the *London Medical and Physical Journal* by 1800. One representative correspondent claimed:

> I have seen a dissector ignorant of the most simple things; ... even while he had the knife in his hand ... Much may be learned in the examination of the dead, without delicate skill and profound knowledge; and physicians, apothecaries, and curious persons, possessed only of a slight knowledge of anatomy, might soon be qualified to perform many useful inspections of the dead.[54]

In other words, there were many types of medical actors present at dissection venues who were skilled, semi-skilled, and medically incompetent working on the same bodies. By 1807 (the date of the quoted letter) the profession was becoming more aware that the brain-stem function differed from the heart-lungs. This was essentially because of what medical science was learning from criminal corpses when checked anatomically for life-signs. For this reason the same correspondent admitted: 'I have myself seen no dissection from which I have not learned considerably; for, to use the remark of [Lawrence] Sterne, *an ounce of a man's own sense, is worth a ton of other people's.*'[55] *Seeing is believing*—was his key theme, and this hands-on trend highlights why more primary research is needed to explore—the meaning *and* timing—of anatomization *and* dissection—since each medico-legal step still merits closer historical scrutiny.

In the provinces and metropolitan centres where surgeons assisted executioners they were starting to appreciate what the brain might be capable of doing to preserve life. There developed a more widespread recognition that 'dying by degrees' happened to a lot of criminals on the gallows, and thus penal surgeons developed a number of other confirmatory tests for loss of vitality in the brain, including by the early nineteenth-century:

- A pressure test on veins to show no refill with blood
- Ammonia was injected to test for inflammation of the subcutaneous tissue—if no inflammation, then no signs of vital life ammonia
- Blunt needle inserted into the heart. If cardiac systole occurred, its motions would be transmitted to flags on the ends of the needle
- Shiny needles placed in the biceps; rusting over time suggested death
- Diaphanous test: loss of scarlet red colour of the finger edges when held to the light
- Cutting through the intercostal space and feeling the heart with the finger tip
- Ligature placed around the finger: failure of distal part to become red or blue
- External pressure to the eyes causes permanent (versus temporary) distortion of the roundness of the pupillage aperture[56]

Contemporary publications like Dr Philip's treatise on *'The Nature of Death'* inspired penal surgeons to start documenting their more prolonged encounters with criminals cut down from the gallows. On arrival at a dissection venue it became *de regueur* for leading penal surgeons behind closed doors to record the dying process at the anatomical stage in minute detail. Some of the best documented examples survive today in the archives of the Royal College of Surgeons in London, Turning to their missing medical perspective, complicates the role of the medical profession in the penal process and its painful subtext over the long duration from 1752 to 1832.

William Clift worked at Surgeon's Hall on bodies executed across London under the Murder Act. He had a long and distinguished career, assisting the famous anatomist John Hunter and working alongside Sir William Blizard. By the early nineteenth-century Clift's record-keeping was meticulous. It reveals that between 1812 and 1830 there are ten out of thirty-five well-documented cases in which he found that the condemned was still alive after judicial hanging—for he privately wrote—'the heart was still beating after the body was received'.[57] In other words, not less than twenty-six per cent had cheated medical death at the gallows and this in a central London location that claimed medico-legal expertise over the English provincial medical marketplace. Self-evidently this must have been a conservative figure of a potentially much bigger logistical problem before the introduction of the 'new drop'. These discrepancies were often highlighted by popular publications, like the *Gentleman's Magazine* throughout 1787 for instance, which reported that 'criminals have sometimes been found alive after hanging near an hour' under the 'short drop'.[58] By March 1815, Clift and Blizzard thus together left detailed dissection notes of 'William Sawyer's *final death*, hanged for murder at Tyburn'[59]:

> Executed 8 o'clock 15th May 1815 in the morning
> Body was received at 9.20 in the morning at Surgeon's Hall
> 9.40am: Body was opened
> 10.05am: Watched the heart dying
> 10.30am: Right auricle still contracting but not the lungs
> 10.40am: The heart and lungs were placed on a dry towel
> 11.00am: The spontaneous action of the right auricle at about a minute and a half intervals, and then remained at rest[60]

At ten minute intervals observations continued until '11.40am' when a final procedure was enacted to ensure the dangerous body was deceased:

> Heart and Lungs were immersed in heated water to about 100 degrees but no motion whatever was produced by the immersion, nor by any future application of the stimulus [by the anatomist's hand].[61]

Only after these careful anatomical checks did those assembled proceed to dissection and indeed Blizard was very careful about keeping the body intact: 'After the parts had been examined by Sir William Blizard they were again returned to the Thorax'. It was noted that the criminal had been on the eve of execution so frightened of the dying process, being very troubled that he might not expire quickly, that 'Mr Box, the surgeon in Newgate called in [to the dissection venue] to say he thought the prisoner tried to poison himself'. Box admitted that he felt merciful towards the suicidal man. To calm him down he 'gave him a cordial medicine to enable him to undergo his sentence'. That action speaks volumes about changing medical perspectives of the execution process. Scientific understanding of a slow medical death in the brain and a related drugs overdose reflected how '*absolute Death*' was still indeterminate. For this reason Clift and Blizard were thorough, and thus a final entry reads: '29 year old, 5 foot 10 or 11, dark hair and remarkably delicate skin, well-grown man' was '*truly dead*'.

Another case from a similar notebook is worth examining briefly because of its remarkable detail and the wider clinical lessons that can be drawn from it.[62] On Monday 24 January 1814 Martin Hogan was executed at Execution Dock in London (the location for the death penalty for serving military men and those found guilty of homicide on the high seas) for the murder of Lieutenant Johnstone. The body was cut down from the gallows and sent to Surgeon's Hall within two hours. This was a fast delivery-service by the blackguards who were paid to transport the criminal corpse from the docks near the shoreline at Wapping, north of the Thames, along the river to where the Old Bailey stands today, across the road from St. Bartholomew's Hospital in the City of London. The dissection notes stated that Sir William Blizard was prompt: 'arrived a few minutes afterward and superintended the dissections.' First however he performed the anatomical checks required by the penal code termed 'anatomization'. This he did by carrying out a number of remarkably detailed and precise confirmatory tests for medical death. To make these engaging for the lay reader the original quotations are cited and in brackets there is a clinical explanation about what he said he was doing, and why, to set medical death protocols in context at the time:

> A needle was immediately introduced through the coats of each eye to the Iris, but no visible effect was produced [checking for brain stem function first and then for signs of waggling in the eyes which confirms blood circulation has stopped].

> On opening the cavity of the Thorax some of the adhesions were observed between the lungs and pleura but no fluid whatsoever [then checks cardio-respiratory failure]

> In the cavity of the pericardium was contained about three drams [it was a winter hanging and he is concerned about extreme hypothermia so he checks the thin double-layered sac which encloses the heart][63]

> No motion was perceivable in either ventricle [of the heart], or the auricle to the left side; but in that of the right, very violent contractions took place, beginning at the apex of the auricle and proceeding regularly towards its base [there were vital signs shutting down but not as yet medical death][64]

The notes explained that from 12 noon when the body came into the dissection room 'the contractions became gradually weaker till about half past 2 o'clock when they entirely ceased'. And this on a criminal corpse that hung for an hour on the gallows—then was transported across London—and for the next two hours still was not in a state of '*absolute* Death'—altogether four hours post-execution. Blizard was sensitive to this medical possibility, and it is noteworthy that by 1814/5 he checked brain function first and then cardio-respiratory failure. Those new procedures reflected how medical understanding of oxygen deprivation and brain vitality were in transition. Thus he decided to carefully double-check the blood flow from the weakening contractions to test their physical function. The notes explained that: 'In the early part of the examination it was necessary to stimulate the auricle; in doing this, a small puncture was made into its cavity by which a considerable quantity of blood issued …made it necessary to compress the heart from time to time, and make the auricle turgid in order to observe the contractions'. Blizard was inquiring, fascinated, patient, meticulous and methodical because medical death had complex organic processes. He learned from a '*living anatomy*' experience in which he also side-stepped human vivisection by performing an act of euthanasia. It is this multi-faceted medical choreography that made the anatomical checking-mechanism of penal punishment termed 'anatomization' distinctive from 'dissection'. For reasons of precision, the lancet was the instrument of medical death and then the implement of post-mortem punishment, in compliance with the sentiment of the Murder Act.

Against this backdrop, disputes and debates emerged about what to do about the medical predicaments of the Murder Act. Surgeons were duty-bound by the legislation to keep secret the conundrum of medical death. They were also being encouraged at the time to take a corporate view of their ethical predicament. The College of Surgeons had reconstituted in 1745, finally distancing itself from the barbers. There was little appetite for stimulating any unwelcome public debate about the capital legislation. Any unsavoury revelations threatened to undermine recent professional gains. The sensible recourse was to focus on the offence of grave-robbing and the perennial problem of low-body supply. These, on balance, were more strategic issues that had the sympathy and support of government. This did not mean that the problem of medical death simply evaporated, quite the reverse. Delving deeper in the archives uncovers important differences between medical actions and attitudes in the capital, compared to those in the provinces. In the metropolis the priority was good public relations, whereas those working in the regions were more prepared to discuss their professional dilemmas in public, often to the consternation of their London counterparts. In many respects an up and coming surgeon in the provinces who did not seek an establishment career in the capital had less to lose and everything financially to gain by being honest and building trust with his host community. The general historical neglect of regional studies of provincial penal surgeons who worked under the Murder Act has therefore resulted in the problem of medical death being neglected. In one representative example 'dying by degrees' was deliberated, disguised and disputed in the Midlands, opening up controversies to public scrutiny by the 1830s.

In the *Leicestershire Chronicle* of 28 July 1832, on the eve of the Anatomy Act, a local surgeon wrote an anonymous letter to the editor in which he detailed how the 'Surgical Profession Hitherto was Injured by the Legislature—Dissection of Murderers'. The extract supports the new findings that form the central argument of this chapter namely that: 'As the murderer's body may be legally dissected, anatomy may still be cultivated, though most imperfectly.' The surgeon went on to explain what he meant by 'imperfectly':

> By making dissection of the murderer's body a condition of the sentence, it makes the surgeon the finisher of the law, and thus stamps upon the source of anatomical knowledge ... They give not over to the surgeon the body of the murderer for the promotion of anatomy, but for completion of the penalty which the crime of murder is doomed to pay. The hangman executes a portion of the sentence which claims the forfeit of the murderer's life [post-execution] ... Were the scalpel of the anatomist employed by the halter of the hangman, some such effect [medical death] might be derived from it...What has anatomy to do with murder or the anatomists to do with the murderer? They are not *participles criminis* [participants in the crime], yet they participate in the punishment. It is the business of anatomy to save life, and the object of anatomy is to teach how to save it; and yet not a murder is committed, or a murderer executed, for which and with whom the surgeon and his science are not severely put upon.[65]

There are two ways to read this source—a traditional one common in the historical literature—versus one that looks in context through anatomical eyes.

On first reading this statement could be interpreted as a straightforward compliant by the surgeon that anatomy and dissection are tarnished by the execution process. Indeed many cultural historians like Jonathan Sawday take the view that: 'As the recipient of the bodies from the execution scaffold, the anatomists was fully implicated in the ambivalent rites of judicial power ...nevertheless working at *only one remove* [author's emphasis] from the grim business of public execution'.[66] If however we factor in the ambiguities of medical death, then there is also an important contemporary argument being made here that is often misconstrued in modern scholarship. On closer reading the extract could be interpreted as complaining that the surgeon is not one step removed from the penal sentence but actually part and parcel of the execution itself because of the problem of medical death. To check if this reading is salient it is necessary through close textual analysis to examine what is being said from a witting and unwitting perspective.

The Leicestershire surgeon begins with a statement of material fact and an ethical concern: 'By making dissection of the murderer's body a condition of the sentence it makes the surgeon the finisher of the law, and thus stamps upon the source of anatomical knowledge'. In this opening sentence it is noteworthy that anatomy and dissection are distinctive parts of the penal process. A first glance, legally they seem to denote post-mortem punishment ('the finisher of the law') but there is also something distinctive about a typical execution process that sullies anatomy ('stamps upon it'). Read from the vantage point of what we have learned from other source material in this chapter about troublesome executions at the scaffold, the unwitting testimony is that anatomy is a separate phase of the penal process and by being so the anatomical sciences are blemished in some way that contemporaries appreciated which modern scholarship has seldom engaged with. The writer then elaborates: 'They give not over to the surgeon the body of the murderer for the promotion of anatomy, but for completion of the penalty which the crime of murder is doomed to pay.' This is a very complex sentence to unpack because it is complicated by the way that anatomy developed as a discipline in the eighteenth-century and briefly looking back to its terms sets the extract properly in context.

At a time when an anatomist got hold of a criminal body he had three ways he could study anatomy—'*speculative anatomy*' (thinking about and studying the body from a theoretical perspective but being hands-off)— '*living-anatomy*' (examining the life-force and living structures of the body by means of an autopsy on the major organ systems in the chest cavity)— and '*morbid anatomy*' (examining dead tissue specimens from dissection preparations before cutting the corpse to its extremities). This was further complicated by an accepted idea in the eighteenth-century that the corpse 'even at the very deepest stages of dissection should be represented as still alive'.[67] An unwitting assumption in the above extract is then that certain types of anatomy were not promoted during the penal process. Seldom was '*speculative anatomy*' a priority at an execution. There was the possibility of exploring '*living anatomy*' because the body was dying not dead. Yet in practical terms the surgeon's priority was to work on the body to get it into a state of '*morbid anatomy*' in order to complete the capital sentence. *Anatomization* was thus primarily about getting the body to become a corpse.

That context helps us to critically analyse the next sentence of the extract which stated that: 'The hangman executes a portion of the sentence which claims the forfeit of the murderer's life'. The executioner does not 'hang the person *until dead*'—he can only guarantee to do a '*portion* of the sentence'—to place the body in a position of legal death which means it is medically *dying*—and this we know from evidence presented in this chapter is one third of the entire capital sentence at best. Notice also the key verb in this sentence—'*claims*'—the executioner cannot guarantee—he can make claim to—but that is not the same as saying he ensures that he can forfeit the murderer's life. Again this is then elaborated: 'Were the scalpel of the anatomist employed by the halter of the hangman, some such effect [medical death] might be derived from it'. In other words, in the surgeon's view, the hangman would need to have the medical man at the gallows rope to guarantee medical death and when not anatomy is 'imperfectly' done because it might involve a mercy-killing. Anatomy should be about saving life, but the anatomist's moral position is the opposite. The logic of that argument is that if he is taking life, then technically the person is not in medical death.

This Midlands surgeon's attitude was not in fact that uncommon. It matched for example what many in the London press had been saying since the 1770s. The *Public Advertiser* on 5 November 1774 for example reflected on medical opinions triggered by famous revivals in the 1760s. It quoted the 'late Dr Barrowby' an experienced barber surgeon who expressed the view: 'That he was less surprised at the Revival of one executed Person, than that many others did not Revive, as he did not think the common Method of Execution was certain Death'.[68] This sets in context why when 'John Howe, alias John Duff' was hanged for murder in May 1788 after the execution was bungled there was a lot of newspaper interest in the biological effects of hanging. John was described as, 'an unhappy man...between 27 and 30...who made a most miserable appearance, being very dirtily dressed, in an old hat and ragged brown great coat' soiled by dirt and disease, 'with a long beard'.[69] On the gallows 'the knot by some accident slipped round to the back of his neck...the consequence of which was, he continued in a kind of convulsive motion for about three minutes'. On being laid out at Surgeon's Hall the medical men acted with precaution, carefully inspecting 'the whole face, the ears, and the neck [which] had a general jet-black appearance, never remembered to have been seen in any subject brought there from the gallows'. Here was precisely the sort of fit young man tainted by poverty that presented a logistical challenge to the executioner and penal surgeons. Only their collective memory could be relied upon when checking for medical death.

Evidence like this points to the need for more detailed medical research to move beyond historical clichés that mislead researchers today into thinking that somehow executions and dissections were seamless punishments; they were not. As one concerned legal correspondent in the letter page of the *Public Advertiser* of the 29 March 1769, exhorted: 'the business of Surgeon's Hall is not to revive and frustrate but to complete the Execution of the Sentence in Cases of Murder'.[70] He said this because the medical stipulations of the Murder Act had little substance until the law was actually applied when the body was cut down and opened up. The legislation's primary purpose was physical retribution, but its wording also reflected deep ambivalence and widespread public anxieties about precisely how to establish medical death and the complicit ethical obligations that followed on. As another letter writer to the *Morning Post* on 2 September 1770 queried: 'Whether a person condemned by the laws of this country, and carried to the place of execution, and is hanged in the usual manner, if that person comes to life, and is seen at large, is not liable to be taken into custody and hanged again, or has suffered the letter of the law by hanging an hour?'[71] In even asking and contemplating that someone could revive enough to walk away, there must have been anxiety around at the time that historians have overlooked.

Taken together then, witting and unwitting testimony, points to a contemporary sense that the death sentence had to have a medico-legal choreography—one third at the gallows—two-thirds beyond the hanging-tree. The question of medical complicity, the degree of moral culpability, and the inherent conflicting emotions there were for designated surgeons were evidently at the heart of some penal experiences. As with all sources each can be read in several ways and there are issues of representativeness that need to be carefully investigated with further research. Nevertheless close textual reading explores the possibility that such extracts can be read anew to facilitate a reassessment of the actual medico-legal workings of the Murder Act. They start to identify with greater medico-scientific precision the penal actions of everyone who actually handled the criminal body on its punishment journey once it had exited the courtroom.

Conclusion: '*I have been* HANGED *and am* ALIVE'[72]

To be 'hung by the neck *until* dead' on the scaffold was never legally guaranteed under the Murder Act. It could not be for the practical reason that death was a medical mystery in early modern England. What happened to the criminal body was that it did not become a corpse until its metabolic processes stopped functioning in the heart-lungs-brain. This biological shut-down then created a cultural sense that there was 'the deep time of the dead'. In the process of which contemporaries appreciated that 'becoming *really dead*...takes time'.[73] Even so, the physical logistics of such a basic medical dilemma at the heart of a capital punishment system ought not to have been neglected by historians of crime.[74] Retrospective medical diagnosis is an unsatisfactory historical tool. It has been necessary therefore to relocate contemporary medical case histories of the condemned in a dying state to investigate continuities and changes in procedures for establishing medical death. For too long these have been neglected in punishment histories that claim to know the theory but neglect to engage with an empirical picture of contemporary medical problems. This chapter has thus set out the early modern terms of medical reference and the working definitions that defined death, dying and vitality for England. The balance of the evidence suggests that those who witnessed an execution saw not a half, but just one third, of the penal process at the eighteenth and early nineteenth-century gallows. There was social and legal death (being condemned and hanged), medical death (being anatomized) and post-mortem punishment (being dissected). It follows that if a missing two-thirds of the punishment rituals came after the scaffold, then there must have been contemporary reasons for this, and their medical choreography is a crucial and neglected aspect of histories of crime and punishment covering the period 1752 to 1832.

Detailed dissection cases recorded at Surgeon's Hall reveal that a considerable number of those who were executed died in the dissection venue, not on the gallows. In London the newly constituted College of Surgeons had a policy of only circulating such sensitive findings within their professional circles. They preferred not to discuss its details publicly because they were self-evidently already tarnished by being connected to penal punishment and dissection. The problem of medical death threatened to intensify cultural revulsion at a pivotal time of early professionalization. Alternative arguments in favour of widening dissection to say patients dying in hospital might moreover be undermined if heightened cultural sensitivities were exacerbated by any discussion of the long-confusion surrounding medical death; something the *Westminster Journal* called for in its edition of 20 December 1746.[75] Against this backdrop it is worth reiterating that of the thirty-six detailed dissections recorded by Williams Blizard and Clift in charge of anatomical procedures on behalf of Surgeon's Hall between 1812 and 1830, there are ten well-documented cases in which the condemned was still alive after judicial hanging.[76] They claimed a level of expertise supposed to be beyond the capacity of provincial surgeons, which implies that this is a very conservative figure of a potentially much greater logistical problem, especially before the introduction of the 'new drop'. Their caseload profile was typically a robust adult (male and female) of average health aged between twenty-nine and sixty with the majority being in their forties at the time of execution, and hanged in, either January (very cold weather), or May (in hotter years) or September (unseasonal Indian summers). Warm and cold weather patterns, coupled with noteworthy physical characteristics like 'a strong neck', and all having been cut down 'within the hour', were common observations in the case notes. Clift also carefully noted the physical strength of the condemned. Most were rough, big men or women, bull-necked, imposing characters, with a strong will to survive. Many were navies, agricultural labourers, men who were blacksmiths, or did a labour intensive trade—and these robust body types appeared to have a higher pain threshold, or quickly stopped jerking on the rope (probably indicating a physical shut down in the body to protect the brain thereby increasing the chances of survival). Overall, in 27.7 per cent (10 of N=36) of Clift and Blizard's anatomical cases before dissection it was 'observed the contractions of the heart', 'the observable vital signs', and when handled the heart 'produced a stimulus': all suggestive of cardiac action with the potential to undergo manual resuscitation provided the blood could be oxygenated enough. That many did die on the gallows should not lessen historical interest in those that did not.

These new findings also set in context a much misunderstood report in 1819 of *The Committee for Investigating the Criminal Law* overseen by central government.[77] In an appendix to these *British Parliamentary Papers* a breakdown of hanging offences committed and their respective death-sentences was recorded.[78] Compiling these and doing so with data from the Sherriff's cravings held at the National Archives makes it feasible to arrive at a conservative estimate of around 5 per cent of convicted cases being sentenced to post-mortem punishment for homicide between 1752 and 1832 in England and Wales.[79] If that contemporary figure is accurate, it begs the uncomfortable question what physically happened to the remaining ninety-five per cent who committed lesser capital offences and were just sent to the gallows? The vast majority were hanged but were not sentenced to undergo 'dissection and anatomization'. Some would have expired on the gallows from fright or shock within the hour. But others could have taken much longer to reach medical death. At present, we have cannot know the precise numbers involved; nonetheless this historical gap is significant given that in the case of those that took longer to expire there was no lancet to commit a mercy-killing. In a lot of contemporary accounts friends and relatives from the crowd assisted the hangman by pulling on the body to help break the neck on the gallows or prayed for the long-drop to work efficiently. Seen afresh in the context of a prolonged medical death those actions are not just explicable, but unequivocally merciful. Hence, the early modern crowd waited, watched, and witnessed the condemned body closing down in preparation for dissection in ways often misconstrued. In the next chapter we encounter just what a bad shape the criminal was in as ordinary people crowded round the co-called corpse as it reached its post-mortem destination.

Notes

1. V. A. C. Gatrell, (1996 edition), *The Hanging Tree: Execution and the English People, 1770–1868*, (Oxford: Oxford University Press), pp. 45–89.
2. P. Linebaugh, (1975 edition), 'The Tyburn Riot Against the Surgeons' in Douglas Hay, Peter Linebaugh, John G. Rule, E. P. Thompson and Cal Winslow eds., *Albion's Fatal Tree: Crime and Society in Eighteenth-Century England* (London: Verso), pp. 65–117, cited pp. 85, 102–3.
3. Ibid., pp. 102–3.
4. Thomas Laqueur (2011), 'The Deep Time of the Dead', *Social Research*, Vol. 78, Fall, III, 799–820, quote at p. 802.
5. A. Cunningham (2010), *The Anatomist Anatomis'd: An Experimental Discipline in Enlightenment Europe* (Aldershot, Hampshire: Ashgate), preface p. xxi, points to a 'seismic' shift in anatomical practices by the last decade of the eighteenth-century.
6. J. Innes (2009), *Inferior Politics: Social Problems and Social Policies in Eighteenth Century Britain*, (Oxford: Oxford University Press), p. 105.
7. Refer, J. M. Beattie (1986), *Crime and the Courts in England, 1660–1800* (Oxford: Oxford University Press), pp. 520–30; Hugh Amory (1971), 'Henry Fielding and the Criminal Legislation of 1751–2,' *Philological*

Quarterly, Vol. 50, 175–92; Richard Connors (2002), 'Parliament and Poverty in Mid–Eighteenth-Century England,' *Parliamentary History*, Vol. 21, 207–31.

8. *Anatomization* has tended to be given a modern spin in standard historical accounts, rather than an early modern focus within a working context of the criminal law of the Murder Act.

9. This is often the case in general cultural histories of the body, like, Helen MacDonald (2005), *Human Remains: Dissection and its Histories* (New Haven: Yale University Press).

10. The term 'euthanasia' is being used here to describe a 'mercy-killing where medical death is an unnatural event in a person's life', see, Ian Dowbiggin (2007), *A Concise History of Euthanasia: Death, God and Medicine*, (New York: Rowman & Littlefield).

11. Roy Porter (2003), *Flesh in the Age of Reason: How the Enlightenment Transformed the Way We See Our Souls and Bodies* (London: W. W. Norton & Co), p. 223.

12. This contemporary eighteenth-century phrase is often repeated in modern science books on the controversial subject of medical death, see, Dick Teresi (2012), *The Undead: Organ Harvesting, the Ice-Water Test, Beating Heart Cadavers—How Medicine is Blurring the line between Life and Death* (New York and London: Vintage).

13. See, Michel Foucault, (French edition, 1979) translated to English by Sheridan, A. (1995), *Discipline and Punish: The Birth of the Prison* (New York: Vintage Books); Colin Jones and Roy Porter eds (1994), *Reassessing Foucault: Power, Medicine, and the Body* (London: Routledge); Jonathan Sawday (1995), *The Body Emblazoned: Dissection and the Human Body in Renaissance Culture* (New York and London: Routledge); Roy Porter (2003), *Flesh in the Age of Reason: How the Enlightenment Transformed the Way We See Our Souls and Bodies* (London: W. W. Norton and Co).

14. Leslie M. Whetstine (2008), 'The History of the Definition(s) of Death: From the Eighteenth-Century to the Twentieth-Century', in David W. Crippen ed., *End-of-Life Communication in the ICU*, (New York: Springer-Verlag), pp. 65–78, quote at p. 65.

15. David J. Powner, Brice M. Ackerman, and Ake Grenvik (1996), 'Medical Diagnosis of Death: Historical Contributions to Recent Controversies', *Lancet*, Nov. issue, panel 2, p. 1220.

16. See, Linebaugh, "The Tyburn Riot", pp. 65–117.

17. 'Report and Trial of Channel and Chalcraft', *The Observer*, 16 August 1818, pp. 1–14, quotes at p. 10.

18. Powner, Ackerman, and Grenvik, 'Medical Diagnosis of Death', p. 1221.

19. *Morning Advertiser*, 29 March 1745; *London Evening Post*, 11 April 1745. I am grateful to Professor Peter King for alerting me to this case, also cited

in Rhiannon Elizabeth Markless (2012), 'Gender, Crime and Discretion in Yorkshire, 1735–1775' (unpublished, MA Dissertation, Department of the Humanities, Roehampton University), p. 146.

20. Andrew Knapp and William Baldwin eds (1828) *The Newgate Calendar, 1824–8* (London: J Robin and Co), 'Case of William Duell, Executed for Murder, Who Came to Life Again While Preparing for Dissection in Surgeon's Hall' e-transcript available on http://www.bl.uk/learning/images/21cc/crime/transcript1595.html.

21. Sir Richard Hoare Esq (1815) *Journal of the Shrivealty in the Years 1740–1*, (Bath: Privately Printed), pp. 37–8.

22. See evidence from contemporary newspapers reproduced in *Notes and Queries*, 1 March 1854 (No 226), p. 174; 13 March, 1854 (No 230), pp. 453–4; 25 March 1854 (No 231), pp. 281; 25 June 1856 (Number 25), p. 490; 26 July 1868 (No 30), p. 73; Refer, also, *London Magazine*, 'The Monthly Chronologer: Dewell [Duell]' (1740), pp. 560–1. Magistrates retained the legal duty of checking on medical death at the gallows even after the 'new drop'. Thus the *Public Advertiser* 20 January 1792 reported that magistrates in Scotland permitted the condemned to 'be applied for by some Surgeons, in order that the usual experiment might be made upon his body for the restoring him to life'. After he was laid out to check his medical death 'a noise was heard' and the magistrates sent for. It turned out to be a dog barking 'who had been locked up in the Hall with the corpse'!

23. See, Jussi Hanska (2001), 'The Hanging of William Cragh: Anatomy of a Miracle', *Journal of Medieval History*, Vol. 27, 121–38, quote at p. 131.

24. Porter, *Flesh in the Age of Reason*, p 223.

25. Roy Porter (1992), 'Medical Journalism in Britain to 1800' in W. F. Bynum, Stephen Lock and Roy Porter eds. *Medical Journals and Medical Knowledge: Historical Essays* (London and New York: Routledge), pp. 6–29. q. at p. 9.

26. Ibid., p. 15.

27. Henry Thomson (1827), "Le Revenant", *Blackwood's Magazine*, Vol. 27, 409–16, quotes at p. 416.

28. See, Leicestershire Record Office, [hereafter LRO], DE3182/1, *The Diary of Thomas Kirkland Surgeon of Ashby and family, c. 1731–1931*, entry 25 January 1796; Kenneth Hillier (1984), *The Book of Ashby–de-la-Zouche* (Buckingham: Barracuda Books Ltd), p. 123. Kirkland would later be an expert witness as the infamous case of Earl Ferrers.

29. Cunningham, *The Anatomist Anatomized*, pp. 6–7 explains this standard sequence was derived from work by the leading Bolognese anatomist Marcello Malphigi—'first the belly, then the chest, then the skull'.

30. Inclusion/exclusion and its practical management are elaborated in Section 2, below.

31. On Aristotle's anatomical views refer, Andrew Cunningham (1997), *The Anatomical Renaissance: The Resurrection of the Anatomical Projects of the Ancients*, (Aldershot, Hampshire: Scolar Press), p. 16.

32. Porter, *Flesh in the Age of Reason*, quotes at p. 222 and p. 216.

33. Ibid., quotes at p. 225 and p. 226.

34. Dr. A. P. W. Philip (1834), 'On the Nature of Death', *Philosophical Transactions of the Royal Society of London*, Vol. 123, 167–198, quote at p. 168. He was a leading doctor, anatomist, and Fellow of both the Royal Society and Royal Society of Literature. See also, Dr. A. P. W. Philip (1831), 'On the Sources and Nature of Powers on which the Circulation of the Blood depends', *Proceedings of the Royal Society of London*, Vol, 3, p. 64.

35. Philip, 'On the Nature of Death', p. 170.

36. Ibid., p. 181.

37. Philip, 'On the Nature of Death', p. 170.

38. Ibid., p. 192.

39. Roger French (1999), *Dissection and Vivisection in the European Renaissance*, (Aldershot, Hampshire: Ashgate), quotes at p. 218, p. 219.

40. Ibid., p. 259 on mechanistic theories, and quote at p. 269.

41. Andrew Cunningham, *Anatomical Renaissance*, p. 128.

42. See, endnote 39 above.

43. Royal College of Surgeons [hereafter RCS], William Clift Collection, Box 67.b.13, Elizabeth Ross case notes and sketch, 1832, anatomized and dissected.

44. These sketches were generally for private consumption by medical men and drawn by artists in the dissection room directed by the Master of Anatomy at the RCS. They are often reproduced in standard histories without checking their provenance and medical circumstances. Illustration, 2.1 ©Royal College of Surgeons Library, William Clift Collection, Box 67b.13, '*Sketch of Elizabeth Ross*' and associated dissection notes on criminal corpses; creative commons license authorised for academic use.

45. Giovanni Aldini was Professor of Physic at Bologna University. His famous experiment is often cited in histories of dissection, like MacDonald, *Human Remains*, pp. 14–17.

46. Philip, 'On the Nature of Death', p. 170.

47. Ibid., pp. 170–2.

48. Teresa E. Hills (2010), 'Determining Brain Death: A Review of Evidence-Based Guidelines' *Nursing* Volume 40, Vol. 12, Dec. Issue, 34–40, quote at p. 37.

49. Ibid.

50. Hills 'Determining Brain Death', p. 37.
51. See for example, D. J. Power (1987), 'The Diagnosis of Brain Death in Adult Patients', *Journal of Intensive Care Medicine*, Vol. 2, 519–25; R. D. Truog & J. C. Fackler (1992), 'Rethinking Brain Death', *Critical Care Medicine*, Vol. 20, 1705–13; J. P. Lizza (1993), 'Persons and Death: What's Metaphysically Wrong with our Statutory Definition of Death?', *Journal of Medical Philosophy*, Vol. 18, 351–74.
52. See, http://www.sca-aware.org/sudden-cardiac-arrest-treatment: 'When a patient undergoes therapeutic hypothermia, it is somewhat startling to feel how cold s/he can be to the touch. This is normal and only temporary. The patient's temperature will be reduced to about 91°F (33°C), approximately 7°F (4°C) lower than normal. The patient will be on a respirator, heavily sedated and unable to move. The therapy typically will last for a maximum of 36 hours: 12 to 24 hours of cooling and up to 12 hours to rewarm slowly back to a normal body temperature of 98.6° (37°C). During the cooling process, the patient will require frequent blood samples to make sure s/he is tolerating the cooling procedure well. Sometimes, additional medications can be given to help control blood pressure and heart rate... you will not know anything about his/her level of consciousness until the therapy is completed'.
53. LRO/13064/13, Leicester Infirmary Minute Books, 23 June 1814–30 June 1819, entry dated 17 November 1815.
54. Letter by an English anatomist and dissector who moved to Maine in America, Editor (1807), *The London Medical and Physical Journal*, Vol. 108, XVII, 155–7, q. at p. 155.
55. Ibid, p. 156.
56. Powner Ackerman, and Grenvik (1996), 'Medical Diagnosis of Death', *Lancet*, Nov., panel 3, p. 1221.
57. The diary of William Clift was first examined by Jesse Dobson, Recorder of the Royal College of Surgeons in the 1950s and her preliminary archive findings were published in, 'Cardiac Action After "Death" by Hanging', *Lancet*, 29 Dec. 1951, 1222–4, quote at p. 1223. I am grateful to Wendy Moore for emailing me about this overlooked secondary source in response to a short article this author published in the *Lancet* in July 2013.
58. *Gentleman's Magazine* (1787), Volume 57, p. 33, letters to the editor, and *Gentleman's Magazine* (1787), Volume 62, p. 673, on 'the escape of criminals from death, after hanging'.
59. Author's emphasis since the precise clinical phrasing is noteworthy.
60. RCS, William Clift Collection, MS0007/1/6/1/3, 1807–32, pp. 6–8.
61. Ibid.
62. RCS, William Clift Collection, MS0007/1/6/1/1, 1800–14, 24 January 1814, pp. 1–2.

63. Ibid; The fluid is contained within the layers and lubricates the constantly rubbing surfaces around the heart's double-layer. The anatomist can also pass a finger posterior to the aorta and pulmonary trunk (arterial end of heart) and anterior to the left atrium and superior vena cava (venous end of heart). This passage is termed the transverse sinus of the pericardium. The pericardium is supplied by several arteries that are in the area (like the internal thoracic), and by the phrenic nerves, which contain vasomotor and sensory fibres. Pericardial pain is felt diffusely behind the sternum but may radiate and was seen as a sign of vitality.

64. RCS, William Clift Collection, MS0007/1/6/1/1, 'Records of the Murderers delivered to the College for Dissection' written inside the front cover of the book. These records have been used in general histories like that of MacDonald, *Human Remains,* pp. 13–27. Their importance regarding the timing of medical death and the precise meaning of the practicalities of the Murder Act has been neglected.

65. Letters to the Editor (1832), 'The Surgical Profession Hitherto Injured by the Legislature—Dissection of Murderers', *Leicestershire Chronicle,* Saturday 28 July, Issue 1140, medical correspondent styled 'PHILANTHROP', pp. 1–5, quotes at p. 3 and 5.

66. Sawday, *Anatomy Emblazoned,* p. 54.

67. Ibid., p. 112.

68. *Public Advertiser,* 5 November, 1774 (A), column 3.

69. *Morning Chronicle,* 5 May 1778, column 2.

70. *Public Advertiser,* 29 March 1769, correspondent styled 'Gracchus' complaining about Jack Ketch; see also SJC, 30 March 1769.

71. 'To the Editor of the Morning Post', *Morning Post,* 2 September 1770, column 2.

72. Thomson, "Le Revenant", p. 409.

73. Laqueur, 'Deep Time of the Dead', p. 802.

74. Ibid., p. 802.

75. *Westminster Journal,* 20 December 1746 (a), column three discussed this at length.

76. See, RSC, William Clift Collection, endnote 57 above.

77. On the workings of the Committee see, Jeremy Horder (2012), *Homicide and the Politics of Law Reform,* (Oxford: Oxford University Press), pp. 7–11.

78. Refer, BPP, *Hansard,* House of Commons Sitting (1819), HC debate, 6 July 1819, Vol. 40, cc 1518–36, Sir James Mackintosh said that the murder rate was 67 murders and 57 executed 1755–1784, 54 murders and 44 executed 1874–1814, with 9 murders per annum in London.

79. These estimated figures have been compiled from BPP (1819), *The Committee for Investigating the Criminal Law* and TNA *Sheriff's Cravings*

(see bibliography). I am grateful to Professor Peter King and Dr Richard Ward for assisting me by calculating that a total of '735 offenders were executed in England, Wales and Scotland in the ten years between 1776 and 1785 for the capital offences covered by the Murder Act 1752'. The project team are very grateful to Professor Simon Devereaux for providing data for the London court circuit.

CHAPTER 3

In Bad Shape: Sensing the Criminal Corpse

The recreation of the gallows scene in the Georgian era by leading crime historians has been deservedly praised for its authentic contemporary picture of eighteenth-century executions.[1] The stimulus for doing this original research was, Vic Gatrell remarked, that 'death by execution may be one of the last agreed obscenities' in modern times but that 'taboo has not served history well'.[2] It was a material reality that 'what the noose really did to people' challenged eighteenth-century sensibilities. Remaining shielded from this physical reality is hence not an option if this book is to retrace all the anatomical things that could potentially happen under the Murder Act. Gatrell's work has now established that on the hanging-tree 'the dead body never failed to betray the nature of its experience however scientifically despatched'.[3] Penal surgeons generally observed four physical traits. It was obvious that it took time to die, and this varied a lot. On arrival at the gallows the typical behaviour of the condemned was described as a combination of convulsive shaking and audible distress, with many paralysed by fear, often fainting. Then, as the convicted murderer climbed the scaffold steps, or were hoisted up, the panic they experienced started to complicate the executioner's course of action. Few prisoners could avoid producing a sharp adrenalin rise, sweating profusely, making it more difficult for the hangman to handle condemned bodies. Historically a hormonal trigger known as a classic 'fight or flight mechanism' stimulated some criminals to try to resist execution, others felt compelled to

© The Author(s) 2016
E.T. Hurren, *Dissecting the Criminal Corpse*, Palgrave
Historical Studies in the Criminal Corpse and its Afterlife,
DOI 10.1057/978-1-137-58249-2_3

escape, and the remainder were rigid with trepidation.[4] Some variant of this involuntary hyper-arousal mechanism was thus inescapable. Normal biological continuities were repeatedly experienced. It followed that a third observation that was commonplace was the awful sense of anticipation caused by an acute stress response. Feeling scared stiff tended to be mentally as terrifying as the physical pain of punishment. In many cases, finally, foul-smelling excrement expelled down the lower limbs as the stimuli of the autonomic nervous system were activated.[5] Reflexive actions like urine incontinence were routine. A lot of condemned bodies were accordingly in a 'bad shape' in the supply chain by the time they reached a criminal dissection location.

When *peri-mortem* (*at, or near the point of death*) the hangman sometimes had to use rougher killing methods to despatch strong-necked murderers. If so, the so-called corpse handed over to the surgeons tended to be more damaged, complicating the working duties of medico-legal officials. Hence, this chapter is all about the sensory nature of execution spectatorship when in a 'bad shape' and 'dying by degrees' (a medical concept introduced in Chapter 2). Everyone present played some part in a stimulating synaesthesia since the physical showcase assaulted the five senses. A nauseous bodily drama, and the immersive theatre of its punishment tragedy, refashioned the 'curious' personality of the early modern crowd too. This context explains why Paul Barber in his studies of eighteenth-century medicine and society has found that the early modern body between life and death has always been visually arresting: 'It can stiffen and relax, bleed at the mouth and nose, grow, shrivel, change colour with dazzling versatility, shed its skin and nails, appear to grow a beard, and even burst open'.[6] All of these physical traits were exacerbated by the execution process. They equally exaggerated what the condemned looked like to the duty penal surgeon post-execution. Yet, seldom have the punishments done to the executed body been considered in-tandem. This chapter is therefore going to retrace their physical relationship at their metabolic connection, known forensically as *peri-mortem*. It analyses the strong symbiosis that there was between the material condition of the body executioners knew and those anatomists actually handled.

If the 'bad body' had been treated clumsily, was diseased by contagious smallpox, or very dirty from being covered in body-lice spreading typhus, then it was unlikely that it would spend a lot of time in a chosen dissection space. This was for the simple reason that it might contaminate the community. The 'good body' by contrast was usually despatched efficiently, did not have dysentery of the bowels, or contain 'bad blood' to taint the penal surgeon on duty. Those that nicked their hands with the lancet were aware of the risk of fatal blood-poisoning (see, also Chapter 6). In many respects, then, the precarious nature of doing general dissection work, which is documented in broad terms by the historical literature, nevertheless remains poorly understood as a penal set of hazards.[7] More needs to be known in crime studies with regard to how the condemned taken from the gallows when *peri-mortem* actually affected the penal circumstances of criminal dissections. Hence, the medical triage of gallows bodies was not done to save lives on the hanging-tree, but to protect surgeons and the crowd from the condemned in a 'bad shape' awaiting post-mortem 'harm' in early modern England under the Murder Act.[8]

Section 1 of this chapter begins by exploring the basic medical problems that the execution process created for penal surgeons and the opportunity costs they presented in terms of a physical set of punishment constraints. Of relevance is: the efficiency (or not) of the short/long drop; the calculation (or lack of it) of body weight to rope length; the careful selection (or clumsy choice) of knot and where to place it around the neck; as well as, the regional customs that endured which were not necessarily medically proficient at the provincial hanging-tree. Above all, it will be shown that historians of crime and medicine need to re-examine what happened physically to the condemned body at two critical metabolic moments— just before—and just after—it stopped shaking on the rope. In this transitional *peri-mortem* phase of punishment certain metabolic processes were commonplace because everyone dies in the same way when put through a traumatic death by a broken neck or strangulation with a rope. These physical characteristics are closely examined since their medical attributes can be identified with some precision despite their historical distance. On arrival at dissection venues, observations were made about what normal symptoms looked like when someone was in capital death (as we saw in Chapter 2). Contemporary accounts of death and dying likewise featured in local newspapers and were retold by ballads-sellers. These sources, used sensitively by carefully sifting sensationalised cases sold to a popular market, can enable researchers to side-step the historical snare of retrospective diagnosis. The external and internal metabolic drama always betrayed itself on the blackened skin, swollen eyes, bruised face, or blood-patched hands, feet, limbs, and so on. Contemporaries thus came to recognise the liminal medical processes of capital death. Accumulating a material knowledge enabled most to learn something about better ways to die on the gallows than others. There was a dreadful, painful side to the procedures when the criminal fought their fate and prolonged the agony. For spectators and perpetrators alike, it was a physical showcase of slippages—in blood, sweat and tears, as well as urine, excreta and semen—even before the lancet was lifted. All of this medical purging recast the condition of the criminal as it became a corpse for public consumption.

In follows that in Sections 2 and 3, this chapter examines several prominent trial narratives and their medical circumstances to illustrate what happened in hanging-towns that dealt with headline-making homicide cases. In the first case we encounter what it meant to kill someone and then be convicted to die by an equivalent set of painful penalties. In the second trial two dangerous men that murdered callously became the public property of penal surgeons, featuring prominently in contemporary publications. These, and associated record linkage work, together reconstruct what contemporaries had an opportunity to see from a medical standpoint once the condemned had been hanged and was about to be handed on. Such scenarios show-case what became known as the '*dangerous dead*'; on delivery to the surgeons these were simultaneously ordered/disordered, attractive/repellent, and mundane/unnerving. In some areas of the country we have no more than a hint of the fusion of these binary ambiguities, whereas in other places they are well-documented.

A rich seam of archival research from Nottingham Assizes is then utilised as an historical prism in Section 4 to compare common body types and their medical conditions. Those delivered for dissection in the heartland of the East Midlands set medico-legal standards for many provincial English towns. Sampling local records introduces a representative selection of criminal bodies that were termed 'dirty', 'diseased', or 'contaminated', as well as the 'good' and 'extras'.[9] All of this human material was said to be 'in limbo', a notion that reflected confusion surrounding the spiritual beliefs and medical sensibilities of the early modern crowd. Since murderers typically shared a history of violence, the equivalent force marked on their bodies left many badly damaged during the execution process itself. This ironically added to the powerfully disturbing bodily sense of criminal potency, even before being supplied for dissection. In so doing when images of the criminal corpse were created by contemporary artists they were often caricatured to try to capture with satire the extent to which execution from a medical standpoint was disturbing in the long eighteenth-century. In the chapter's final Section 5, therefore, several of the most iconic images featuring renowned characters condemned to die that were once famously sketched by William Hogarth and Thomas Rowlandson are re-appraised. Re-examining notorious criminal corpses reveals a surprising set of medico-legal perspectives that call into question certain art historical clichés. The overall intention in this chapter is then to envisage medical men waiting to get hold of the *peri-mortem* body, to grasp the next step on a highly symbolic punishment journey.

THE CONDITION OF THE CRIMINAL CORPSE: CONCESSIONS AND OPPORTUNITY COSTS

On 7 September 1771 a convicted murderer called John Chapman was hanged at Chester.[10] He was executed in one of the main 'hanging-towns' in the North West of England. His execution rites soon proved to be controversial in the locality. Chapman resisted the hangman with considerable force. His body had therefore to be very roughly-handled to complete the death sentence. This potentially damaged the outward shell of his corpse before it was handed on to local surgeons. The controversial situation thus exposed common logistical problems that often damaged the criminal body badly before it reached its dissection destination. The courtroom facts were that Chapman was a sailor. He was convicted of homicide after a port brawl. Throughout the capital hearing he protested his innocence. It was rumoured that the trial evidence that convicted him was decidedly circumstantial. His fellow sailors thought him innocent and were angry at the death-sentence verdict. They consequently threatened to interfere with the Chester hangman. Chapman was equally indomitable. In prison he declared that he was determined to go on the offensive and resist execution. It was a priority therefore at the gallows for the legal officials to try to conclude matters speedily.

The standard execution procedure at Chester was that a clergyman was appointed to accompany the condemned to the hanging-tree. A spiritual counsellor was supposed to keep the convicted man or woman calm. They were also there to reassure the crowd that any repentance speech was genuine. The clergy had to attempt to bring about a final confession, being duty-bound to persuade the prisoner to atone murder to cleanse their soul. That confessional was written down and then printed as an intended moral lesson. Due legal process had to be seen to attend to the prisoner's spiritual needs and protect the well-being of provincial society. Chapman nevertheless refused to be consoled or counselled. On getting into the waggon with the appointed clergyman to travel to the gallows, the convicted man charged at his confessor's belly. The churchman was thrown violently out of the conveyance. The sheriff's officers seized the prisoner and 'tied him with ropes to the cart'.[11] Still he struggled. On arrival at the hanging-tree he had to be strapped round the base of the gallows. Then the execution rope was prepared. The hangman 'had to try to lift him up to tie the rope about his neck'. Chapman reacted even more forcefully and lurching forward again 'got the hangman's thumb in his mouth, which he almost separated from the hand; he was at last tied up, but with great difficulty.' Ideally he should have been hanged for just over an hour. On this occasion he was cut down within a 'short time' to avoid a riot by the sailors and resentful crowd. It was a public order situation that could easily have got out of control. The executioner judged it necessary to react to the controversy by opting for a standard medico-legal concession. The condemned body on this occasion would be more badly damaged to finish the penal process without further delay at around the forty minute mark. Suffering was an objectionable but intrinsic part of the capital code. If the rope got tied in the wrong place or was not tight enough then the prisoner must suffer the painful penalty of being obdurate.

Generally medical men did not intervene to improve the condition of the body supplied to them because it was the business of the executioner to decide how and when to execute his duties. When therefore John Chapman was handed over to the surgeons his body had already been 'blemished' (by the initial tussle), 'bruised' (by the rope marks and manacles fastened very tightly in the cart), and 'battered' (having been strapped to the bottom of the gallows, gripped on the rope, and forcefully swung). He was despatched in haste, too hasty in fact; symbolic of the sorts of 'bad' bodies that surgeons often received which could vary a lot in terms of their physical condition even before post-mortem 'harm' commenced. Jon Lawrence indeed has indicated that anatomists were often not consulted about the ideal length or placement of the rope to ensure a quick death on the gallows.[12] And this despite the fact that Simon Devereaux has found that English executioners generally lacked basic medical expertise compared to their European counterparts.[13] To appreciate properly then the damage done to the criminal corpse and why at the hand-over this mattered medically-speaking to penal surgeons, it is necessary to briefly re-envisage the condemned dangling from the hanging-tree. The aim is to first build up a general historical picture of the sensorial elements before engaging with trial narratives that confirm the specific material circumstances *in situ*.

Under the Murder Act, a common occurrence, hence, was that the hangman's rope generally marked the skin badly for surface anatomy. It was either tied behind the ear (a standard but inefficient technique as it often slipped) or to the front of the voice box (preventing the criminal from screaming out but usually prolonging death). Another medical short-coming was that it was usual to measure the height, but not the weight of the body before execution in the eighteenth-century. This meant that strong necks which took a lot longer to break than the maximum period of one hour's hanging-time, troubled executioners especially those that remained unbroken even when cut down.[14] A robust 'bull neck' allowed the vertebral and spinal arteries to function, maintaining continual blood-supply to the brain. Even when the carotid artery in the neck did get cut off by the rope marks, the major systems did not drain of blood in-tandem. The force of gravity often enabled the rest of the body to draw what blood supply it could from the major organs and this was a power-ful survival mechanism. Typically, the contemporary picture was that of a criminal starting to die by strangulation with their skin being stretched under their body weight and a dislocated but not broken neck. Most pris-oners passed out from the excruciating pain, but they were still technically *peri-mortem*.[15] If this lack of a neck break continued, and the rope was kept tightened, the cause of death was execrable. Blood stopped circulat-ing in and out of the head. The heart meantime still tried to pump blood to the brain even though it could not go anywhere because the tautened rope restricted blood-flow to and from the head. There was equally too much blood inside the brain with nowhere to escape. An enormous build-up of head pressure occurred. Catheters in the brain became so pressurised they started to haemorrhage. The criminal brain in the process was turned to a bloody mush.[16] The visible signs of this happening included eyes pro-truding like stalks, and the face turning from vivid purple to black: both were often reported in provincial newspapers. Then the criminal was in the most confusing medical state of all, '*the living dead*' and few could legally determine whether this painful spectacle was torture or not. It was now that the executioner had the discretion to make concessions to appease the crowd.

The capital code decreed that the hangman did not have the option of using a sword or dagger to finish off the criminal. That penal authority had been revoked as uncivilised, by the early modern period.[17] It was routine however to find other ways to speed things up. As a matter of course many hangmen elected to give the surgeon(s) on duty a rough-handled body.

This had one advantage and one downside. If the execution method was rough-and-ready, the murderer would die when being mishandled. It thus averted the ethical issue of human vivisection. A 'bad body' could potentially limit how much original research and teaching the surgeon might want to do, but the criminal would in all likelihood become a corpse on the gallows. If, alternatively, a surgeon got a 'good' body that was less damaged by the execution, then he had gained a worthwhile research and teaching commodity. Yet, it might present him with an ethical predicament on arrival in the supply-chain. The tying of the hangman's knot was not therefore simply about an efficient execution. Its placement influenced the post-execution scenario too: a medical perspective often missing in crime studies of the period. In early modern England there was hence ongoing medico-legal debates about how, where, and for how long, to tie the hangman's rope on a criminal neck. Just how adept the medico-scientific thinking was on location was often disputed. Standards tended to differ in provincial and metropolitan settings. They were usually dictated by local and regional customs, as well as the executioner's preferred working practices. At Chester, the hangman always hoisted up the condemned prisoners facing each other because it was customary to accumulate criminals to stage double-hangings at each Assizes.[18] The executioner used a black hood to cover each prisoner's face, with an underhand knot on the neck. This only worked smoothly from a medical standpoint if the condemned were co-operative and resigned to their legal fate. In cases where the convicted resisted execution to protest their innocence, like John Chapman, or had much stronger physical constitutions, medically-speaking methods of punishment ranged from the clumsy to inept. This contrasted with rituals in Leicester where the hangman worked faster than that his counterpart at Chester. Medically he prioritised getting the execution done to a precise medico-legal tempo.[19] He always started proceedings at the gallows as the city clock struck noon and then ended them when the hour chimed at one o'clock. The crowd therefore could countdown life ebbing away. The length of such hanging-times was nevertheless disputed, and again this mattered when it came to the timing of medical death and the general conditions under which surgeons first handled bodies.

Official eighteenth-century sources usually said that the standard execution time was 'up to an hour' but conflicting newspaper accounts often stressed that around 'twenty to forty minutes' tended to be the norm.[20] At Tyburn for example in April 1733 four men were hanged for murder and their executions reported in the *London Magazine*. One of the felons, William Gordon, 'was said to have procured his Windpipe to be opened by a Skilful surgeon...'Tis imputed that though he was a heavy Man, he was observed to be alive after the rest were dead'. He was hanged for '3 Quarters of an Hour' but on being carried into a 'House on the Tyburn Road, he open'd his Mouth several times and groan'd'. In the opinion of the medical men present, most felons, even after being hanged for forty minutes, could fully recover.[21] At Bristol in September 1736 the *Historical Chronicle* reported a similar outcome.[22] Joshua Harding and John Vernham were hanged together, cut down once they stopped jerking on the rope and displayed in two open coffins. Later they 'came back to life' in the house of the duty-surgeon. He bled them and had the foresight to declare to those assembled that they had a 50:50 chance of survival— Harding recovered but Vernham expired. After 1752, another contemporary commentator wrote of the execution rites of the Murder Act that: 'cases of semi-hanging are not a few, and, if those which are unknown, on account of the secret having been well kept, were made public, the list, I believe, would contain some score of names'.[23] He claimed that 'tying or placing the rope in a particular way [generally to back or front of the neck instead of to the left-side of the windpipe] or by cutting it down sooner than usual, the hangman could render the subsequent revival of the condemned a comparatively easy matter'. In a medical sense then executions tainted by '*semi-hanging*' and the '*half-hanged*' betrayed the hangman's legal options to rough up the body, for reasons of corruption, to speed matters up, or to cover up basic medical ineptitude. The post-mortem sequel of events this created for the duty-surgeon mattered a lot less than the legal priority to carefully handle a public order situation that threatened social stability in a locality. The contemporary attitude was that the medical profession were lucky to get the criminal body even though it should have been guaranteed by the capital legislation. An odd medico-legal status quo therefore came about.

Once convicted in court the murderer no longer had any legal rights. Yet, the Murder Act had described the sorts of punishments that ought to happen next. This meant that in some controversial cases events at the execution could contravene the expectations of everyone. Gatrell has shown that there tended to be a spectrum of sentiments expressed on a hanging-day.[24] These ranged from the blood-thirsty in the case of a brutal murder, to considerable sympathy for the condemned if they seemed the innocent-party. Empathy was sometimes pronounced for harmless or gullible characters too. The sort of unlucky characters caught up in an unfortunate set of personal circumstances also won public sympathy. Any emotional outbursts during domestic violence that had escalated to culpable homicide resonated with the early modern Biblical belief: 'There, but for the Grace of God, go I'.[25] Regardless, however, of the public mood in the execution crowd or the personal capacity of spectators to engage with an emotive storyline, there was a customary expectation that when the condemned was cut down they would at least 'look dead'. In general the attitude was that the person should still resemble the human being seen in the courtroom as they dangled from the hanging-tree and it was up to the hangman to be dignified and skilled about it. They could not therefore yet be less than human when taken by the surgeons. For the punitive steps to have moral authority and function as a capital deterrent, the executed body could be marked, mishandled, even broken, but not yet defaced or despoiled. The condition of the criminal corpse under these complex circumstances was hence subjected to the medical gaze of a lot of interested parties. They each claimed to wield official power or have moral agency over it: as Figure 3.1 illustrates (see, next page). This fluid contemporary context raises a fundamental historical question about whether it is possible to be more precise about the medico-legal damage being done to the *peri-mortem* body on what were three open-ended punishment pathways that happened concurrently and therefore are very difficult to disaggregate. They served the popular imagination, painful spectatorship, and criminal infamy (refer, discussion Chapter 1, page 11). At the handover one way to think again about these puzzling physical circumstances that involved medical steps to achieve legal ends, is to adopt Peter King's model of crime and justice for early modern England.

King envisages the English crime and justice system as a corridor of 'connected rooms' in the long eighteenth-century.[26] Each 'room', he explains, contained decision-makers, motivated by a desire to realise their concepts of justice. King emphasises that each 'space' had a different 'shape' according to 'different legal constraints and customary expectations' behind a hallway of doors. There was a classic triangle of social relations present in each 'room' too. It was generally peopled by—the medical fraternity—executive personnel that staffed the legal system—and the crowd—who all pulled and pushed for agency (refer, again, Figure 3.1). This judicial framework was overseen by state power, but it too functioned through its client representatives (medical and legal) and could not ignore the way that the criminal corpse was viewed culturally (by the crowd). Hence different social groups had different roles and their influence varied depending on

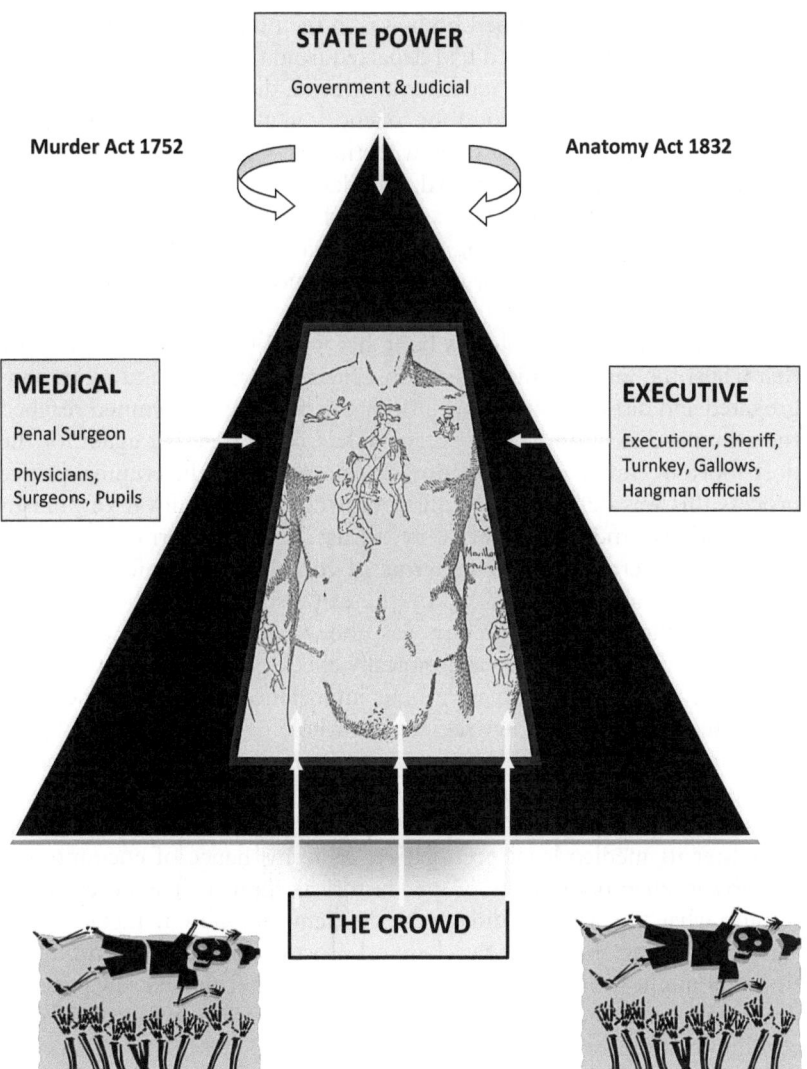

Figure 3.1   The medical gaze & the embodied criminal corpse, 1752–1832.

how far along the condemned body was in the punishment 'corridor'. By the time that the condemned had departed from the 'courtroom' space, it was moving into a medical territory in which a diverse audience was now assembled. The surgeons stood for 'science' on a path to full professional recognition but they had to work with the law-makers who had to make physical concessions to the crowd. The law and what could lawfully be done to the convicted murderer's body had therefore 'different meanings for different people' depending on the homicide case, logistical situation at the gallows, and the performance of post-mortem 'harm' the community expected to see.

Extending King's model, this book has found that it was also the case that whilst the various official and non-official actors came and went, congregated and dispersed at executions, by the time the condemned reached the criminal dissection 'room' everyone had come together again for the post-mortem encore. In a 'corridor' of 'rooms' inside the criminal justice process this was definitively *the* most crowded in the hallway of punishment options. And this is why to be 'thrust on up the corridor', as King puts it,[27] mattered, not just in terms of completing a capital sentence, but what was about to be done by the exit also circumscribed how the entire punishment choreography was about to be viewed morally and physically. In which case, to anatomically check on medical death and its unstable condition, took on a symbolic importance, not just for medicine *per se* but eighteenth-century society as a whole. If then the medical gaze in Figure 3.1 was powerful, it was also continuously seeking to become power-laden. Surgeons thus relied on things going wrong at executions because these were opportunity costs that medicine needed to exploit to bolster its medico-legal prerogative. In terms hence of encountering the *peri-mortem* body in its proper historical context, it is necessary to rethink what the weather did to the condemned—how it metabolically shut-down—and the ways that chemical processes like *peri-mortal* digestion—all might spoil what was being handed over. The next few pages of discussion detail these essential physical circumstances before then proceeding to compare them to actual trial narratives.

It was not difficult to find things going wrong because executions often took place in places with logistical problems, as we saw in the last chapter. Winter cold sometimes caused extreme-hypothermic bodies, but summer executions presented equivalent difficulties too. The timing of the Assizes was in fact crucial. It coincided with the Quarter Sessions courts at the four religious festivals of Epiphany, Easter, Midsummer and

Michaelmas. Mary Shelley thus wrote in *Frankenstein*, 'The season of the assizes approached' when setting out her famous storyline about resuscitation.[28] Generally hangings staged after Easter were fore-shortened compared to winter executions, less than an hour; the recommended time span. This was because bodies rotted very quickly in the summer heat. So there were always local disputes about ideal rope lengths and optimum hanging-times. The official pressure tended to be more intense in summer to make a declaration of legal death to release the body before medical death. Samuel Hey of Leeds explains why this was so.[29] Young Samuel when undertaking his medical training at Leeds Infirmary, where he did anatomy and dissection work, lamented to his brother the foul smell of decaying human flesh after execution during the Indian summer of September 1831. Samuel later elaborated in his private correspondence: 'I rather expect a post-mortem tomorrow, which I understand is the most horrible thing possible, on account of the intolerable stench'.[30] In a follow up letter he recounted: 'I had to trudge about a mile in the rain with William [his uncle] and Dr Hobson a physician in Leeds, about 9 o'clock at night and work at the chap for two hours, and then march home again carrying bits of him in my pocket, and a stinking case under my arm'. His clothes smelled of the stench of death and 'the worst part of the problem was that I pricked my fingers sewing him up but I sucked them as hard that it did me no harm'. He ended with: 'I certainly shall never wear the clothes I had on then, for no other purpose; no stink can compare to a body stink'.[31]

The Newgate Calendar likewise records how hard it was to execute in warmer weather. This outcome happened for instance at the double-hanging of William Proudlove and George Glover on 28 May 1809 at Chester.[32] In the warm weather, the ordinary clerk related how the platform was dropped away 'but, alas, horrid to relate both ropes snapped a few inches from their necks, and the poor sufferers fell upon the terrace'. It was noteworthy that 'the miserable men appeared to feel little either in body or mind from the shock they had received'. In other words, they were 'in the name of Death'—their vital functions had started to shut down—but they were not in 'absolute Death'. The hanging time was too hurried, the substandard rope snapped, the murderers sweated too much, and so they revived 'and spoke of it as a disappointment' believing they were optimistically 'going instantly to heaven'. In this case, there was no mercy because they revived under the gallows and were therefore not technically legally dead. The executions were restaged the next day with much stronger ropes, a bigger drop and longer hanging time. The *Historical Magazine* similarly claimed in 1789 that at Exeter highwaymen were often suspended from a gallows and slipped the knot in summer weather when sweating with fear in the excessive heat.[33] In the case of 'William Snow alias Skitch, for burglary, and James Wayborn, for highway robbery' both men 'fell to the ground' after the ropes started to burn under the heated friction and snapped. To the amusement of the crowd, Skitch cried out: 'Good people, do not be hurried: I am not hurried: I can await a little'. This time the executioner decided to 'lengthen the rope' and the men were hanged one at a time in front of a 'crowd of thousands'. Skitch was made to watch until Wayborn was cut down. The stronger rope was then placed on Skitch's neck and he remained on the gallows for several hours to make sure he was beyond resuscitation in the summer heat. The key point to appreciate in all these weather scenarios is that in homicide cases surgeons worked in grey medical territory. They had to somehow find ways to exploit the fact that criminal bodies would be damaged and mishandled as legal concessions were made in front of crowds when basic procedures became controversial or were sub-standard.

Envisaging then how the condemned body looked on arrival, began to visually decay, and started to smell badly, means historically rethinking a sensorial process and its biological events. Taking this practical approach involves revisiting the medical circumstances that have endured in death (rather than seeking in vain detailed medical accounts written by the crowd in an illiterate culture). For it is a material fact that the condemned in a history of the body has been a stable subject of biological continuity. *Knowing how this works* was an implied cornerstone of the Murder Act. That material recognition makes it feasible to retrieve what was an arresting experiential encounter that surgeons had to try to exploit. They were generally advantaged if it was a very gruesome execution with people being repelled to turn away earlier than expected. At the other extreme, an unspectacular, quick, send-off meant they could usually claim the body without being challenged. This sets in context why it was that nearly all contemporary newspaper accounts featured the condition of the hands and neck as notable indications of death by hanging.[34] First-hand accounts provided a short commentary on the black bruising from the rope marks. The face was likewise commented on when contorted by blood rushing to the surface of the skin. The pallor, reporters recounted, generally looked clammy and then waxen-yellow once rigor mortis set in about four hours after the legs stopped jerking on the rope. Everyone, in all periods, looked like this at a hanging. It was a sensorial encounter that legitimised the actions of the surgeons at the hand-over and which historians can better appreciate.

The metabolic shut-down can be more accurately understood too.[35] Medically-speaking chemical changes in the muscle had to have occurred after cellular respiration halted in the deep tissue of the condemned. Historically this has happened to every human being at about three to four hours after a fatal trauma. This timing explains when exactly the arms and limbs started to stiffen after being hanged on the gallows and at what point the surgeons hoped to exploit that physical situation. Hardening in limb-muscles would have been at their most rigid by twelve hours after metabolic death. Then twenty-four to thirty-six hours later the stiffness began to wear off, but slowly: it always depended on the ambient body temperature and this would have influenced the rate of tissue putrefaction. Early modern sources confirm that it was always much easier to handle a corpse immediately after an execution when the body was in a physical condition known as primary flaccidity.[36] Since rigor mortis generally wears off around a deadline of twenty-four hours in everyone, this explains why many surgeons preferred to leave the body until the next day when the corpse could be easily handled. Sometimes they cut the tendons around the Achilles heel to stop the feet and legs stiffening, but in general there was a time lapse between being laid out for inspection to confirm medical death and being dissected to allow rigor mortis to wear off, as we saw in the case of Elizabeth Ross in Chapter 2. In winter, the surgeons worked more slowly on the body because rigor mortis in the cold took about six hours to take hold, rather than the standard three to four hours. In summer, the dissector knew that in the intense heat rigor mortis often happened much faster at around two hours. Whatever the seasonal time-span, the foul smell of decay was unavoidable.[37] In terms then of the general condition of the executed, bodies hanged at the winter Assizes worked much better for surgeons everywhere.

There were physical aspects of the execution that hangmen and the executive personnel preferred to keep silent about, and which penal surgeons dealt with swiftly by shaving and washing the body on its arrival at a dissection venue (see also, Chapter 4). Again, these actions reveal opportunity costs to get the criminal body into a medical ambit more speedily to avoid public embarrassment and yet that logistical context has often been overlooked. Whereas for instance chemical changes and the problem of rigor mortis could be stage-managed, other body functions represented more of a taboo immoral issue. Legal officials refused to speak about how often the rope constricting the neck triggered an auto-erotic response in men during the summer heat when mishandling the corpse was a heightened problem.[38] In summer flies and wasps buzzed around the expelling body fluids attracted by the sugary scent of sweat. Those nearest to the gallows got a strong whiff of urine ketones (smelling of cat pee) mixed with hydrogen sulphide (stinking faeces stools) because many criminals literally wet or defecated at the fatal time; others smelt of alkaline bases of chlorine (ejaculated semen) too, aroused by the imagined terror. Grabbing hold of a male body in these slippery conditions and manipulating the sexual organs before public display at a dissection venue was a test of a surgeon's dexterity. They could generally work behind the scenes assisted by the executioner because handling such tasks was considered distasteful in front of a crowd containing boisterous men and receptive females said to be impressionable. When women were hanged the physical purging was vivid red too, since an unpleasant side-effect was that females tended to menstruate spontaneously. The pull of gravity on the lining of the womb generally caused a prolapse of the sexual organs as the torso stretched downwards.[39] Barber surgeons trained in naval and military warfare were familiar with the scents of blood products, but the fishy odour of menstrual evacuation mixed with bacteria in the vagina when exposed to fresh air was still a disquieting side of their penal work. A cultural of denial also surrounded the side-effects of the basic functioning of the involuntary hyper-arousal mechanism (mentioned earlier and discussed in Chapter 2). Essentially the sympathetic nervous system activated the adrenal medulla in the brain to secrete catecholamine's norepinephrine and epinephrine.[40] These have four unavoidable outcomes relevant to an executioner's handling/mishandling of the condemned as they expired: blood flow increases, blood pressure rises, blood-clotting speeds up, and muscle tension stiffens to provide the body with extra speed and strength. Three of these together better oxygenate the blood, and provide a very powerful survival mechanism when in a traumatic situation. Yet, few historical accounts trace such a commonplace biological chain-reaction and its chemical impact on the criminal body. Common medical symptoms beyond historical dispute have thus tended to be overlooked or misconstrued. Surgeons that sensibly elected to clean or manipulate the sexual organs and trim off all the body hair at the hand-over took necessary sanitary precautions. Yet, the processes of preparation for dissection also covered up the unsavoury side of execution that they were exploiting to bolster their medical prerogative too.

Digestion adds another dimension to biochemical death by execution and it likewise could be exploited by the surgical community. It was a biological fact that 'different parts of the body decayed at different rates' to complicate criminal dissections.[41] The bacteria in the gut did not die when the criminal was hanged. In all human beings microbes keep on functioning in everyone after death (regardless of how the person expires). They feed off the last meal and are then expelled as noxious gas post-mortem. The smell lingered for those working in close proximity to an execution because gaseous exchange goes on occurring for up to six hours until involuntary digestion ceases. If someone was young—and most of those executed for capital offences were males aged between 20 and 40[42]—then their body tended to be healthier. This meant that their flesh decayed more slowly than the corpse of a diseased person that was older.[43] In the past, if the body was placed under a flowing water tap—common in the basement of some provincial dissection rooms—then again this could slow down decay.[44] In general, a criminal body 'decomposes in air twice as quickly as in water and eight times as rapidly as in earth'.[45] An unpleasant aspect of this digestive shut-down was that when gases accumulated in the lower abdomen a side-effect was that the lungs were forced upwards in the chest cavity. Decomposing blood had to escape from the mouth and nostrils. Once this happened it was easier for the surgeon to be in a position to claim the body. It 'looked dead', still resembled a human being, but was self-evidently moving into a new punishment phase in which they could claim more expertise than the hangman. A related physical characteristic of this bloody purge was eerie. It adds to our historical appreciation of why there was often a subdued mood amongst the crowd commented on in contemporary newspapers. Once rigor mortis happened in the dying body the wind pipe in the neck area stiffened. The corpse was now filled with gas that needed to escape but not all of it could leave the anus, and since escaping blood blocked the nose and nostrils, some air had to be expelled by passing over the windpipe. It could be heard as moans, groans, squeaks and rasps by those witnessing the death. This biological soundscape might sometimes have been lost in a noisy crowd at the execution, but once the legs stopped jerking another feature of the punishment ritual seems to have been to listen out for the death-rattle.

In terms then of the body's basic condition when a medical official wanted to take over legal responsibility, to the touch it would have been grey, flaccid, often waxen, and the skin was bruised a lot, fragile, paper-like, especially around the iron-manacles or rope-marks. Hand and toe nails looked longer because the skin retracts when circulation slows with the last shallow breathes; likewise the hair seems lengthier. The jaw dropped in criminals with teeth in a poor condition. This was not unusual in the eighteenth-century when decayed gums receded a lot; some also sold their teeth to dentists to pay for a good send-off.[46] Altogether it was commonplace to see mouths sunk into faces. It would have been an arresting sensorial experience to look at such criminals, either close-hand, or when trying to handle them. Criminal histories do highlight that there were 'degrees of death' and 'numerous ways to die at the execution site' in the eighteenth-century, but few examine the longevity of their sensorial or medical dimensions in-depth.[47] As one English commentator wrote: *"Tis well known there are some kinds of Death more sharp and terrifying than others"*.[48] This reflected how for those crowding round the corpse there were 'two kinds of dead: one in Nature, the other in Culture' as it lay on the ground.[49] It meant that in terms of bodies leaving the execution scene for punishment accompanied by large crowds 'the corpse and the person [were] not irrevocably sundered'.[50]

Although therefore Georgians lived with the foul smells of humanity in their over-crowded neighbourhoods lacking basic sanitation, it was more desirable for surgeons to pay the price of obtaining bodies by letting them take over. Yet, this very important medical context seldom features in standard cultural or criminal histories with the exception of Gatrell's admirable work. Essentially, as Esther Cohen remarks: 'The ritual was worthless unless people knew and understood its symbolism': to be given access to a body cut down for a criminal dissection close-hand was all about the immediacy of its material reality and its unsavoury nature became the business of the surgeon.[51] Resented many may have been pre-execution, but increasingly not post-execution, when the condemned was in a 'bloody mess'. Making the punishment 'alive' or 'fresh' in people's minds was about slippages—shedding, smelling, swelling, and shrinking. A representative selection of trial narratives brings these ubiquitous sensorial experiences into view and sets the opportunity costs they implied for provincial surgeons in their historical context. Two high-profile trials highlight medical narratives that were well-publicised in the retelling, revealing how medico-legal officials worked under challenging conditions, in Sections 2 and 3. Later in Section 4 will see that much of the headline detail was also duplicated in the East Midlands. In this way, the substance of the detailed trials in Sussex and Surrey outlined next, also relates to a spectrum of bodies handled by surgeons elsewhere under the Murder Act in provincial English life.

BULL-NECKED BAD BODIES: MEDICAL NARRATIVES
ON TRIAL IN SUSSEX

In August 1831 a shocking murder was to make national headlines. The brutal killing and the gruesome disposal of the human remains of Celia Holloway raised distressing questions. The homicide was a textbook copy-cat killing. It mirrored how retribution by penal surgeons was done to the criminal corpse in death. Newspaper reporters investigated what motivated John William Holloway to murder his wife Celia. They wanted to know where he learned to duplicate anatomy and dissection techniques. John confessed to local magistrates in Brighton that he tried to strangle Celia with a cord, then hanged her with a rope, quickly cut down the corpse, and afterwards dissected and dismembered his estranged wife. Celia was pregnant with John's child (even though they had been residing apart before the fatal day). He was currently living in Brighton with his accomplice and mistress. Celia lived nearby but John had deserted her. He was determined to kill his liability. In three remarkably detailed confessions Holloway described how he adopted standard execution methods to his deadly-design:

> I then tied the cord as tight as I was able and then dragged her into the cupboard, and hung her up to some nails that were placed there before. As I dragged her into the cupboard, I felt the poor dear infant struggle in its mother's womb, surprisingly strong indeed. Oh what a shocking sight! I shall never forget it. I did not remove the cord from Celia's neck, but took an over-handed knot, and I made the ends fast to the nails, so that she was hanging by the neck. I proposed then cutting her up; but Ann Kennett [his lover] *told me to wait until the blood was settled.*[52]

In Horsham jail John Holloway elaborated to his confessor what it looked and felt like to kill his former wife, an act of pre-meditated murder:

> When I called Ann Kennett [accomplice] from the cupboard, I believe that poor Celia was able to see the person whom I called, although her eyes were nearly turned into her head; yet, when Ann Kennett appeared, her wild appearance increased; the cold sweat in big drops on her pale cheeks; and almost as soon as Ann Kennett took hold of the cord [to pull it one last time], poor Celia dropped. Her wild appearance seemed to terrify Ann Kennett, as well as myself; for Kennett said to me, 'Oh how wild she looks'. I made no answer that I can remember. At the time that I committed the horrid deed, I am not able to describe my feelings; but I think you will understand me when I say that I had none.[53]

John admitted that he resented his wife for compelling him to marry when he was '18 or 19 years old...she was with child by him after their intimacy, and almost seven years ago the poor law overseers compelled him to marry her. They lived for some time pretty comfortable, and might have continued to do so, but for the ill usage of her relatives'.[54] He claimed that he was always being criticised and then was shunned by his in-laws for his lack of financial acumen. Eventually John ran off to sea, before returning to co-habit with his lover Ann Kennett (sometimes called Kinnard). The Preston parish overseers, near Brighton, his former wife's relatives, and others in the Sussex neighbourhood, meanwhile continued to induce him to pay Celia a weekly maintenance allowance. His fury, he claimed, turned to paranoia that they all wanted to 'destroy his peace of mind'. Young love had evidently turned to hate. John thus planned to briefly co-habit with Celia once more to regain her trust (this accounted for her pregnant condition at the time of the murder) before luring her 'to a private place and assassinate her'. Having 'cut the body up', he confessed to putting the parts 'in a trunk, dug a hole and buried them' in a local park. In point of fact a local constable rediscovered the dismembered head and limbs in a privy. Meanwhile John's lover was arrested too for being an accessory to murder.[55] The sensational elements of the case gave rise to a lot of speculation as to why John was so impressionable. He had not just observed, but admitted adapting methods of execution and dissection he had seen to suit his deadly purpose. Evidently the capital punishment process of the Murder Act was well-known and had enticed someone susceptible to commit a violent crime: the opposite effect intended in the legislation.

The unwitting testimony contained in Holloway's confessions is instructive. It reflects more broadly how the body's fate on the gallows and its post-mortem punishment entered popular culture. John confessed that by 1831 it was generally known that it took an 'over-hand' knot to secure a body on the end of a cord or rope because a lot of executions had been bungled before the crowd. He tells us that he attempted to strangle his wife but found it very hard to kill her quickly by the 'ordinary method of strangulation'. This is why he decided to tighten the cord, elevate the body under a strong knot (he used a trefoil the second time), and got an accomplice to help him tighten the noose: techniques adapted by executioners on the gallows. He knew there had to be a 'drop' from which the corpse was suspended but he could only create a 'short-drop' when there clearly needed to be a 'long-drop' to hasten death. John related his impulse to cut up the body to get rid of the physical evidence but his accomplice understood that the body works with gravity and drains of blood from the head to the toes when dying—Anne *'told me to wait until the blood was settled'*. As we have seen in Chapter 2, new scientific ideas were in circulation that challenged the out-dated idea that cardio-respiratory failure was medical death. That explanation was old-fashioned because of anatomical research on the brain's ability to survive head trauma (refer, also Chapter 6). By 1831 it was known that the brain drained of blood in death, and not until then would the person truly expire. Indeed in jail Holloway confessed that Celia hanging on the end of the tightened cord was sensible of her murderers despite the excruciating pain. Holloway noted her physical reactions—a 'cold sweat', 'big tears', 'pale cheeks', 'eyes nearly turned into her head', and a 'wild appearance'. Being hanged by the neck looked physically like this for everyone and because it was well-known this part of his evidence was not contested in court.

Under cross-examination John Holloway conceded that once he started to dissect the limbs the fresh 'blood went everywhere'.[56] The prisoner remarked that the practicalities of dismembering a corpse were far more gruesome than he had anticipated. In his arrogance, he revealed that he knew the basics of how to do anatomy (cutting open the body and looking) but not dissection (dismembering it into body parts). He soon discovered the latter's dreadful side. It is worth keeping in mind that dissectors had no latex gloves, suction tube, or electric light to work under in the long eighteenth-century. Fresh blood could spurt all over the candle-lit dissection table, especially if the anatomist was inexperienced and cut the stomach or intestines containing the prisoner's last meal. If they did, the smell of food-decay could knock them out. This is one reason why the condemned was usually just fed bread and water at their last meal: again, a seemingly trivial but important physical part of the execution ritual often overlooked in criminal history. Naturally contemporary courtrooms were in the thrall of such gruesome confessions. Newspapers in Holloway's case made a number of medical observations relevant to the discussion in this chapter. All noted that the death penalty had a dehumanising side. Even experienced hangmen in high-profile cases preferred to have surgical assistance. Not all necks broke on the rope, and a prisoner's willpower could defy the procedures of capital death. Medically-speaking surgeons took advantage of this situation. Yet, the crowd increasingly had an accumulated knowledge about why it was desirable that they did so. The unsavoury aspects of cutting up a warm body, revealed by Holloway in court, confirmed that it was best to let a surgeon handle those that 'looked dead' in reaching post-mortem 'harm'.

The authorities in Brighton were so appalled by the brutality of Celia Holloway's murder that they sent for an expert executioner from London. In court John Holloway stated that it was his intention to make a dramatic entrance at the gallows by dressing in a black-cloak. The trial judge dismissed such theatrics as self-serving. According to local newspaper reports, the prisoner's bravado held up until he faced the actual gallows. A local account described his countenance as 'grief-worn, pale and cadaverous' with a 'restless oscillation in his eye'.[57] The dramatic adjectives and building of theatrical tension did of course serve to increase newspaper sales. Nonetheless, it was common to witness a prisoner with agitated eyes and a pale face, usually caused by stress. Insomnia was also a physical factor since few slept well before the fatal day. In John's case, the hangman 'tied his arms and hands' and it was reported that for the first time the prisoner 'exhibited signs of fear, trembled excessively, and expressed in a low tone of voice a hope that he would not suffer excessively'. Local records stated that the rope was placed around Holloway's neck and was 'at least three feet in length'. The hangman now stood in front of the chest of the condemned whose 'heartbeat was racing'. In all executions, the autonomic nervous system in every human being was reflexive like this.

The court records confirm that the executioner of John Holloway was highly skilled. Newspapers likewise stressed that 'a medical gentleman was present' to monitor death. Procedures were meticulously managed for publicity reasons. Given the notoriety of the prisoner, full justice had to be seen to be done, and efficiently. We thus learn from local accounts that Holloway's 'chest heaved' as he grasped for breath; 'his clasped hands quivered, and dropped'; there was a 'convulsive movement of his lips'. It was then reported that:

> After the body had hung about a quarter of an hour, a man ascended the scaffold with the executioner, and seating himself on its edge, took off his hat. The hangman then loosened Holloway's hands, the palms of which he rubbed on the forehead of the countryman under the absurd notion that the death sweat would remove the excrescence [monstrosity]. The executioner (the Newgate hangman) was to be paid half a crown for this disgusting spectacle. The under-sheriff, however, very properly gave him to understand, that he would not suffer a repetition of such proceedings until after the body was cut down.[58]

The account described how a local parson was permitted to also climb on top of the scaffold and give the last rites after fifteen minutes as a precaution because local opinion held that medically-speaking death was indeterminate. It was popularly held by evangelicals in Sussex that the soul darkened by sin needed to have spiritual help to disconnect from the bad body. Equally there were those present who believed in a version of the superstition of the *'dead hand'*: in this case wiping the forehead of the 'countryman' with the prisoner's sweaty palm was said to release the spectre of monstrosity to the judgement of God. The local newspaper reporter took a more Enlightenment attitude. He questioned 'the fitness or unfitness' that 'such public religious harangues at the moment when a fellow creature hangs suspended from the gallows'. He also noted that a fifteen minute duration was too short to declare medical death and that as a further precaution the executioner checked again after 'one hour, the 'usual time', when the hangman 'ascended the platform, and untying, not cutting, as was formerly the custom, the halter, the body fell into the arms of the turnkey, who stood beneath'. The reason for carefully undoing the knot was even after sixty minutes a faint pulse was sometimes felt. Holloway could revive by the very act of releasing the rope and this made him dangerous. The corpse was thus removed into a room nearest the scaffold, where the hangman took the rope from the neck, which on the knot to the left side was deeply cut by the ligature. He then stripped off his clothes, his prerequisites, and the body lay naked on the floor. On examination by the surgeon, the neck 'proved not to be broken, from the extreme strength of the muscles, the deceased being what is termed bull-necked'.[59]

John Holloway's condemned body in capital death was not yet a corpse more than an hour after being hanged. It had been inscribed by the symbolic processes of social and legal death, and the metabolic procedures of medical death were underway but not yet finished. These required a number of interventions even before a dissection venue was decided. The body was stripped of its clothing to start to erode the prisoner's identity. The naked corpse was laid on the bare ground to demarcate it as 'the other' in society and, yet, it had a medical potency. The court record confirms that a young man from Lewes who assisted the under-sheriff was given the hangman's rope to sell to 'for a gratuity'. This folk-belief reflected local customs about the potency of the gallows-rope as a quack cure or cabinet-curiosity in Northern Europe. The body now had to be made safe by the surgeon. This was after all someone considered to be dangerous. There was an unbroken neck and it was essential to cut the carotid artery. It was therefore not a coincidence that John's body was left on the ground after execution. Raising the head slightly to cut a body lying in a prone position (face-up in this case) was the best way to do a mercy-killing in the neck area to speed up the dying process: a new medical perspective of criminal justice. Severing the carotid artery had the added advantage of keeping the torso blood-free because body fluids drained behind the head before rigor mortis set in. These basic procedures were ethically questionable and an opportunity cost for the surgeon, as well as seemingly merited in a murder case that had inspired a public outcry. One local newspaper reporter who followed the case throughout thus described the work of the appointed surgeon on duty as: *'like a tempest in the natural world, which although it might inflict a partial evil, is in the end productive of the general good'*.[60] In its scientific performance, medico-legal procedures going wrong in this case are instructive about the shifting historical standards of crime, justice, and punishment rites in provincial England under the Murder Act.

John Holloway's corpse was brought into the dissection room of the Brighton voluntary hospital. It was said in local newspapers that it would be 'publicly displayed the following day'. Later reports stated that it was a 'spectacle...to which hundreds resorted'. Afterwards it was dissected 'to its extremities'. In practical terms this meant that the surgeons on duty did three key things. First, they double-checked for life-signs. The body was then secondly washed down with camphor to try to kill the rotten smell. Thirdly, it was shaven all over to prepare it for dissection. This third procedure marked the condemned as the ultimate outcast. The razor cut the final connection they had in mainstream eighteenth-century society because having hair all over when alive had real significance. Most people were hirsute at the time. Beards were bushy, bodily hair was seldom shaved; pubic, back, leg, and under-arm hair grew prolifically and was very fashionable. Thus for instance when John Hogdkinson from Hereford visited London in 1794 he wrote in his diary about how surprised he was to see the close-shaven Turkish Ambassador and his retinue: 'They had no beards' he exclaimed, 'except what grew on their upper lips!'[61] His surprise reflects the fact that most Englishmen were long-haired. Both men and women's hair was at least collar-length at a time when tall powdered-wigs were in vogue. It was rare in Georgian society for any person to be depicted without hair unless they had taken the King's shilling to join the army or navy. Shaving the criminal corpse bald thus publicised a dangerous social pariah.

This common observation was a central feature of perhaps one the most famous murder cases in criminal studies, namely that of William Corder executed for the infamous Red Barn Murder at Bury St. Edmunds in 1827.[62] Once shaven local newspaper reporters claimed that a lot of young women came several times to see his coarse head laid out in the dissection room (refer, Chapter 6). Primeval notions of survival of the fittest evidently existed before Darwin labelled them. The human psychology of coarse appearances may have held some sort of basic physical attraction between early modern men and women. As Albert Mannes' recent work explains in past societies 'men with shaved heads were rated as more dominant than similar men with full heads of hair... men whose hair was ... removed were perceived as more dominant, taller, and stronger than their authentic selves'.[63] It is an intriguing possibility that stripping John Holloway's corpse, shaving his head, and washing him for post-mortem punishment, might have enhanced his powerful attraction for those that crowded round. Holloway was not simply an exhibition of terrible retribution; he could be an exciting primitive encounter too. Taken together, sources depicting John Holloway's case set in context why there was a dangerous frisson in the air on arrival at the dissection venue. His body debauched by immorality, strangely shaven, exuded a masculinity that had a bad criminal stench.[64] It was this context too that arrested crowds of people in Surrey at two notorious executions when the medical condition of the condemned bodies became the focus of popular attention in our second detailed trial narrative of the early nineteenth-century.

PROVERBIAL BAD MEN WITH BAD BODIES: MEDICAL NARRATIVES ON TRIAL IN SURREY

In the early morning of 10 November 1817 two bodies were found dead in a town-house in Godalming, Surrey.[65] William Parsons, a local surgeon, examined the corpses. He informed the local coroner that he found two murder victims: 'He saw Elizabeth Wilson on the floor, with her throat cut and other marks of violence on her head. She was cold and stiff. The wounds appeared to have been afflicted a considerable time, and the wounds were flaccid and cold.'[66] Upstairs Mr Chennel, the property-owner, was dead in bed:

> reclining on his right side, with his throat cut and his skull fractured [He] thought life was suspended by the blows of the hammer, and his throat afterwards cut, and the reason this was the course of the proceeding was, that there was not such a flow of blood as there would have appeared if the throat had been cut first, while the pulsations of the heart were still active.[67]

Extensive enquiries were instigated door-to-door to establish the estimated time of death. Sometime between 9 and 10 pm there was a short
killing-spree. The coroner concluded that there was probably more than
one perpetrator. He informed the Inquest jury that the murderers were
either arrogant, over-confident, or had little remorse, because they left
behind a blood-stained carving knife and large hammer. The forensic
evidence indicated that the two homicides were pre-meditated, carefully
orchestrated, and this enabled the killers to make a silent getaway.

In the coroner's court witnesses testified that John Chennel junior was
in debt. He had recently separated from his wife and was desperate for
ready cash to entice a new lover. His father refused to fund his son's dissolute ways. On the fatal night John robbed him. Elizabeth Wilson, the
family housekeeper, tried to prevent John stealing two pounds from his
father's coat-pockets. He thus murdered her downstairs. Upstairs meanwhile an unscrupulous man-servant called John Chalcraft killed John
Chennel senior. The Inquest jury were told by a dozen witnesses that
Chalcraft 'was an infamous character, and ought not to be believed unless
confirmed by other testimony'.[68] Parricide and the murder of a master by his servant were perfidious charges. To secure a conviction there
needed to be meticulous evidence-gathering. It would take nearly nine
months to bring the case to trial. In due course, on 14 August 1818 at
the Summer Assizes in Guildford John Chennel and John Chalcraft were
found guilty of the Godalming murders.[69] The judge used his sentencing
powers to decree that the penal punishment must be a very public affair.
The Godalming murder story is then in many respects an historical prism
that illuminates the ongoing medical position of surgeons in the legal processes of punishment for homicide under the Murder Act by the early
nineteenth-century.

On 15 August 1818 a large crowd congregated in the centre of Godalming
to witness the execution of Chennel and Chalcraft.[70] An editorial in the
local press reported that the number of people was 'immense and lined the
road as far as the eye could see…In the narrower places they were pressed
together so closely as to be endangered by the horses and raised clouds of
dust that literally enveloped them'. Most were dressed as 'farm servants' for
the reporter noted that he had never witnessed 'so many smock-frocks and
straw-hats' assembled in 'mournful procession'. The Under-Sheriff for the
county of Surrey ordered a gallows to be constructed on Harvest Saturday
near the murder spot. After execution the bodies were to be returned to the
fatal premises where the original homicides were committed to undergo a

symbolic dissection. This public recourse to physical retribution was not in fact unusual but the local record of what happened next was more detailed. The trial accounts state that the 'bodies were cut down and given to the two surgeons of Godalming, Mr Parsons and Mr Hayes'. August, as we have already seen, was the worst month of the year to obtain an executed body because putrefaction happened much faster than in winter or spring. After twenty-four hours the human material had started to smell badly. Another practical problem was that in summer the bodies could only be left on the gallows for the minimum-time. By convention the court record usually said that each hanging took an hour. But these official records were sometimes written before the day of execution so that official accounts of how the prisoner repented and confessed their sins could be distributed to the crowd to increase the moral value of the execution spectacle. Local newspaper reports tended to be more historically accurate about actual hanging times, even if the general tone was sensational to sell copies. The key point to appreciate is that Surrey journalists did report much shorter hanging times of around forty minutes depending on the season. Hanging from the rope like this was practicable but medically suspect. Hence the Godalming records state that 'the two prisoners were brought out, with irons on their feet, hands pinioned, and the rope with which they were to be hung round their waists'. The executioner carried a 'drawn sword' but he only used this to keep public order. He did not deploy his weapon of office to stab the heart or behead the corpse on the gallows, like in Tudor times. Making sure the criminal *'looked dead'* was a short-run performance of rough retributive justice. Everyone was in a hurry to get the hanging-stage done at the height of August's harvest heatwave.

Surrey court records confirm that the bodies of John Chennel and John Chalcraft were 'cut down...received into the waggon' which doubled up as 'the elevated stage for execution'.[71] Provincial newspapers claimed that contemporaries held that there might still be a sign of life after hanging on the rope for a 'shorter timespan in summer'. This is confirmed by the 'curious way' that the crowd now acted in Godalming. The bodies were 'conveyed in slow and awful silence' accompanied by those present walking alongside the wagon in procession. They listened and watched over the bodies for any visible life-signs. The mood was not just poignant, it also reflected medical practicalities. The crowd proceeded with precaution. The surgeons were not given any special access to the bodies at the gallows. Their opportunity cost to obtain a medical prerogative therefore came later in the punishment choreography than those cases discussed earlier in this chapter. Local opinion nevertheless held that some 'murderers had a stronger life-force' such was their determination to evade natural justice. There was hence a 'solemn, slow, and symbolic procession' that did not sensationalise the spectacle of the corpse now that they were '*in the name of Death*'. The crowd instead stood watchful at the threshold of life and death. The surgeons now took centre-stage as they dealt with the medical condition of the *peri-mortem* bodies.

Mr Parsons and Mr Haynes were waiting on the criminal corpses: strictly speaking they were both body and corpse, as they were in a condition of medical 'limbo'. In the kitchen of the Chennel townhouse the post-mortem 'harm' got underway. The court record and local newspapers reported that Mr Parsons 'performed the *first office*' and this was an anatomical check of the sort seen in the last chapter.[72] He took a lancet and opened the torso from the navel to the neck, in turn, laid out on two kitchen tables horizontally. As the court record states, 'the bodies in this state were left to the gaze of thousands, who throughout the day eagerly rushed to see them'. The reason that they did so was because the anatomical cuts exposed the major organs, including the heart. If there was any sign of life, then the surgeon's first duty was to use the lancet to commit a merciful act. In other words, the '*first office*' (to use the exact contemporary phrase) made the surgeon's lancet the executioner's medical accomplice. Only then, would he proceed to the '*the second office*' of post-mortem punishment by dissection. Local newspapers in Godalming described the latter as 'awful… they may be imagined but not described'.[73] It was understandable why contemporaries would take this editorial line. Post-mortem punishment was supposed to be a deterrent. If the opening of the corpse was restricted by custom to the public view of an anatomical exposure to check on medical death, then dissection had the potential to instil more fear. From the surgeon's viewpoint these local sensibilities also gave them the opportunity cost they had been waiting on. It was now feasible to arrange that as there was not enough physical space for the execution crowd to be part of the dissection in the kitchen, they should exit before the extremities were cut from the torso. In Chennel and Chalcraft's case, and numerous accounts elsewhere that we will encounter later in this book, very few tried to stay behind. The condition of the corpse that ordinary people wanted to crowd around was generally one with recognisable human features. It was a complete body in the sense that the limbs were not yet severed and the head was not decapitated. In terms of Northern European sensibilities surrounding body-integrity in death, at criminal dissections the essential wholeness of condemned bodies had enduring appeal (a theme elaborated in Chapter 6). Meantime, many murderers in 'jail' were said to be in 'limbo'—an eighteenth-century term for sinful souls that had crossed a threshold of immorality requiring the redemption of capital death.[74] Typically, these sorts of condemned bodies were handled by the Nottingham Assizes and they highlight the common logistical problems of provincial surgeons in many locations outside of London.

CRIMINAL BODIES IN *LIMBO*: AN EAST MIDLANDS MEDICAL PERSPECTIVE

Samuel Haywood was a very experienced executioner from Appelby Magna who carried out forty-four executions in the East Midlands, covering Leicestershire, Derbyshire and Nottinghamshire during his tenure from 1820 to 1847.[75] He was also hired by Gloucester, Lancaster, Liverpool and Warwick Assizes in recognition of his proficient skills. Newspapers followed his adeptness with alacrity. Reporters highlighted the logistical difficulties he overcame that confounded others. A detailed record of his final execution at Derby in April 1847 reveals for instance how in the case of a condemned murderer named John Platts his light-weight body caused a lot of logistical medical problems. Haywood had to place extra irons tied on the feet of Platts because the condemned was just 'four foot eleven inches tall', 'stocky', and 'trembling'.[76] It was no easy matter to ensure that he died on the gallows. Such a small man took a lot longer to die in a fraught state because the force of gravity did not work so well. Local reporters stressed that 'even with the extra irons secured to his feet, his neck did not break'. His hands and feet 'were seen to rapidly jerk to and fro as far as his irons would allow, and then slowly as if a tipped bottle was emptying, all motion ceased'. He did not try to run away since this was impossible but fear (the classic 'fright' response) was seen as a factor in his painful execution. Hence, Platts was hanged for 'over an hour', cut down, and checked anatomically by Haywood and a surgeon. They did so very carefully and the crowd did not disperse but were described as 'very orderly', 'quiet', and 'subdued' by newspaper reporters at the scene. Pathos was a powerful experiential part of the medical rituals. It also gave surgeons the medical prerogative to get more hands-on when things became emotive as death's dominion was contested. Even a skilled hangman relied on a medical official to make sure that the condemned was lifeless. John Platts' physical short-comings were an opportunity cost not to be missed or indeed mistimed.

The position at Nottingham, where research material is rich in detail, was that surgeons needed the body to establish their professional status as soon as it was timely to do so in the execution process. They were always confronted with one logistical problem. A dutiful executioner might rough-handle the condemned for a lengthier time period. This placed the duration and extent of post-mortem 'harm' at risk. Strategically therefore for the penal surgeon the ideal situation was to be in a medico-legal position to exploit the condemned *peri-mortem*. The criminal 'looked dead' in a near-death situation, but was not strictly-speaking in chemical putrefaction (in death, cell membranes become permeable and break-down releasing cytoplasm; this chemical medium contains bacterial enzymes that cause putrefactive changes in the deceased). In the East Midlands, and elsewhere, most penal surgeons got involved as close to the hanging-tree as they dared. This backdrop, emphasises why anatomization developed into a distinctive stage in the punishment choreography. In the last chapter we saw that it became a standard checking-mechanism for life-signs on arrival at a dissection venue. It was also a clever penal device to let the surgeon deal with robust bodies. Many near-death situations might otherwise have been delayed coming to them. There was no need to inflict more damage on the hanging-tree if medical expertise was prepared to finish the task. That tactic agreement meant that surgeons learned to exploit whatever concessions executioners were prepared to make to them. Extensive damage to the body to complete the death sentence of a dangerous murderer was never outlawed (quite the reverse). It was different timings and understandings of discretionary justice that shaped the actual moral standing and practical working relationships of medico-legal officials across the East Midlands. In the end, the penal surgeon's medical prerogative was decided by body types and the timing of capital death on the Nottingham gallows: the medical logistics are therefore illuminating in the archives.

The Stranger's Guide to Nottingham (1827) described how local executions had traditionally taken place at Gallows Hill on the outskirts of the town. The hanging-tree was then moved to the front of the Shire Hall, which was rebuilt around the time of the Murder Act. It was: 'A good brick building, faced with stucco. In the front is a clock, which strikes the hour upon a bell on the roof of the building'. Its purpose was to count down each execution, for, as one contemporary wrote, the clock-bell has 'a very fine mellifluous sound and was heard at a considerable distance.'[77] To control the large crowds, 'The west front is guarded by iron palisades, through which we pass by a flight of stone steps into the hall, where the assizes and sessions are held, and also the Mayor and Sheriffs' court once a fortnight, and the Sheriffs' county court once a month.' The gaol was conveniently located here too and a door in the high-pavement opened to take prisoners down into cells cut out of caves, a natural occurrence, in the subterranean underworld of the town-centre. Situated opposite was St. Mary's Church with a convenient graveyard in which to bury the condemned, but not before local surgeons had an opportunity to punish the corpse post-mortem. From the 1760s they received all sorts of bodies in a variety of physical conditions and medical circumstances. This reflected how Nottingham developed a number of execution traditions that were influential across the Midlands.[78] It had for instance been accepted practice to compel the condemned to attend a final sermon at St. Mary's Church on the day after sentencing but before their send-off. The Biblical theme of the 'wages of sin are death' was preached from the pulpit by a clergyman exhorting repentance. If prisoners were not penitent then a second tradition was to take the condemned out into the churchyard and get them to lie down in their newly dug grave to make sure it fitted their body size: a misleading public relations exercise since after dissection there would not be much left to inter. A third way to prepare the prisoner for atonement was to dress them in a shroud and lead them through the town-centre in a public procession to the execution site. This meant that when a penal surgeon checked for life-signs the condemned was fully clothed in a rough woollen garment from head to toe. Quickly stripping the body was essential before monitoring medical death since the shroud was covered in excrement, sweat and urine. Some prisoners had thus a 'good body (young and muscular)', whereas others were 'contaminated bodies (dirty and diseased)', with the majority in a 'bad shape (after hurried or mishandled executions)'.

At Nottingham on for example 17 April 1784 when two culprits called 'Henfrey and Rider' were hanged 'the executioner fixing the rope too far back on Henfrey's neck' had to ask the prisoners to assist him in getting the rope into the correct position.[79] Newspaper reporters noticed that 'the caps provided' to cover their faces were 'too little'. Things were so badly run that 'Henfrey called on Rider to make a spring [escape]' because he thought they had a good chance of making a getaway. Ironically the act of pulling away on the misplaced rope actually tautened it around Henfrey's neck as he leapt forward and 'doing for himself at the same instant, by which he broke his neck and never stirred afterwards'. In a young and muscular man such a quick break left behind a 'good body' to be argued over because as a highwayman he did not automatically go for dissection but was a valuable research tool that might be an 'extra'. Rider meanwhile had lost faith that he would be killed efficiently. The local press noted that 'he asked a gentleman for his watch, to look at what o'clock it was' to count down his deadline.[80] In such a state of heightened anxiety and fear his corpse was unsurprisingly in 'a bad condition' at the end.

Other bodies were handled much better like that of Anne Castledine on 17 March 1784 convicted of child murder. At a double-hanging she was executed with 'Robert Rushton the fellow murderer' and both were despatched with efficiency. Yet it was Anne's body that aroused intense medico-legal interest in the Midlands. The *General Evening Post* recorded that 'both bodies were 'taken to county hall in order to be publicly exposed and dissected'.[81] Further source material uncovers however how gender dictated the precise medico-legal steps. Robert's body was muscular and therefore valuable. He was opened up to be anatomically checked and later dissected in Nottingham town. Anne's corpse was initially opened up with a 'crucial incision', the cross-like cut on her torso, to establish her medical death. Then it was 'exposed on boards and tressels [sic] in front of County Hall for two days' so that ordinary people could walk around it and see that a child killer was *'truly dead'*.[82] In the next chapter we will encounter this basic equipment in more detail. For now what is important to appreciate is that the table was mobile, it could be levered up and down to take in and out of County Hall each night, and had to be erected twice on two separate days to satisfy the large crowds filing past over a forty-eight hour period. Meantime there was considerable local discussion about where to dissect such a 'good body'. She was a fertile young woman and corpses like it attracted a lot of medical competition. In the end a decision was taken by a judge in consultation with the local medical fraternity to send her body to 'a surgeon in Derby'. Nearby, men like Erasmus Darwin were considered better qualified to make full use of such a valuable medical research opportunity (we will be revisiting the Derby medical scene in more detail in Chapter 4). The matter however did not rest there.

A lot of urban myths soon circulated about post-mortem 'harm' in controversial cases across the Midlands. The 'good body' might be morally contaminated but it was believed to have romantic connotations and material potency that endured *'in the name of Death'*. Local gossip thus held that just before the Derby dissection of Anne Castledine got underway a 'strange gentleman took up the heart, kissed and shed tears upon it, squeezed a drop of blood on a handkerchief and rode off'.[83] Such was the power of the dissection story of this criminal corpse that almost a hundred years later local histories repeated the heartfelt tale verbatim. It is a keen reminder that bloodletting to check vitality was also bound up with popular notions of the wounded heart, being broken-hearted, and the heart as an emblem of the emotional centre of humanity. As Louisa Young in her recent reappraisal of the history of the heart reminds us: 'It is painful for the heart to be written on, engraved, branded: and painful to be opened up and read'; if moreover 'the heart is full of blood, then blood can be ink and the heart an inkwell ...from the 17th century heart-shaped inkwells abounded' in popular culture.[84] The deterrence value of this type of heart-narrative is self-evident and was exploited by legal, medical and religious writers. Yet, the blood-stained handkerchief dipped in the heart-drops of a child murderess by a stranger, who may have loved and lost, also keyed into popular notions about the vital nature of bodies and their organs sent for criminal dissection. Surgeons had to work with such traditions or risk losing bodies altogether across the Midlands region.

Another set of executed bodies was far more troublesome because getting them undamaged was hazardous. On 10 August 1803 William Hill was hanged at Nottingham for the 'rape (attended with great brutality) of Mrs Sarah Justice, the wife of a respectable farmer at Bole, near Gainsborough'. Local newspapers reported that Hill had been brought up 'very imperfectly, – an associate of bad men, dissolute, and grossly licentious'. He confessed to the prison-clergyman that: 'He had made criminal attempts upon other females, one of whom was only 12 years of age, but without success'. Newspaper sources suggest that there was little local sympathy for such a violent character and few locals cared what happened to his body after being hanged. The Murder Act did not send convicted rapists to be dissected but few doubted that he deserved the harshest post-mortem sentence. Sentiments hardened when Hill tried to escape on the way to Gallows-hill. On the morning of the execution it took six men to get him out of his cell, he managed to leap from the cart transporting him, and the crowd was said to be in 'great astonishment and agitation'. In the end

> Several of the Sheriff's men to hand immediately struck him with the blunt end of [their] javelins, and in spite of his horrid language and struggles, he was again forced to ascend the cart, and compelled to meet his fate. This hardened malefactor was about five feet eight inches high, and extremely robust and muscular.[85]

This was the sort of 'extra body' that surgeons hoped to get, but being bludgeoned before execution did damage the surface anatomy. So 'extras' tended to be about opportunity costs too. They often came with surprises as well. Thomas Dover who had been a physician in practice for fifty-eight years wrote in 1762 that he once found a large quantity of 'mercury... in the Perineum of a Subject I took from the Gallows for Dissection' which had 'rotten bones' caused by the poison that did not show up on the skin's surface before execution.[86] This standard treatment for venereal disease made his contaminated body doubly dangerous in death. When indeed the executioner was really inept 'extras' often turned out to be 'lost causes'. This was the case on 16 August 1791 when John Milner was executed for 'cow stealing' and the local press reported:

> He had hung a few seconds, when the knot of the rope gave way, and the wretched man fell heavily to the ground. Upon being raised, and the preparations for his suspension commencing afresh, he seemed painfully conscious of his situation and exclaimed "My God, this is hard work". Either from the inefficiency of the hangman, or some cause unexplained, several minutes elapsed ere the preliminaries were re-adjusted ; and it was not until a ropemaker named Godber pushed the blundering functionary aside, and tied the man up himself, that the cart was a second time driven away. The execrations of the spectators at the executioner were very loud and general. Milner was a native of Eakring, and was a very stout, broad-set man, nearly six feet in height. His body was buried the same day, in St-Mary's church yard.[87]

There was of course the option to disinter for 'extras' not covered by the Murder Act but again local surgeons knew that acting illegally was tricky. In the well-known case of Thomas Hallam a convicted thief sentenced to death in 1738 he was

> Interred in the churchyard of St-Mary, Nottingham, and a surgeon of the town, anxious to possess the body for the purpose of dissection, engaged a man named Rolleston to clandestinely remove it, from its resting-place. Accordingly, Rolleston and a confederate disinterred the body the night after burial, and placing it in a sack, the former took it away on his back... the undertaking, thus far, was comparatively easy. But on passing along Stoney-street, Rolleston fancied he heard the dead man pant, and his terror was so great, that he gladly left hold of his burden, and with his assistant, ran as if for life. Becoming, however, somewhat reassured, they, after an interval, deliberated on the course best to be pursued. If they took poor Hammond to a surgeon, it was their conviction, uncharitable as it might

seem that he would cut him up, dead or alive; and if he really should not be dead, it would be equally dangerous to leave him in the street as to take him away. They therefore took the corpse, and placed it in a barn of a field. It was found there in the morning by a woman who came to milk, and was subsequently reinterred by friends, at Sutton-in-Ashfield.[88]

The physical condition of the corpse—whether it was in a *good* or *bad shape*, *contaminated* or an *extra*—reflected a wide range of medical opportunism in Nottingham, across the Midlands and other English counties too. This seems to explain why it was that some engravings of the Murder Act became so infamous in contemporary culture. Behind the imagery is a disturbing set of surgical encounters that merit reconsideration too in the final section of this chapter.

THE ARTFUL CARICATURE OF THE CRIMINAL CORPSE

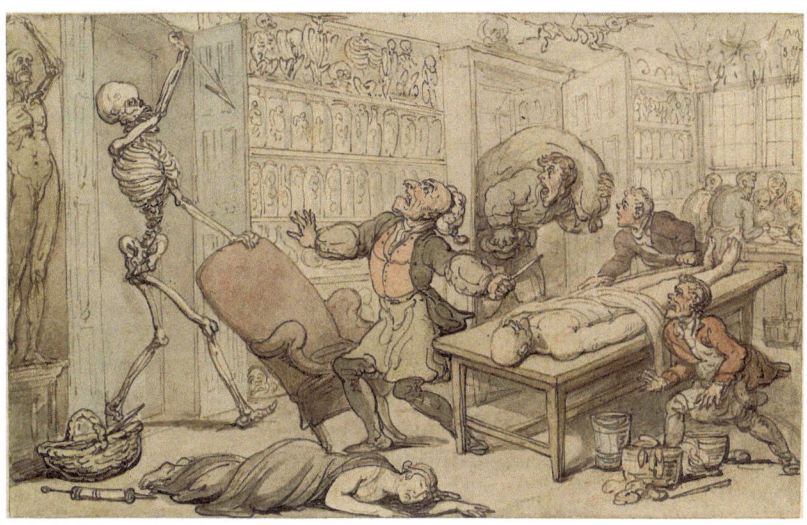

Illustration 3.1   © New York Public Library, Digital Collections, NYPL: b16830809, taken from Spencer Collection, Spencer Coll. Eng. 1815–16, Thomas Rowlandson (1815), '*Death in the Dissection Room*', original sketch published in '*The English Dance of Death*', *from the designs by Thomas Rowlandson with metrical illustrations* by J. Diggens published by R. Ackermann London 1814–16; Creative Commons Attribution-NonCommercial-ShareAlike 4.0 International License (CC BY-NC-SA 4.0)

In 1815 Thomas Rowlandson produced a cartoon described in art history books as '*Death in the Dissection Room*' (above Illustration 3.1). A central skeleton figure threatens with violence medical men and their resurrection-suppliers. It looks as though two anatomists and an assistant are about to dissect a cadaver, or does it?[89] Art historians have traditionally interpreted the giant skeleton to the left as *Death* (he is well over 6 foot in height). This angry and aggressive embodiment of the Grim Reaper seems to be seeking revenge on anatomists who have paid thieves to disturb graves for medical-profit. This book argues that a conventional reading of sack-men pillaging cemeteries is taken out of its medical history context. It is all too easy to read a picture caption, and look at what it says, rather than what is actually depicted.

Strictly-speaking the giant skeleton in the cartoon is not *Death* but *Medical Death* standing-by to kill a body on the dissection table: an important distinction. This is why the medical men and sack-men are caught by surprise. The skeleton pulls aside a large red chair (satirizing perhaps this most tradition symbol, a chair in medicine) to stop the anatomist cutting the corpse. Medical death has dominion here and he is poised to take the plunge. Notice for instance just how carefully Rowlandson has drawn the slanting perspective of the anatomist's knife. It is positioned in a diagonal-line with medical death's fatal arrow. The effect is to create a viewing point that draws the audience's eye to the question of who has anatomical authority over 'absolute Death'. In this duel for medico-legal supremacy, the combat is not one the anatomist can win. What is needed is the presence of medical death in the room to proceed to dissection. Even though the lifeless looking body looks ready for a clean cut to the carotid artery in the neck, or along the chest cavity on the pencil-line of the breast-bone, it is self-evidently not '*truly dead*'. Medical death is roused because the pale cadaver has vitality that is still in his purview. This general subtext is repeated in the placement of the objects on the floor where there is a basket of medical tools and a syringe to the side. These had two functional uses: to first inject hot water into the vital organs to test for vitality and then to squeeze wax into the body to preserve any tissue for anatomical specimen-taking. Looking more closely at the cadaver laid out is informative about body conditions in the room.

This is a body that has been cleaned up because it is unblemished and bloodless, a spectre of a human being but one still identifiable to the crowd: a crucial distinction. It has been shaved, is hairless, naked, and seems to have been washed down with camphor by the anatomist's assistant—notice his foot in the bucket, water spilling out, normally used to clean and then swill down the corpse. The skin is very pale and therefore drained of blood in the epidermis. So the person appears to be according to eighteenth-century sensibilities on the boundaries of life and death in the worst state of all 'the Name of Death'. It is very difficult to decipher whether there is any indication of a rope mark around the neck, but it is a tall man of considerable stature, partially draped with a cotton winding-sheet. If he has been dug up then he is remarkably unsullied by the wooden spade, the strong pull of the grave-robbers out of the top of the coffin, and there is no sign of any bruising, mud-stains or straw-marks, on his exposed flesh from the exhumation. The winding sheet looks skimpy emphasizing that he may have been a poor man, his family unable to afford a secure burial, or a criminal cut down from the gallows not yet prepared for burial since he must undergo post-mortem punishment. Whatever his supply-source although he is lying in a dissection room, no anatomical cuts have been made as yet to his torso. We are seeing what the crowd would have looked at as they accompanied the 'good body' on the punishment journey of the condemned.

Perhaps the most intriguing thing about the image is the lack of a response in the medical-men huddled around another body in the background. Self-evidently medical death has no dominion in that part of the picture because a dissection and dismemberment is already underway on someone in 'absolute Death'. Nor does the skeleton figure of medical death threaten the woman lying-down on the floor in the foreground. She is a curious figure and to modern eyes looks superfluous and somewhat melodramatic. She cannot be dead since her limbs are not in rigor mortis. There is fluidity in her bodily expression. Her artistic placement seems therefore to have an experiential purpose. She has fainted just as the body is about to be cut open. Maybe she was the type of female that was often mentioned in newspaper reports who were enticed to the dissection room by the spectacle of an alpha-male, dangerous and damned. In this scenario, the reality of seeing a dissection was a fainting prospect. Another possibility is that in private medical museums along Fleet Street women were sometimes trained to prepare specimens and act as a guide to anatomical collections.[90] Rowlandson seems to be mocking the weaker sex whose desire to be involved with notorious criminals had a material downside at criminal dissections: a theme we will be returning to in Chapter 6 when we consider how difficult it was to endure a full-scale dissection. There are of course many ways to read satire, but the balance of the visual and medical evidence surrounding the image suggests that the criminal corpse was a complex medico-legal commodity when it came into the dissection venue. It is important not to over-interpret visual sources since they often contain a certain artistic integrity that we may never grasp or which we have to learn more about since Rowlandson drew for private and popular consumption. Yet images like this also command our historical attention by encouraging us to look afresh at the complexity of medical death, anatomy and dissection, especially when we consider that this interior scene is basically what many provincial anatomy spaces looked like after the Murder Act (see Part II). They also alert us to lots of ambiguities and contradictions stimulated by the Bloody Code. Of these, the basic condition of corpses was subject to medico-legal modifications and new anatomical thinking over time. Those trends bring us to the famous images drawn by William Hogarth of criminal dissections and what they reveal about bodies being punished in the presence of the crowd.

In the history of eighteenth-century crime, Hogarth's four engravings *The Four Stages of Cruelty* (1751) are iconic. The final image in the series depicts a criminal archetype called Tom Nero who has been tried and found guilty of murder. He is about to meet the *Reward of Cruelty* by being dissected in an anatomical theatre. The engraving anticipated the Murder Act (1752) by twelve months. It depicted Tom's corpse opened by surgeons and about to be hauled up high on a pulley so his skeleton can be seen by those assembled under the auspices of a leading professor from the Royal College of Physicians. Looking anew at this symbolic representation from the perspective of the archival material presented thus far in this book, there are three things that are striking (see Illustration 3.2 below).

Illustration 3.2   © Trustees of the British Museum, William Hogarth (1751), '*The Reward of Cruelty*', Image Reference, S, 2. 126, Digital Image Number AN16677001, original engraving; Creative Commons Attribution-NonCommercial-ShareAlike 4.0 International License (CC BY-NC-SA 4.0)

The first is that this is arguably one of *the* most disturbing images of the eighteenth-century capital code because Tom Nero looks more alive than dead. Vic Gatrell first commented on the emotional appeal of the pained face. It is difficult to avoid the twisted visage that seems to express genuine feeling. The pincers on his nose are props for Hogarth's satirical angle that maybe Nero can smell his own body stink. Can he experience being in a bad shape, a contaminated state, with a muscular frame that is a medical prize? There is an unmistakable human impulse that pain however justified can be filled with pathos too. A second observation is that those surrounding the corpse do not flinch in the face of this pain and this is normal. This is the price of murder and the upside of medical advancement; or is it? These medical men according to the Hippocratic Oath are healers. Here they have become agents of a criminal justice system that involves them in taking life. Again, Hogarth's satire is bitter tasting. Tom Nero's story is about a cruel society that has bred a cruel boy who embraced cruel criminality and meets a cruel end—even so, it would be cruel indeed if the anti-hero had not yet reached medical death! Either Hogarth exploited his artistic license to dramatize the scene's medical melodrama to increase sales of his engraving or he had seen from his visits to St Bartholomew's Hospital dissection rooms that not everyone died on the gallows.[91] It is important to be alert to such alternative readings because in many respects the image is so familiar that it has become almost an art historical cliché.

Another, third, striking feature in the engraving is that the image made sense to those at the time: a point Vic Gatrell first made. Hogarth seems to be encouraging the viewer to see the rule of law as contradictory and contrived. At the heart of the *Reward of Cruelty* are shifting concepts of medical authority that are powerfully in transition, just as the criminal body is a powerful instable metaphor being deployed by the State. It's medico-legal character will go on being modified for the next thirty years and so an image that looks finished is actually about to be continually refashioned. It resonates powerfully to viewers because the engraving was produced when the Murder Act was imminent but still lacked legal bite (unlike the dog eating the intestine sausages!). Hogarth knew broadly what the new capital punishment entailed and he was familiar with dissection sessions, but only after 1752 were these brought together for wider public consumption. So this image is a modified version of an artistic vision of what is coming into force. The crowd are outsiders, but within a year they will be insiders, seldom excluded from the post-mortem rites. The terms of the new legislation's authority are therefore still being refined, reworked and reviewed. The deliberate headcount of 21 medical men given a privileged access hints at this instability since the post-mortem punishment peopled by medical actors is fluid. Historical sources confirm that the physicians, surgeons, and barbers who look united in the image were the opposite. Their authority was a divisive one and could be acrimonious. It is notable that modified reforms would result in the barbers and surgeons splitting: the latter forming the Company of Surgeons in 1745, later awarded a Royal Charter by 1800. They would change premises too: the surgeons moving from the Old Bailey to Lincoln's Inn Fields in Holborn by 1797. So Hogarth's chosen medical space, its staging, the predicted audience, and the medico-legal scenery, are under constant negotiation because of the multiplicity of meanings attached to the central figure of the criminal corpse, and medically-speaking, he is not a corpse, but *peri-mortem*. The image is a performance of a large-scale drama and this is Act One. The 'bad shape' of the main character promises lots of scene-changing, Tom Nero arrests attention because what cannot be denied is that his body is in a pained condition and he is a medico-legal creation. Ironically this revitalises his disempowered corpse as life ebbs from his body: the reverse effect to the official one intended.

We end this brief discussion on the condition of criminal corpses at a theatrical time in imaging-making. The *Drury Lane Journal* featured this artistic trend on 20 February 1786. A satirical play had just opened called '*Punishment of Felonies, Burglaries and the like*'. In advance of opening night a poster invited theatre-goers to read a synopsis of its script. The contemporary mockery was evident for the extension of even harsher capital legislation:

SCHEME

FIRST, That one or more surgeons be appointed for every jail, to make ottomies [anatomies] of all the condemn'd bodies.

SECOND, That all the malefactors, within two days after the death sentence is pass'd upon them, be cut up alive in the prison yard; and that every one confin'd there for capital offences be oblig'd to stand by and see it done.

THIRD, That, while they are thus ottomising [anatomising] they be tyed hand and foot to prevent their struggling and that their mouths be gagg'd, to hinder their horrible shrieking and groans.[92]

If, as the satire of Rowlandson, Hogarth, and their circle of artists that attended criminal dissections seems to suggest in their image-making, the melodrama of human vivisection had a strong basis in medical reality, then their artistic integrity deserves better recognition by criminal historians.

The crowd that gathered round the corpse in a 'bad condition' came to know its medical potency in the dissection and playhouse theatres of Georgian England.

CONCLUSION

Penal surgeons had to exploit those opportunity costs that came within the scope of a working-choreography of punishment rites at the gallows. They tried to assert their medical prerogative when the condemned became *peri-mortem*. It was essential that those convicted to die were handed-on with as little damage as possible from the execution process nevertheless the material condition of the human material was inconsistent. Surgeons could try to intervene to speed things up but this was often decided by two local factors. If there was a lot of 'natural curiosity' about the condemned, then it was better to stand back and wait to be called forward by the hangman. On other occasions it was possible to be more assertive when resentment against the convicted was intense in the vicinity of a violent homicide. The varying nature of the emotional engagement and the changing tempo of the penal rites meant that the *peri-mortem* stage was a medical dance. It had to be estimated strategically from a material and metabolic standpoint.[93] This was then a liminal penal space filled with synaesthesia and of stimulating *emotives*. These aroused the archaeology of emotional expressions the early modern crowd had an opportunity to give voice to, in the sorts of ways that William Reddy has envisaged (introduced in Chapter 1).[94] Although it has not been possible to hear every conversation in oral cultures at this historical distance, the fact that a lot of conversing was going on was however a consistent feature of the average penal surgeon's working-life: a theme Chapter 6 returns to. The actual bodies about to be punished smelled foul and were hard to handle. They slipped and slid through surgical fingers, wet with urine, faeces, semen, blood and sweat. Even experienced medical men like those of the famous Hey family of surgeons in Leeds admitted privately that: 'No stink can compare to a body stink'. Strong prisoners capable of strangling murder victims with their bare hands had powerful willpowers to survive. They also had muscular bull-necks that did not break on the rope. Their shaven heads and tattooed torsos made them in visual terms striking physically from head to toe. This made them mightily attractive and pungently repellent. Most spectators therefore experienced mixed-emotions about *peri-mortem* bodies and surgeons navigated that spectrum *in situ*.

On being cut down from the gallows some convicted murderers were strictly speaking still the '*dangerous dead*'. Potentially many threatened

the public good of the local community by being contaminated by common diseases like head lice, dangerous summer fevers, and life-threatening
conditions such as dysentery of the bowls. Lying on the floor or on the
dissection table some were pock-marked by smallpox. A nick of the dissector's knife was a well-known health risk to the penal surgeon on duty
(see Chapters 5 and 6 for more details). Bad blood could then literally
blood-poison the post-execution rituals. Against this hazardous backdrop
it is striking how many ordinary people wanted to be in close proximity to
the criminal corpse oozing body fluids and spirited danger. That outcome
expresses the disturbingly powerful nature of the post-execution punishment rites for many ordinary early modern people. Trial narratives suggest
that the five senses of each spectator were arrested as they jostled to get a
good view of the body laid out for their inspection; and this, regardless of
the scale of underlying emotions, whether apathetic, mildly interested, or
thrill-seeking. Above all, it was the condition of the criminal corpse that
shaped what was about to happen next after the gallows. Neglecting to
trace the onwards passage of the body through the punishment process
is therefore a significant historical omission in eighteenth-century studies.

A wide variety of contemporary sources suggest that the power of the
criminal corpse was mutable and pluralistic. It was the subject of a sort
of 'cultural compost'.[95] The body degraded physically in the dissection
venue but the processes of disintegration and their timings had a powerful symbolism too. The pre and post-executed body naturally worked in-
tandem and had a medical symbiosis that has been neglected. This explains
why anatomization was about multi-tasking: to check lifelessness, manage
rigor mortis and slow down 'harm' done on the gallows. It got the condemned into a medical sphere of discretionary justice before putrefaction
speeded up. Artists like Thomas Rowlandson and William Hogarth consequently saw the criminal corpse as a compelling subject for satire. The
challenge for historians is to look anew at what their artistry reveals about
a cross-section of public opinion. It was not obvious or inevitable that
penal surgeons would open the corpse cleanly. They could be caught in
a medico-legal duel for the condemned in the cartoons of Rowlandson.
Often they were depicted in an unedifying ethical situation surrounding
human vivisection illustrated by Hogarth. In turn, both anatomical trends
were mocked in melodramas on the London stage. Enacting the Murder
Act was never easy, nor palatable, to early modern sensibilities confronted
with different types of medical opportunism. Spectators might relish an
execution but be unable to stomach its aftermath however much they
were determined to do so. Even the most blood-thirsty found it hard to

bear the olfactory showcase. A worrying logistical problem was how to manage the types of unsavoury anatomical information (like spontaneous menstruation and ejaculation) that came into the public domain from those that crowded round the criminal corpse. Copy-cat killings were likewise unnerving for a legal fraternity charged with maintaining public safety. In notorious cases, dangerous criminals did use execution methods and dissection techniques in circulation to murder their victims. Opening up the criminal justice system to try to increase the deterrence value of capital punishment had therefore a troubling downside. To get closer to the dilemmas contemporaries encountered, Part II of this book now journeys onwards with the criminal corpse that captured widespread 'public curiosity' in English provincial life.

NOTES

1. V. A. C. Gatrell (1994) *The Hanging Tree: Execution and the English People, 1770–1868* (Oxford: Oxford University Press).
2. Ibid., p. 29.
3. Gatrell, *The Hanging Tree*, p. 46.
4. On the longevity of medical definitions, see, Arthur S. P. Jansen, Xay Van Nguyen,Vladimir Karpitskiy, Thomas C. Mettenleiter, and Arthur D. Loewy (1995), 'Central Command Neurons of the Sympathetic Nervous System: Basis of the Fight-or-Flight Response', *Science*, Oct. Vol. 27, 644–6.
5. The autonomic nervous system (ANS) comprises two branches: the sympathetic nervous system (SNS) and the parasympathetic nervous system (PSNS). The SNS is the "fight or flight" system, while the PSNS controls heart rate, reflex actions, and brain input. The SNS is a quick response system that mobilizes the body into action whereas the PSNS slows down strong physical reactions for emotional well-being: the two branches work by balancing each other under normal conditions, but tend to act independently in response to fear and trauma.
6. Paul Barber (1998), *Vampires, Burial and Death: Folklore and Reality* (New Haven: Yale), p. 119.
7. See, M. M. Shoia, B. Benninger, P. Aqutter, M. Loukas, and R. S. Tubbs (2013), 'A Historical Perspective: Infection from Cadaveric Dissection from the Eighteenth to Twentieth Centuries', *Clinical Anatomy*, Vol. 26, II, 154–60.
8. Some historical demographers have seen the long eighteenth-century as an age when the triage of social cleansing typified English society, see, for instance, Richard L. Rubenstein (1983), *The Age of Triage: Fear and Hope in an Over-Crowded World*, (Michigan, USA: Beacon Press, University of Michigan).

9. 'Extras' were those hanged for a lesser capital offence like highway robbery. Some sold their body before death to pay for a good send-off, or their family were paid to supply the surgeons.

10. *Chester Annual Register*, (Chester, 1771), entry 7 September, execution of John Chapman.

11. Ibid.

12. J. Lawrence (1983), *The History of Capital Punishment* (London: Citadel Press), pp. 44–8.

13. Simon Devereaux (2009), 'Recasting the Theatre of Execution: the Abolition of the Tyburn Ritual', *Past and Present*, Vol. 202, 127–74, argues that this increased the 'theatrics of the execution' staging and processes.

14. Gatrell, *The Hanging Tree*, p. 54.

15. Physical details are outlined in, A. Koestler and C. H. Rolph, (1961, 1st edition) *Hanged by the Neck* (London: Penguin) and Dr Harold Hillman, *Guardian newspaper*, 15 December 1990, both discussed in Gatrell, *The Hanging Tree*, p. 45, endnote 54.

16. Thomas Stuttaford, "Swift end rests with skill of the hangman," *Times* newspaper online, (1 January, 2007), http://www.timesonline.co.uk/article/0,,3-2526006,00.html.

17. Robert Hutchinson (2007), *Thomas Cromwell: The Rise and Fall of Henry VIII's Most Notorious Minister* (London: Weidendfeld and Nicholson), p. 64, explains that Cromwell was in charge of the first executions of the Reformation under Henry VIII. It was he who permitted using a dagger or sword to cut the head off the criminal body whilst it was still jerking on the rope, which the Murder Act repealed.

18. *Chester Annual Register*, (Chester and London: Baldwin, Craddick and Joy, 1771), entry 7 September, execution of John Chapman.

19. On Leicester executions, see, Alfred Temple Patterson (1954), *Radical Leicester: A History of Leicester, 1780–1850* (Michigan: University of Michigan Press), p. 160.

20. On eighteenth-century capital punishment timings and their disputes, see, Jerry White (2013), *A Great and Monstrous Thing: London in the Eighteenth Century*, (London: Random).

21. *London Magazine*, (April 1733), pp. 213–3.

22. *Historical Chronicle, Gentleman's Magazine*, Volume 6, (September, 1736), p. 549.

23. See, list of eighteenth-century sources cited in *Notes and Queries* (January-June, 1861), 2nd series, Volume II, p. 314.

24. Refer endnote 1.

25. This Biblical phrase is attributed to the English evangelical preacher and martyr, John Bradford (1510–55). He is quoted as saying, 'There but for the grace of God, goes John Bradford', when witnessing criminals being led to the scaffold. Later he was burned at the stake in 1555 and went to the gallows

pronouncing, 'We shall have a merry supper with the Lord this night'. The sentiment inspired early modern Christians, like, Edward Bickersteth (1822) *A Treatise on Prayer* (London: private publication), p. 60.

26. Peter King (2000) *Crime, Justice and Discretion in England, 1740–1820*, (Oxford: Oxford UP, 2000), p. 1.

27. Ibid.

28. Mary Shelley (2003 edition), *Two Works by Mary Shelley* (London: Peverell Press), chapter 21, Frankenstein.

29. He was the nephew of William Hey senior, a leading Leeds surgeon, who received gallows bodies in the city.

30. Leeds University Special Collections, MS/1990/5, Hey Family Correspondence, 1828–42, Samuel Hey wrote to his elder brother William at St John's College Cambridge about his medical training experiences.

31. I am grateful to Dr Richard Ward for altering me to the fact that William Hey senior also discussed his preference for bodies in winter and spring rather than summer; see, Richard Ward (2015), 'Wilberforce, Anatomists, and the Criminal Corpse: Parliamentary Attempts to Extend the Dissection of Offenders in Late Eighteenth-Century England', *Journal of British Studies*, January, Volume 54, Issue 1, 63–87; copy available at Leicester University, pp. 1–28, quote at p. 12, taken from Spencer-Stanhope MSS, SpSt/11/5/1/2, WYAS.

32. Andrew Knapp and William Baldwin eds (1828) *The Newgate Calendar, 1824–8* (London: J Robin and Co), 'Case of William Proudlove and George Glover executed at Chester, 28 May, 1809, for salt-stealing, after a First Attempt to hang them had Failed' e-transcript available at http://www.exclassics.com/newgate/ng502.htm.

33. 'Remarkable Domestic Events: Fortitude', *The Historical Magazine*, (April, 1789), p. 271.

34. These attributes are now well established, as a result, in crime histories, see, for a recent example, Richard M. Ward (2014), *Print Culture, Crime and Justice in Eighteenth-Century London* (London: Bloomsbury).

35. See, historic-standing in, Carol Heron, John Hunter, Geoffrey Knupfer, Antony Martin, and Charlotte Roberts (1995), *Studies in Crime: An introduction to Forensic Archaeology* (London and New York: Routledge).

36. Refer, James Copland (1833), *A Dictionary of Practical Medicine, Vols. 1–3* (London: Messrs Longman).

37. This is often noted in diaries about hospice care or dying at home. Hon. Jane Clark, wife of Alan Clark MP for instance records that Alan's dead body laid out for burial in their bedroom was 'I must say, had made our room a tiny bit high, and the window had to be shut because of the flies' in late-summer, A Clark, transcribed and edited by Ion Trewin (2002), *The Last Diaries: In and Out of the Wilderness: Alan Clark Diaries, Volume 3*, (London: Phoenix Press) entry Monday 6 September 1999, p. 491.

38. Refer, for example, W. Lambert, medical commentator (1831), 'Letter to Lancet: Effects of Hanging Upon the Organs of Secretions & so on', *Lancet* dated 14 Sept., published in the Sept. edition, p. 808, stated that: 'penal erection of the male sex by the infliction of death by hanging, [it] is a well-established fact'.

39. See, Edward H Campbell, Surgeon, Royal Navy (1831), 'Letter to Lancet: Occurrence of Menstruation during Hanging', *Lancet*, letter dated 20 August 1831 but published in the Sept. edition, p. 704 and discussion with its taboo nature in Charles Cooke, Surgeon, Holloway Prison (1831), 'Letter to Lancet: Occurrence of Menstruation during Hanging', *Lancet*, letter dated 1 September, published in the Sept. edition, pp. 751–2.

40. Refer, Rick Harrington (2013), *Stress, Health and Well-being: Thriving in the Twenty-First Century* (Belmont, USA: Wadsworth Publishers), p. 83.

41. A general point made by Barber, *Vampires, Burial and Death,* p, 105 but seldom with regard to criminal dissections.

42. Refer, John M. Beattie (2001), *Policing and Punishment in London, 1660–1750: Urban Crime and the Limits of Terror Part II,* (Oxford: Oxford University Press).

43. Normal life expectancy was around forty-eight years old over the course of this book's chronological focus.

44. E T Hurren (2011), *Dying for Victorian Medicine: English Anatomy and its Trade in the Dead Poor, 1832 to 1929,* (Basingstoke, Palgrave), chapter 3, details the nineteenth-century use of water baths for body preservation.

45. Barber, *Vampires, Burial and Death,* p. 110.

46. See, Colin Jones (2014), *The Smile Revolution in Eighteenth-Century Paris* (Oxford: Oxford University Press).

47. See, notably, David Garland (2011) 'Modes of Capital Punishment: The Death Penalty in Historical Perspective', in David Garland, Randy McGowan and Michael Meranze eds. *America's Death Penalty: Past and Present* (New York: New York University Press), chapter 2, pp. 30–71, quote at p. 42.

48. Stuart Banner (2003), *The Death Penalty: An American History* (New Haven: Harvard University Press), chapter 3, 'Degrees of Death' pp. 33–87, quotation at p. 70 and endnote 33 citing original English sources.

49. Thomas Laqueur (2011), 'The Deep Time of the Dead', *Social Research,* Vol. Fall, III, 799–820, quote at p. 802.

50. Ibid., p. 805.

51. Esther Cohen (1989), 'Symbols of Culpability and the Universal Language of Justice: The Rituals of Public Execution in Late Medieval Europe', *History of European Ideas,* Vol. 11, I-VI, 407–16 quote at p. 410.

52. John William Holloway (1831), *An Authentic and Faithful History of the Atrocious Murder of Celia Holloway*, (London: W. Clowes and Son Ltd), p. 235.
53. Ibid., p. 237.
54. Evidence compiled from contemporary reports in the 'Confession of Holloway', *The Observer*, 22 August 1831, and *Brighton Herald*, 20 August 1831.
55. In court John defended that his lover was innocent and the constables failed to find enough reliable evidence to convict Ann Kennet as his accomplice. She was subsequently released and disappeared from the local area.
56. Holloway, *An Authentic and Faithful History*, p. 269.
57. Ibid., pp. 287–88.
58. Holloway, *An Authentic and Faithful History* p. 291.
59. Ibid, p. 291.
60. Holloway, *An Authentic and Faithful History* p. 292.
61. F. and K. Wood eds (1992), *A Lancashire Gentleman: The Letters and Journals of Richard Hodgkinson, 1763–1847* (Dover: Alan Sutton Press). p. 50.
62. I am grateful for an advance copy of Shane McCorrestine (2015), *William Corder and the Red Barn Murder: Journeys of the Criminal Body* (Basingstoke: Palgrave Pivot).
63. Albert Manns (2012), 'Shorn Scalps and Perceptions of Male Dominance', *Social Psychology and Personality Science*, July, VII, p. 1, led to him concluding that 'men experiencing natural hair loss, may improve their interpersonal standing by shaving [more]'.
64. A gender theme explored by social scientists like Richard Collier (1998), *Masculinities, Crime, and Criminology: Corporeality and Criminal(ised)*, (London: Sage Publications).
65. It was a small town in Surrey, 30 miles south of central London.
66. 'Report and Trial of Chennel and Chalcraft', *The Observer*, 16 August 1818, pp. 1–14, quotes at p. 1.
67. Ibid, p. 2.
68. 'Report and Trial of Chennel and Chalcraft', p. 2.
69. Ibid. p. 3.
70. 'Report and Trial of Chennel and Chalcraft', quotes at p. 1.
71. Ibid., pp. 13–14.
72. Reconstructed from record linkage work on court records in, Andrew Knapp and William Baldwin eds (1828) *The Newgate Calendar, 1824–8* (London: J Robin and Co); 'GEORGE CHENNEL AND J. CHALCRAFT, *Executed August, 1818 for the atrocious murder of Chennel's father and his Housekeeper, at Godalming,*' e-transcript available on http://www.exclassics.com/newgate/ng577.htm; and detailed contemporary newspaper reports in *The Examiner*, No 555 (16 August 1818), p. 519; *Leicester*

Journal and Midlands County Advertiser, Volume LXVII (21 August 1818), Issue 3346, pp. 1–4; *Edinburgh Advertiser,* (21 August 1818), p. 4.

73. See, *The Observer,* 16 August 1818 and a subsequent 'Report of the Trial of Chennell and Chalcraft' published as a pamphlet by Duncombe, 19 Little Queen Street, Holborn, 1818.

74. Morton D. Paley (1996), *Coleridge's Later Poetry,* (Oxford: Clarendon Press), p. 56 points out that there was a strong eighteenth-century connection between the word 'jail' and the concept of 'Limbo' for sinful souls.

75. He carried out 44 hangings in the East Midlands and seems to have been responsible for another 22 elsewhere, totalling 66 between 1820 and 1847. When he died aged 70 the *Law Times and Journal of Property,* (1848) Vol. 10, p. 510, said his eldest son was to succeed him as hangman.

76. Ron Knight (2009), *Murder in the Shambles,* (Milton Keynes: Author House UK Ltd), pp. 83–6, quote at p. 86.

77. Anon (1827), *The Strangers Guide through the town of Nottingham being a description of the principle buildings and objects of curiosity in the ancient town* (Nottingham: Sutton and Sons), relevant entries can be accessed online at http://www.nottshistory.org.uk/books/nottingham1827/guide6.htm.

78. These are recorded widely in local histories like John Potter Briscoe (1895), *The Old Guild Hall and Prison of Nottingham* (Nottingham: Sutton and Sons); James Granger (1907), 'The Old Streets of Nottingham', *Transactions of the Thornton Society, Volumes III and IV,* 3 and 7 February issues.

79. 'Nottingham executions', *Felix Farley's Bristol Journal,* 17 April 1784, Issue 1851, p. 1.

80. Ibid., p. 1.

81. 'Nottingham executions', *General Evening Post,* 23 March 1784 – 25 March 1784, Issue 7813, p. 2.

82. William Stevenson [of Hull] (1893), *Bygone Nottinghamshire* (Nottingham: Hard Press Publishing), p. 183.

83. James Orange (1840), *History and antiquities of Nottingham,* (Nottingham: Nabu Press), p. 447.

84. Louise Young (2002), *The Book of the Heart* (New York: Doubleday, Random Books), p. 225.

85. See, http://nottinghamhiddenhistoryteam.wordpress.com/page/11/for contemporary records access.

86. Thomas Dover (1762 edition), *The ancient physician's legacy to his country. Being what he collected himself in fifty-eight years of practice, and so on* (London: H. Kent, for C. Hitch; J. Brotherton; and R. Minors), p. 84.

87. See, http://nottinghamhiddenhistoryteam.wordpress.com/page/11/for contemporary records access.

88. Ibid., p. 11.

89. Image depicted at http://www.18thconnect.org/news/?p=80yet.

90. Matthew Craske (2011), 'Unwholesome' and 'pornographic': A reassessment of the place of Rackstrow's Museum in the story of eighteenth-century anatomical collection and exhibition', *Journal of the History of Collections*, Vol. 23, I, 75–99.

91. It has recently been established that the dissection might have taken place at St Bartholomew's Hospital, where all three identifiable surgeons in the image were based. It does however also feature the Cutlerian Theatre of the Royal College of Physicians, near Newgate. The throne is the same and bears their arms; and its curved wall did resemble a cockpit. The niches of the Barber-Surgeons' Hall were not used for dissection displays of skeletons until after the surgeons split away to form the Company of Surgeons in 1745. On the context, refer, Ronald Paulson (1992), *Hogarth*, (London: James Clarke & Co), p. 35.

92. 'Punishment of Felonies, Burglaries and the Like', *Drury Lane Journal*, 20 February 1786.

93. Building on the approach of, Fay Bound Alberti (2009), 'Bodies, Hearts and Minds: Why the History of Emotions Matters to Historians of Science and Medicine', *Isis (Chicago Journals, The History of Science Society)*, Vol. 100, Dec. IV, 798–810.

94. See, notably, William Reddy (1997), 'Against Constructionism: The Historical Ethnography of Emotions', *Current Anthropology*, Vol. 38, 327–51.

95. I owe this apt phrasing to A. Nicolson, (2011), *The Gentry: Stories of the English* (London: Harper Press).

PART II

Preamble

The Beauties of the Magazines (1762)

Do not ye prudes pretend to dread reading any further, lest you should meet with some wanton descriptions that might alarm your sensibilities; and you should be so shocked at such obscene writing, that you could not think of anything else...But to satisfy, or speak more honestly, to dissatisfy you, in this work there will not be those common place pictures, or descriptions, that only tend to make weak minds yet weaker; this work being intended a DISSECTION OF THE MIND, to lay nature naked to view.[1]

Charles Bissett MD on Putrefaction, Newcastle-Upon-Tyne, (1766)

I put the blood contained in the right auricle and ventricle of the heart, the bile that was taken out of the gall bladder, and the tainted bilious humour that collected in the duodenum into separate galley-pots; these I covered with a loose paper; and placed them on a shelf in the surgery...in order to observe their progress to putrefaction...In two hours a very small proportion of blood serum separated...At the sixth hour, several maggots were moving upon its surface...At the eighteenth from the...death, it had a disagreeable smell, and fetid putrefaction...[2]

A Colloquy between the Gallows and the Hangman:
The Evils of Execution (eighteenth century)

Museums enriched and collections enlarged:
With relics of murder and blood overcharged
And often these scenes are dressed up for the stage
And the votaries of pleasure and fashion engage
There, in all its grim horrors, the tragical sight,
Is enacted, afforded unbounded delight![3]

PREAMBLE PART II

In the winter of 1815, John Keats, an aspiring Romantic poet, was a young medical student in his early twenties doing the rounds of the dissection rooms of London. At night he wrote about a secretive anatomical world where all living things were cut open:

> [...] skeletons of man,
> Of beast, behemoth, and leviathan,
> And elephant, and eagle, and huge jaw
> Of nameless monster. [...]
> The gulphing whale was like a dot in the spell,
> Yet look upon it, and'twould size and swell
> To its huge self; and the minutest fish
> Would pass the very hardest gazer's wish,
> And shew his little eye's anatomy.
> *Endymion, IIII.*[4]

These private pursuits were costly on his emotional well-being and financial capabilities. The young poet could not fund the medical fees from his writings. Keats's mother had to pay on his behalf Guy's hospital fees, a '£1 2s administration charge' and another '£25 4s' required to 'register for twelve months as a surgeon's pupil'.[5] This would enable him to train for six months on the hospital wards to obtain enough experience to be examined by the Royal Society of Apothecaries. The longer term plan was to then study with the Royal College of Surgeons as an apprentice. Keats was hence one of '159 students that paid ten guineas for a course on *Anatomy and the Operations of Surgery* taught by Mr Astley Cooper and Mr Henry Cline' in 1815–6. Both were very skilled surgeons yet experts in different forms of dissection, as Robert Gittings explains: 'Cline was reckoned the soundest and most mechanically ingenious of operating surgeons, Cooper the boldest, most dashing and experimental' when working on the living and the dead.[6] Often each gave evidence in high profile murder cases in which expert medical testimony was required to convict the prisoner of homicide at the Old Bailey courtroom, next door to the Company of Barbers and Surgeons in the City of London. Keats kept a detailed medical notebook of the available dissections he paid to see.[7] In it, he wrote, that he was struck by Cooper's audacious speculations about whether or not 'Blood possesses Vitality [and how arteries] expel Blood in the last struggles of Life'.[8] He also drew two skulls, almost certainly criminal ones, in the margins of his note-taking. Later what he learnt from those *peri-mortem* bodies delivered for dissection emerged in his poetic musings as: 'This living hand, now warm and capable'. After Keats became a qualified apothecary his nocturnal adventures stood him in good stead. He took up the position of surgical dresser in the operating theatre of Guy's hospital. There, he learnt that human vivisection was commonplace.

In eighteenth–century England, surgical operations were harrowing experiences because alcohol and laudanum could only dull the pain for the patient before the discovery of modern anaesthetics. On returning home at night exhausted Keats wrote about how: 'Full many a dreary hour have I past/ My Brain bewildered/ and Mind o'er cast/ With heaviness'.[9] It was a cheerless task to work long hours on the wards and at nightfall have to venture out to pay hard earned cash to watch penal surgeons working on rotting bodies taken from the gallows. The medical fraternity tried to take advantage of how contemporaries thought that the condemned polluted early modern society with a sinfulness that the redemptive nature of social justice could not remedy alone. The English state, anxious and guilt-ridden by its inability to prevent the 'horrible crime of murder', had to rely on medicine to lance—literally and figuratively—the canker of homicide.[10] Part II of this book is consequently all about the sorts of anatomical settings this happened in, the surgical men present in the dissection room, and the medico-legal circumstances surrounding criminal corpses actually delivered and then cut up to 'harm' them under the Murder Act.

Each chapter that follows reflects one of three vantage points that have been neglected in the historical literature. Their themes have been identified from the source material cited on the opening page of Part II. The key gaps in our knowledge are—the shock to the sensibilities of seeing actual dissections—the medical reality of decaying flesh putrefying to be cut up—and the material collections that were created from macabre criminal work. Hence, Chapter 4 examines how the condemned was actually cut open and the extent to which procedures changed over time under the capital legislation Then, Chapter 5 investigates the locations where criminal cadavers actually became available for dissection and how punishment venues differed in the capital compared to provincial life. The punishment journey of the condemned concludes with Chapter 6's assessment of the types of original research that took place and how specimen-taking created a medical museum culture displaying criminal afterlives.

Essentially, then, *Dissecting the Criminal Corpse* utilises John Keats's experiences at St Guy's dissection room in London and others like it in

English regional society. The overall aim is to rediscover whether first-hand accounts written by famous surgeons like Sir William Osler who cut open criminal corpses in front of the would-be poet, can be relied on, or not:

> On entering the room, the stink was most abominable...The pupils carved them [limbs and bodies] apparently, with as much pleasure, as they would carve their dinners. One, was pouring Ol.Terebinth [oil of turpentine used normally as a purgative but here as a crude preservative] on his subject, & amused himself with striking his scalpel at the maggots, as they issued from their retreats; here, were five or six who had served but a three years apprenticeship, most vehemently exclaiming against that regulation in the Apothecary's bill, which obliges everyone to serve five years.[11]

To what extent this medical commentary was representative of punishment experiences, ritual methods, and spatial settings, in *all* of early modern England, remains open to historical dispute. The time has then come to delve inside dissection room doors opened up to the public by the Murder Act, still awaiting their rediscovery in the archives.

NOTES

1. Editorial (1762) *The Beauties of the Magazines selected including the several original comic pieces selected to be continued the Middle of every Month* (London: Waller publishers), p. 8.
2. Charles Bissett (1766), *Medical Essays and Observations* (Newcastle-Upon-Tyne: Thompson Publishers) pp. 60–1.
3. *A Colloquy between the Gallows and the Hangman: The Evils of Execution* was an anti-capital punishment poem published by Albert Mildane in London in 1851 looking back at the long eighteenth-century practices.
4. Jack Stillinger ed. (1978), *John Keats: The Complete Poems* (Cambridge, MA, USA: Harvard University Press), pp. 116–8.
5. Nicholas Roe (2012), *John Keats* (New Haven: Yale University Press), p. 74.
6. Robert Gittings (1973), 'John Keats, Physician and Poet', *Journal of the American Medical Association*, Vol. 223, 51–5, quote at p. 53.
7. John Keats (1934, 1st edition), *Anatomical and Physiological Notebook* (New York: Haskell House Publishers).
8. Roe, *Keats*, p. 80.
9. Ibid., p. 95.

10. See, for context, A. McKenna (2006), 'God's Tribunal: Guilt, Innocence, and Execution in England, 1675–1775', *Cultural and Social History*, III, 121–44.
11. Gittings, 'John Keats, Physician and Poet', p. 54.

CHAPTER 4

Delivering Post-Mortem 'Harm': Cutting the Corpse

INTRODUCTION

The iconic image of the criminal corpse has been closely associated in historical accounts with one legendary dissection room in early modern England. Section 1 of this fourth chapter revisits that well-known venue by joining the audience looking at the condemned laid out on the celebrated stage of Surgeon's Hall in London. It does so because this central location has been seen by historians of crime and medicine as a standard-bearer for criminal dissections covering all of Georgian society over the course of the long–eighteenth and early–nineteenth centuries. It is unde-niable that inside the main anatomical building in the capital an 'old style' of anatomy teaching took place on a regular basis under the Murder Act. This however soon proved to be a medico-legal shortcoming once a 'new style' of anatomy came into vogue during the 1790s. By then leading surgeons that did criminal dissections were being tarnished with a lacklus-tre reputation, even amongst rank and file members of the London Company. This meant that their medico-legal authority was increasingly dubious. It transpired that their traditions were too conservative at a time when anatomy was blossoming across Europe. As it burst its disciplinary boundaries, embracing morbid pathology with its associated new research thrust, London surgeons started to look lacklustre. A prime location of post-execution 'harm' that has dominated the historical literature does not then on closer inspection merit its long-term reputation for teaching

© The Author(s) 2016

125

E.T. Hurren, *Dissecting the Criminal Corpse*, Palgrave
Historical Studies in the Criminal Corpse and its Afterlife,
DOI 10.1057/978-1-137-58249-2_4

and research excellence. Progressively, Surgeon's Hall was over-shadowed by the rising prominence of provincial theatres in the North, South and Midlands of England too. There, criminal dissections served an expanding medical sector by 1800.

A selection of bodies distributed along this complex supply chain, presented in Section 2, illustrates the sorts of penal surgeons that actually handled the criminal corpse in the provinces. To establish a good business reputation for medical innovation it was important to be seen to receive bodies from the hangman in a local area on a concerted basis. Career-standing was more and more dependent on the publication in the medical press of cutting-edge post-mortem work. As that sector of newsworthy information expanded, the medical establishment started to change its views with regard to the anatomical value of criminal dissections staged outside of London. They were no longer seen as necessarily second-rate. At the same time, a conjunction of socio-economic factors slowly altered the financial calculations of surgeons that worked from provincial business premises. The fiscal situation was that those who had trained in the capital and became officially licensed as surgeons were still obliged to serve at Surgeon's Hall in London. On a rotational basis company members had to take their turn about once every five years to act as either Master of Anatomy, or perform a supporting role, for a dissecting season. Many however elected to pay a substantial annual fine, rather than temporarily relocating their households to the capital. Keeping the loyal custom of wealthier consumers meant that many penal surgeons were reluctant to move far from the vicinity of home in a competitive medical marketplace. Few wished to neglect the local hangman's tree either since they relied on that supply to publish original findings. Those that remained *in situ* avoided the expense of a locum and established their credentials in the neighbourhood. They were in a more positive business position to provide a bespoke service that nurtured the goodwill of their fickle patients. It was then serendipitous that a lot of provincial penal surgeons found themselves advantaged by the fact that by the early nineteenth-century more condemned bodies were being supplied from the local gallows rather than execution sites in the capital: a sentencing trend that justified them making a business decision to stay in the provinces (see, Chapter 5 for timings and supply figures). This complex commercial backdrop complicated the medico-legal duties and official reach of penal surgeons in practical terms. Hence, the historical prism of criminal dissections reveals the changing surgical nature of central-local relations understudied in eighteenth-century histories.

How then to cut the corpse to make maximum use of its research opportunities, is the focus of Section 3's discussion. The career path of Sir William Blizzard, introduced in Chapter 2 and expanded on here, illustrates how a leading figure that worked from Surgeon's Hall was very critical of the criminal code's underlying ethos. He, like many other penal surgeons, started to question the nature of the discretionary justice in their hands, and how exactly to cut up the criminal corpse to dissect and dismember it. This discussion mattered because it symbolised changing attitudes to medicine and society inside and outside the surgical community. The medico-legal purpose of post-mortem 'harm' was redefined in practical terms. It will thus be shown that from the 1760s there were a lot of medical debates about what 'anatomization' as a legal duty actually entailed and how it should differ from dissection. These private discussions were revealed in the press as a result of one of the most infamous murders of the period committed by Earl Ferrers of Staunton Harrold in Leicestershire. He was tried in a high-profile murder case in London. As a peer of the realm the anatomical fate of his body gave rise to considerable public speculation about how much each criminal corpse should be punished by the lancet. In the course of which, working methods were clarified, particularly in relation to class. Altogether, seven anatomical methods were described under the Murder Act for the first time, and these related to agreed guidelines about cutting up the condemned.

At the heart of all of these material reveries, novel anatomical angles were exposed—outside/inside—dorsal side/ventral side—supine/prone. In terms of public consumption early modern audiences found new ways of seeing the '*dangerous dead*'. It was the promise of engaging with the material demise of the deviant that captured the attention of many diarists of the period too. Their recollections frame this chapter's focus on first-hand and hands-on experiences of dissection. Often commentators admitted in private how much 'public curiosity' they observed. It appears to be what motivated many ordinary people to enter Surgeon's Hall. In time, those with 'natural curiosity' went further afield as well. Elsewhere, new, and sometimes, more intense, emotional experiences were being staged. By 1800, compelling home-grown murders, and the strong reactions they generated, shifted press attention from London reporting to the English regions. These contemporary developments reflected how much, as Fay Bound-Alberti observes: 'as objects of scientific knowledge, emotions were (and are) unstable and transient experiences' that nonetheless are no less deserving of historical attention since all human beings encounter 'emotions as sensory, embodied experiences' especially when confronted by a fresh corpse that reminds them of their mortality.[1]

UNDER-DOOR AT SURGEON'S HALL

15th September 1773: Saw two men hanged for murder. I should not have gone if it had not been reported that they intended to make some resistance. Was afterwards at the College [of Surgeon's Hall] when the bodies were received for dissection. They bled on the jugular being opened, but not at the arm.[2]

Silas Neville in his private diary styled himself a radical. As a medical man of fashion he also followed the anatomical entertainments in the capital. During the London season from 1767 to 1773 his diary entries were all about the new sensation of seeing criminals dissected. Silas obtained his MD at Edinburgh and then he moved down to London, where he walked the wards of St. Thomas's Hospital as a pupil. This was on the recommendation of his friend and mentor, the Scottish professor of medicine, William Cullen (1710–90).[3] His theatrical taste for medical dramas often reflected how his working life blurred with his private tastes. In early September 1767, for instance, he wrote that he suffered from painful toothache, bought a quack remedy from Elizabeth Miller of Whitechapel, and drank at the Chapter Coffee house in Paternoster Row near St. Paul's Cathedral. Here he mixed with penal surgeons that peopled Child's Coffee-House and attended Surgeon's Hall close to the Old Bailey criminal court.[4] An inveterate gossip, Silas gleaned privileged medical news, and was given tickets for the latest criminal dissections like that of Elizabeth Brownrigg found guilty of murder:

Wednesday 16 September 1767: After waiting an hour in the Lobby of Surgeon's Hall, got by with great difficulty (the crowd being great and the screw stairs very narrow) to see the body of Mrs. Brownrigg, which, cut as it is, is a most shocking sight. I wish I had not seen it. How loathsome our vile bodies are, when separated from the soul! It is surprising what crowds of women and girls run to see what usually frightens them so much. The Hall is circular with niches in which are placed skeletons.[5]

Silas was 'curious', pushing up a narrow spiral staircase. He claimed to be shocked by the bloody scene. This private admission is striking, given his medical training in basic human anatomy at Edinburgh. Either he was being disingenuous writing for posterity in his diary or his surprise was genuine. Like many medical students he had studied '*living-anatomy*' which involved looking at the major organs in the body but not '*extensive dissection*'.[6] He was also used to a male-privileged anatomical training, and this explains why it was disturbing for him to see women and girls running to partake of the post-execution spectacle. Helpfully he recorded in some detail the competing 'entertainments' on offer in the vicinity of Surgeon's Hall in 1767. These included a 'collection of curiosities' that he paid to go round at Pimlico featuring 'Birds' that had been dissected. They were part of a travelling exhibition of 'animals preserved in spirits' that he had first seen 'in the Haymarket'. In theatre-land he likewise bought a ticket for the 'Pit' to see Mary Ann Yates (one of the greatest tragic actress of her day) in a 'Pantomime' called 'Harlequin Skeleton'. She was, he remarked, an expressive actress. Her eyes he thought 'particularly affecting' even if the storyline was in his opinion 'foolish'.[7] We have already seen how Georgian theatrical shows often linked the world of medicine to that of dramatic storytelling on the London stage. The meaning of the word 'theatre', as Andrew Cunningham observes, meant 'literally a *place for seeing*'.[8] Surgeon's Hall was thus conveniently close to the main playhouses of the capital. If audiences were eager to pack out dissection venues, then why not exploit their macabre taste for shocking out of body experiences by featuring the dancing skeletons of infamous criminals on the stage.

Silas Neville said he disapproved of this macabre theatrical consumption: a predictable attitude perhaps for a 'gentleman' expected to act with 'good sense' and 'decorum' in Georgian society. Even so, his private musings are in many respects an historical prism of broader cultural trends. Like many contemporaries, Silas had a 'natural curiosity' and this overrode his personal misgivings. Few missed out on the anatomy theatre's fare that everyone was talking about in the Coffee Houses. The gossips speculated about how best to cut the corpse open and whether penal surgeons could revive the condemned before proceeding to post-mortem punishment. These medical conundrums were likewise debated in the provinces: a perspective often neglected in crime studies. John Baker in 1773, an attorney from Horsham, and like Silas Neville a diarist, noted carefully how the penal surgeons in Sussex generally bled the executed man before a full-scale dissection: 'After Cannon had hung half an hour, he and two others were cut down when Mr Reid, the older, and Dr Smith and three others of the faculty bled him and carried him to Mr Reid's and tried blowing and other means to recover him, but all ineffectual'.[9] The fact that it was standard practice to do this, sets in context that there were medical fashions at criminal dissections adopted everywhere.

The eighteenth–century was seen by many commentators as an era of conspicuous medical consumption. This allowed diarists to justify their theatrical tastes as social commentary. Frequently, they featured the architectural scaffold of punishment venues and to delve inside we need to follow suit in the capital before comparing conditions in the provinces. It happened then that a history of London written in 1790 praised the central location and convenient setting of Surgeon's Hall (see, Figure 4.1):

> On the outside of *Ludgate*, the street called the *Old Bailey* runs parallel with the walls [of London] as far as *Newgate*...The *Sessions House*, in which criminals from the county of *Middlesex* and the whole capital are tried, is a very elegant building, erected within these few years. The entrance into it is narrow so as to prevent a sudden ingress of the mob... By a sort of second sight, the *Surgeon's Hall* was built near this court of conviction and *Newgate*, the concluding stage of the lives forfeited to the justice of their country, several years before the fatal tree was removed from *Tyburn* to its present site. It is a handsome building, ornamental with iconic pilasters; and a double flight of stairs to the first floor. Beneath is a door for admission of the bodies of murderers and other felons; who noxious in their lives make a sort of reparation to their fellow creatures by being useful in death.[10]

Most diarists visiting from the provinces devoted time to seeing the impressive scale of the medical architecture and their theatrical enticements inside. When Richard Hodgkinson steward to the wealthy Hesketh family (major landowners in Hereford and Leicestershire) came to London on business in March 1794, he wrote that taking a medical tour of the capital was very fashionable. His carriage drove past St. Bartholomew's Hospital, the Blue Coat Hospital nearby, and then surveyed the Old Bailey on a morning's outing: 'This [the courthouse] is an immense piece of a Building being as I conjecture about 160 yards in front'. Surgeon's Hall he said was renowned as a major tourist attraction for the *beau monde*; together the courtroom and theatre next door occupied a distinctive urban space. By the 15 March, Hodgkinson had obtained tickets for the most popular lecturers on anatomy: 'Mr Johnson called upon me and took me to the lectures of Dr [George] Fordyce's'. The theme was '*The Death of the Patient*' and Hodgkinson followed the crowd avid for more information about resuscitation methods.[11] Surgeons, he observed in his letters

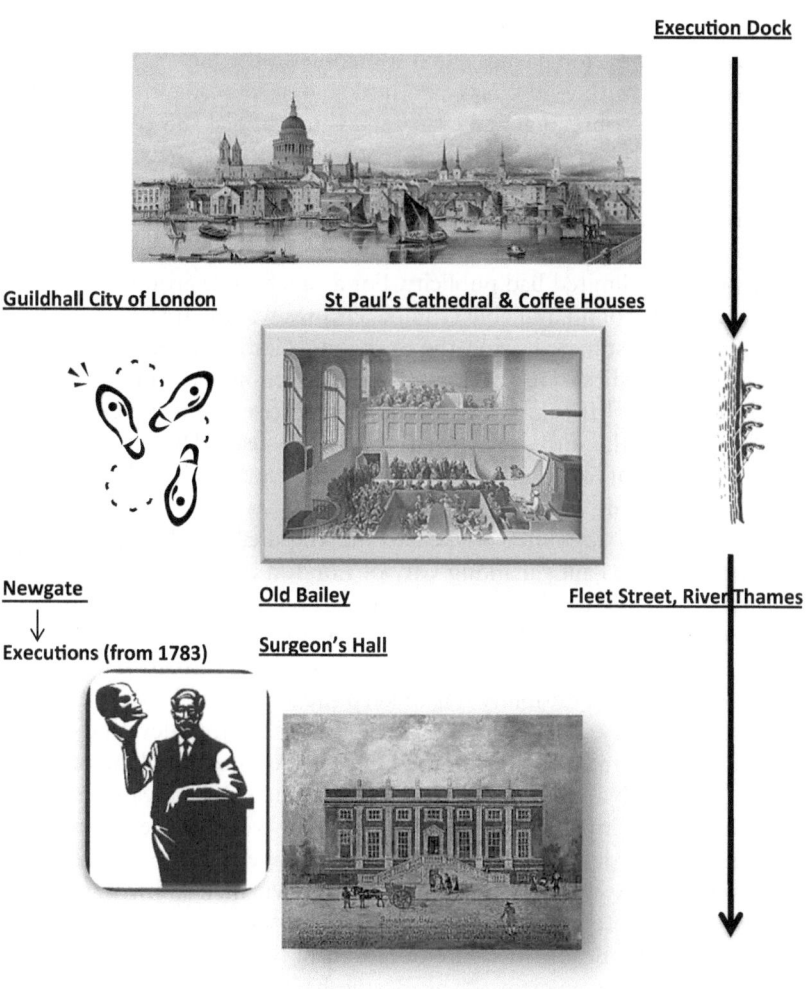

Figure 4.1   Geography of buildings and places associated with capital punishment in the City of London after the Murder Act 1752.

home, had three key sources of supply: they obtained bodies that died within the City of London area and were handed over by customary right. By tradition there were also permitted to acquire corpses retrieved from Execution Dock. These were the bodies of those sailors that murdered which came under the Admiralty court jurisdiction or they were a civilian convicted of homicide having killed someone on the high seas and therefore were dealt with by the Navy. A third source was cadavers sent via the criminal justice courts, hanged either at Tyburn until 1783 (on the site of Marble Arch today) or thereafter at Newgate prison (next to the Old Bailey). This basic geography (refer Figure 4.1) confirms that anatomists favoured a door-step business of supply: a trend under the Murder Act that continues today. It ensured that corpses were fresh on arrival and the close proximity limited bad publicity. For a square mile around Surgeon's Hall shuffling around available criminal corpses was the norm.

It has been estimated using the Surgeon's Company financial records that some 80 bodies were sent for criminal dissection in the fifty years after the Murder Act.[12] It is in fact difficult to be precise about the exact numbers because of the surreptitious nature of body deals. The man in charge was the Beadle of Surgeon's Hall. He was pivotal to the criminal corpse's punishment. Being Beadle to the surgeons was an important ceremonial role. The person appointed walked to the Guildhall on feast or religious days or attend St Paul's marking City of London celebrations. Figure 4.1 illustrates the close geographical networks served by the Company Beadle; for it was he who scheduled all dissection work. In the histories of crime and medicine his role has often been understudied. Perhaps the Beadle's most important responsibility was to keep the key and unlock the '*Under-door*' at Surgeon's Hall. This entrance was at the street-level below the main double staircase in front of the building. As Silas Neville wrote, visitors had to climb a spiral staircase up into the anatomical theatre. So bodies had to come in via a trap-door because they were too heavy to lift up the narrow crowded stairwell. Generally a carriage drew up on execution days and a body was carried in by rough-hewn men known as '*black-guards*'. They were said to be of robust stature, 'as tall as they were wide', employed by the Beadle to secure bodies from the gallows. The *Annals of the Barber Surgeons* explains: 'The Beadle has always had his "house" at the Hall.'[13] He lived on the premises to maximise supply opportunities and so call the '*blackguards*' out. In typical petty cash entries in the Company of Surgeons accounts, 'Mr George Search' (an alias to denote his body-finding duties as *blackguard*) was paid '£1 16s 0d' on 21 May 1767, and in January 1768 '£0 5s 6d' for 'bringing the bodies' to 'John Wells the Beadle'. Another important discretionary activity was to distribute a petty cash purse. By tradition the hangman was 'entitled to the dead man's clothes' at executions. But in the scramble this could damage the body and result in the mob carrying it off in the fray. So the Beadle gave out gratuities giving 'him [the executioner] compensation' for clothing and belongings, and he invited the hangman 'to the Hall regularly for his Christmas Box'.[14] The Company recognised that there was no substitute for a personal connection. This ensured that each time a body became available it could potentially be made full use of on the premises.

The first task of the Beadle on hearing about an available body was to call upon 'Mr Bates the Constable' who helped to make the deals with the legal officials.[15] Thus for instance petty cash of '£0 15s 0d' was paid on 9 October 1765 for 'three Murderers' and in the same month, 'Paid Bates for wine for Anatomical Officers £0 3s 10d'. Often the Beadle called at the Coffee Houses inside the precincts of St Paul's Cathedral and settled the bill for entertaining the support staff like that on 13 May 1766 'Paid a Bill at Child's Coffee House on the Anatomical Officers £1 15s 3d'. In hotter weather getting a fresh body was thirsty work. Associated bills reveal that Bates, his fellow constables (usually no more than four at a time) and the *blackguards* on duty wrapped the bodies in a winding sheet to carry them aloft. Thus on 1 May 1762 the Beadle 'Paid for Linen used by the Anatomical Officers £1 10s 11d' covering transportation and swabs. Others in the chain of supply included those that were brought in a 'Shell'—this was an elm coffin, its waterproof wood resistant to leaking bodily fluids. It was usually hinged at one end to be recycled—typically, the Beadle 'Paid Marks the Undertaker on April 27 1765' for supplying a 'Body from Tyburn'. Generally such dissections were advertised in advance in the London press as a public relations exercise. The Beadle then called by Child's Coffee House to collect equipment needed for the dissection twenty-four hours later. On 18 September 1761 he thus paid for food and drink of '4s 6d' and 'Paid Edward Stanton the Cutter his Bill £1 8s'.

Record linkage work reveals that Stanton ran a lucrative business through the Saw and Crown public house on Lombard Street up the road from Surgeon's Hall. A surviving business card publicised his services as 'London Surgeons Instrument Makers' (see, Illustration 4.1).[16] He sharpened the dissection knives made blunt by sawing through bone, ribs and the skull. He also made bespoke lancets and midwifery kit. His exclusive contract with Surgeon's Hall was however renegotiated soon after Easter 1761 because Edward died. In his will he left a lucrative business to his sister Mary and her husband William Sparrow who then traded under the family name from St Paul's Churchyard conveniently next to Child's Coffee-House where the surgeons congregated after anatomical sessions.[17]

Illustration 4.1   ©Wellcome Trust Image Collection, Slide Number M0015855, *'Edward Stanton at the Saw and Crown in Lombard Street London'* (1754–61): 'lancet-maker: maketh and selleth all sorts of surgeons instruments likewise razors scissors penknives knives & forks... note: lancets and other instruments carefully ground and sett', business card; Creative Commons Attribution-NonCommercial-ShareAlike 4.0 International License (CC BY-NC-SA 4.0)

Thus on 22 September 1756 the Beadle 'Paid for grinding of dissecting knives' some '4s'. He also ordered that the Hall itself had to be kept clean and so on 11 December 1756 'a woman was 'Paid £3 3s' the annual fee for sweeping up and washing the theatre. Another unnamed cleaner of lower status was given '8s for taking away the Dust 2 years to Xmas' in February 1762.[18] In fact the material waste at Surgeon's Hall was extensive, so much so, that it caused a local public health crisis. On 16 July 1766 the Beadle 'Paid a Sewer tax being the Company's proportion for drainage and cleansing the Common Sewer' down which they swilled with cold water, blood, tissue and, associated human waste in a culvert under Surgeon's Hall: we return to this theme later in this chapter when we explore how the body was cut to 'harm' it.

Inside Surgeon's Hall there were also expenses to be covered for the building fabric that made dissections feasible.[19] After the Murder Act on 9 July 1752 'Bowman the Smith' was paid to erect 'Iron Railings in the Theatre' anticipating the need for greater crowd control at a cost of '£20 12s 6d'. The Beadle also had fixed in position better 'Lighting Lamps' outside and a man called 'Nash' charged '£6 8s at Michaelmas' for keeping the '*Under door*' well lit at night to receive criminal corpses. There are numerous petty cash payments for tallow candles made from beef or mutton fat too. They were cheaper to make and used a lot to light the theatre during long winter sessions. In a pre-refrigeration era it was also helpful that they could be stored for longer than wax candles in sealed containers. Coloured hot wax tended to be used as an anatomical preparation to make models. An ongoing expense from 1755 was glaziers' bills for 'mending and cleaning the windows' to ensure maximum daylight. The repair bill also covered broken glass because the crowd did sometimes stone the building to protest about a controversial criminal dissection. This sets in context a bill by October 1755 of '£2 7s 'for wire work to Iron rails' to better control the crowds determined to press forward inside the theatre. Most were eager to get closer to the body. In winter the room temperature was kept lower to try to counteract the body-heat of the audience and keep corpses fresh. In December 1753 it was so cold however that the Beadle decided to pay 'for Chocolate for Masters and Stewards of Anatomy'. A hot drink must have been welcome because later that month a decision was taken to then pay 'for a Stove for the Theatre' costing '7s'. In the bitter cold of December 1760 tellingly it was deemed necessary to obtain 'a Brazier for the Theatre 6s 6d', and this despite the ambient temperatures audiences generated on public days. The average quarterly

'Bill for Coals' by 15 January 1762 was an expensive '£23 12s 6d'. Additional features found in the petty cash accounts facilitate a reconstruction of the interior circular platform where the criminal corpse was displayed before dissection.

To enhance visibility a 'Horseshoe Table with Black Leather' was placed centre-stage, designed for '£5 5s' in 1767.[20] China and wooden dishes to collect organs and tissue specimens were placed on this main table, as regular bills for 'Turnery Ware' attest of '£3 9s 8d in 1768'. Helpfully, a bill survives for the 'Jack' that the blacksmith made for the Company to hoist up the corpses on 4 August 1752 costing '5s 6d'. The purchase suggests the company expected to be busier once the Murder Act came into force (Illustration 4.2).

Illustration 4.2   ©Science Museum, Science and Society Picture Library, Image Number 10572107, *'Set of dissecting chain hooks, steel, by Savigny and Co. of London, 1810–1850'*; Creative Commons Attribution-NonCommercial-ShareAlike 4.0 International License (CC BY-NC-SA 4.0)

Indeed the design-concept of the central table and pulley evidently worked because it was copied elsewhere, notably at Cambridge (see Illustration 4.3 below of the rotund). Likewise the Beadle paid for a large bundle of 'towels and sheets' costing '£3 12s 6d' and he popped along to the 'skeleton maker'. In the case of Thomas Wilford, the first corpse under the new capital legislation, the supply bills were:

Paid Bill for expenses of Thomas Wilford executed   £1   6s 2d...
for murder
[hangman and constable-in-charge's supply fee]
Paid Skelton Maker for Making Wilford's Skeleton   £1   5s 0d
[cleaning bones, boiling them, returned a month later]
Paid for mounting Wilford's Skeleton, etc   £4 14s 0d
[famous murderers were set in circular niches
with nameplates]

Total cost: £7   5s 2d[21]

On hand, was a sharp razor 'for shaving the body', generally done by the duty Master of Anatomy on arrival of the criminal corpse. Although a pencil sketch of an oval dissection table in use up to the 1760s was drawn by a visitor to the Hall, the horseshoe table design of 1767 soon became *de rigeur*. The Company made the change because most leading anatomy schools were introducing revolving tables to improve visibility, so a horse-shoe-design was seen as a distinctive innovation. The basic equipment at Surgeon's Hall looked a lot like that in the contemporary sketch which survives in the Royal College of Surgeons Museum Collection today.[22] These were arresting details that Hogarth had the foresight to satirise in his famous cartoon *The Reward of Cruelty* (1751) in Chapter 3. He how-ever never knew how bodies in the supply chain arrived because the Company preferred to do its dealings in secret across London and use the Beadle to co-ordinate body trafficking long before the Anatomy Act.

In general, the Company discouraged individual surgeons from making supply deals at Newgate prison next door. It was easy to be tricked. The Beadle was far more worldly-wise in the subterfuge of body-dealing. One anecdote published in 1819 recalled the sort of double-dealing that could catch surgeons out under the Murder Act.[23] A convicted man described as a 'hardened villain was given a capital sentence and 'contrived to send for

Illustration 4.3 ©Wellcome Trust Image Collection, Slide Number M0010176, J. C. Stadler (1815), '*The Anatomical Theatre at Cambridge*', (Cambridge: R. Ackermann's History of Cambridge), original sketch; Creative Commons Attribution-NonCommercial-ShareAlike 4.0 International License (CC BY-NC-SA 4.0)

a surgeon...he offered his body for dissection after his execution for a specified sum'.[24] He wanted the surgeon 'to advance him the money immediately, that he might make himself while he lived, as comfortable as circumstances would allow'. The surgeon fell for the trick. He decided that since 'no person could present a better title to the body than the wretch who offered to sell it' the proposal was profitable, but as a precaution he made the convict place 'his signature to a written article, which he thought would be legal' in exchange for the money. The surgeon then told a fellow member of the Company that he had made a great deal. But his friend was sceptical: 'He shook his head saying: *I am very apprehensive that he has tricked you, even under sentence of death'*. The criminal in question was so notorious that the judge had sentenced him to be '*hung in chains*'. There would be no body to collect because it was destined for the gibbet. The duped surgeon was furious and confronted the prisoner who 'confessed'. Laughing, he pointed out that since nearly all the money had been spent and he was already 'placed beyond the dominion of the law' by being condemned to die within days, it was hard luck. The Beadle by contrast knew by long experience and close personal ties which bodies to target and which not.

The Company members had the advantage of fostering corporate ties with the City of London Guilds. These personal connections facilitated the smooth running of body supply, display and disposal. It was the Carpenter's Company that provided the most support staff to Surgeon's Hall: a recent archival finding. John Hopper for instance was a carpenter who resided 'near the George' public house on Drury Lane in 1777 and he came on a regular basis to assist on dissection days.[25] So did Thomas Pacey and Thomas Mansell, fellow carpenters. Since coroner's records survive of these craftsmen serving together as jury-members at inquests into suspicious deaths, it is perhaps not surprising to discover that they were familiar with the 'view' of dead bodies in a dishevelled state.[26] Carpenters had two hands-on encounters with criminal corpses at Surgeon's Hall. They made the recycled coffin shells that were used to take what was left for burial after being '*dissected to the extremities*'. If the corpse could be kept for longer in the coldest winter months sometimes a carpenter's backboard was inserted under the spinal cord to keep the torso intact to protect the integrity of the human material for another dissection day. Once the surgeons had cut down to the bones, a carpenter's wooden cross was occasionally used to pin limbs to. That is, until the skeleton-maker came to collect them to be boiled down, wire them up, and then four weeks later brought them back to be displayed. There has been some historical dispute about how often bodies were dissected at Surgeon's Hall and whether extensive use was made of the potential teaching material, or not. The medical press and newspapers did consider some surgeons to be lack-lustre. Yet, payments to the carpenters for work on the premises, seems to have denoted a busy working-session. When less cutting was done it was because bodies tended to be in a bad shape on arrival at Surgeon's Hall, as we have already seen in Chapter 3. Another factor was how the theatre space at Surgeon's Hall was peopled. It determined expectations about how to dissect; something there had been a lot of ongoing discussion about amongst company members since the 1730s. It is then fortuitous that a major proposal to reform internal procedures has survived in the Halford collection at Leicestershire Record Office. There was it reveals an open-ended policy of continually modifying duties, reflecting the forward-thinking ethos of the Company until the Murder Act was passed. It then became somewhat conservative as a medical institution concerned to be seen as part of a new 'scientific' establishment. This stagnation meant that it lost credibility by the 1790s when it did not embrace 'new anatomy' with gusto. Revisiting therefore the reform proposals dating from the 1730s kept amongst the family records of Henry Halford pinpoints debates that did not abate about how exactly to cut the criminal corpse: this chapter's main focus.

Sir Henry Halford (1766–1844) was an ambitious medical man. In 1795, he made a fortuitous marriage to the Honourable Elizabeth Barbara St. John Bletsoe of Wistow Hall in Leicestershire. This propelled him into aristocratic circles. He used those connections to become medical adviser to the Royal family. Considered handsome, discrete, and talented, he was appointed as Regus Physician to four monarchs from George III to Queen Victoria. The young Henry evidently had innate surgical skills. He also benefitted from having a number of renowned surgeons on his paternal and maternal family-lines. These connections helped him to navigate a competitive medical market-place; training in Edinburgh, moving to London, but keeping his surgical links by marriage, with the Midlands. His grandfather, Henry Vaughan, ran a lucrative medical practice on the corner of New Street and Friar Lane in central Leicester in 1763. From here, he helped to found the Leicester Royal Infirmary in 1766. Strategically, this family background placed Henry at the centre of medical debates in provincial society and the capital. Amongst his collection of surviving family papers it is therefore instructive to rediscover that surgeons in his family had contributed to debates about the role of the Royal College of Physicians and its relationship with Surgeon's Hall. Starting in the 1730s, of particular interest, was how the London Company should be staffed in the decades running up to the Murder Act.

Henry Halford's surgical relatives took an avid interest in proposals to restructure anatomical teaching on criminal corpses in the capital. These have survived in draft form and in a final edited version. Their handwritten testimony permits us to gaze in through the windows of Surgeon's Hall at a pivotal time in the Company's internal restructuring and rebuilding work.[27] It should be stressed that the Halford surgical papers were never intended for public consumption. They attest instead to the internal debates there had been about the teaching function of the Company and the format for public anatomy over the long eighteenth-century, but especially around the time of the Bloody Code. According then to the draft notes, at criminal dissections the corporate ambition was to reform the working-day as follows: the 'first professor of anatomy' to examine 'the parts' of criminal corpses 'during 2 hours not less every day so long as those Bodyes [sic] can be kept sweet' and 'afterwards' to dissect 'human preparations...[to] show where he could not upon the said Bodyes[sic]'. A second professor of anatomy was meanwhile to:

> Give 2 courses of all ye operating practical upon human Body (Those of the Bones excepted) with your instruments, operations, and dressing properly by belonging to every Respective operation of the Bodyes – [if] they cannot be kept sweet long enough, that he shall shew them in the best manner he can.[28]

A third professor was then to take responsibility for 'the ligaments of the Bones and other parts useful in the case of fractures and dislocations, and during the summer season give 2 courses on the human skeleton'. This together with instruction on 'dressing...& distemper and all ye Bandages observed in practice for your distempers of the human body'. If the Company acquired a female body then a fourth professor 'shall every year complete a course of midwifery viz two of these at Surgeon's Hall and 2 others in different parts of London and for the instruction of Midwives'. A fifth and sixth professor were then given the task of demonstrating 'all the other parts of surgery and compression under the foregoing head viz *Principia Chirurgrie*'. They were to give additional instruction in 'The Doctrine of Tumours, of Ulcers, of Wounds, and the apparatus and method for the cure of Distempers, the *material medica* and all the Chirurgical instruments'. These men were hence in ordered ranks to stand around the dissection table in the theatrical space (see, Figure 4.2). Their career standing and desire to reform working practices were together pivotal in the development of the sort of experiential routine this book has been recovering in the archives.

The draft notes make it clear that there were to be '3 demonstrators of anatomy and surgery and they shall be coequal, but to prevent confusion in the Discharge of their respective duties that Mr John Douglas and Mr Abraham Chovet shall prepare one private and one public Body and shall make the very best use that can be made of such Bodies'. John Douglas was a surgeon attached to the Westminster Infirmary, a Fellow of the Royal

Society from 1720, and praised for lithotomy. He was expert at the removal
of stones from the kidneys and bladder, a dangerous and highly-skilled
surgical procedure in the eighteenth-century. Abraham Chovet meanwhile
was a surgeon who lived in the central London parish of St-Martins-in-the-
Field.[29] He was renowned in medical circles for displaying in 1733 an
Anatomical Venus, the wax model of 'a woman...suppos'd open alive'
which was used to reveal the circulation of the blood in pregnancy.[30] Their
respective expertise explains why Chovet did the public anatomy days and
Douglas the private dissection ones: each specialised in different types of
post-mortem work. Douglas was tasked with checking on the medical status of the lifeless looking criminal corpse and then doing a dissection on the
extremities. Chovet injected with coloured wax at which he was highly
skilled and talked at public sessions on the general mechanics of the body.
Behind Douglas and Chovet stood 'Peter Macculough [sic] shall in like
manner...shew...such ligaments and joints of the Bones'. In several Old
Bailey murder trials dated 20 April and 6 September 1737 he was a surgeon
working in 'Westminster'.[31] He generally appeared as an expert witness in
homicides, serving at coroner's hearings in the vicinity where he had conducted an autopsy into unexplained deaths. He was thus in an ideal position to journey with the criminal through the judicial system into the
'*Under-door*' of Surgeon's Hall. And he did not do so alone.

This ritualised set-up depended on the assemblage of a large parade of
Company members. They walked into and out of the premises in ranked
formation. *The Annals* contain reports of over seventy carriages travelling from the Hall down to St. Paul's Cathedral and its Coffee Houses
to celebrate the end of dissection sessions in years when executed bodies
were more plentiful.[32] There was an expensive feast, plenty of fine wines,
and a public parade of medico-legal officials to convey that this set of
ritual punishments was endorsed by the City of London. All this testified
to the broader cultural role of the Hall in creating normative standards
of natural justice. Yet despite its iconic status more still needs to be
known about how exactly the corpse was opened-up to a 'public curiosity' in lots of theatrical settings. Medico-legal procedures everywhere
were fluid and this meant that additional reform proposals threatened to
cloud what it really meant to attend a criminal dissection by the 1790s.
Comparing and contrasting then what happened in the capital with procedures at other prominent venues around the country is instructive
about common predicaments. The twin focus in what follows next is the
neglected provincial scene and how it came to predominate by the early

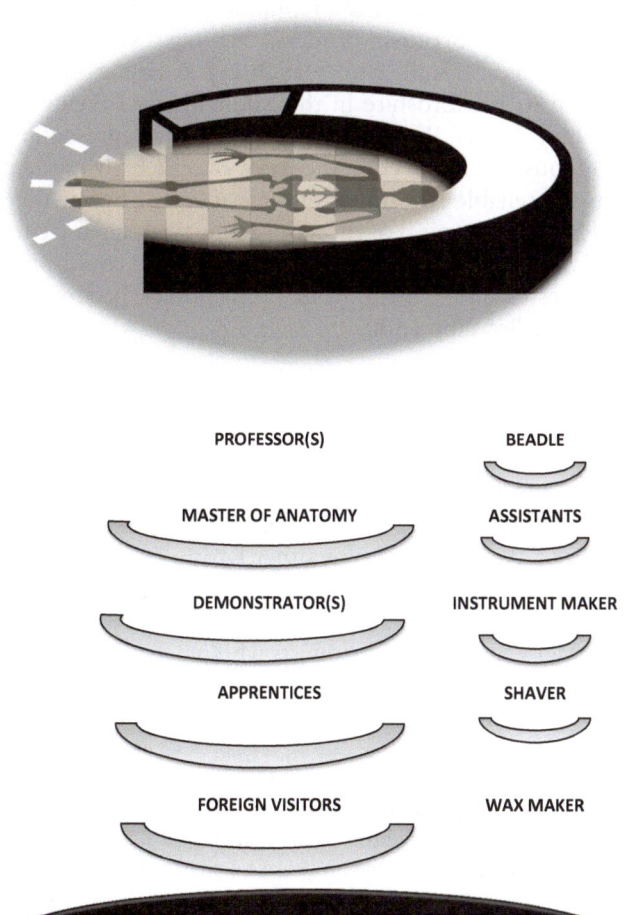

Figure 4.2 Ranking of the Company staff and visitors at Surgeon's Hall, London, circa 1734.

nineteenth-century because that perspective is omitted in criminal histories that have often overstated the reputation of the London Company. Travelling from the Scottish borders, the next section begins with Newcastle's Surgeon's Hall, before alighting at Exeter in Devon, then going back up to Bedfordshire in the Midlands, and finally heading to Lincoln in East Anglia. The chosen examples are representative of the sorts of surgeons that received bodies on a regular basis in the provinces taken from the reliable records of the Sheriffs Cravings at the National Archives. These were typical of those surgeons that paid a fee to London Surgeon's Hall to avoid serving there so that they could protect their volatile businesses and benefit from more bodies becoming available from the local gallows by 1800.

NORTH, SOUTH, EAST & WEST: SENDING BODIES TO THE SURGEONS

In 1723 Daniel Defoe toured Newcastle: 'here is a Hall for the Surgeons where they meet, where they have two skeletons of Human Bodies, one a Man the other a Woman, and some other Rarities'.[33] He observed that it was a place dominated by bad weather, dusty coal, and endemic poverty. By the time of the Murder Act, an infirmary for the sick poor had been established. Many prominent anatomists who practiced surgery walked its wards. By nightfall they undertook criminal dissections. Thus on 19 August 1754 it was reported that:

> Dorothy Catinby of Love Lane on the Quay Newcastle was executed on the town moor, there, for the murder of her bastard child. She behaved in a very penitent manner but persisted to the last that she did not murder her child. The body, after hanging the usual time, was taken to Surgeon's Hall where it was dissected, and lectured upon by Mr Hallowell, Mr Stodart, Mr Lambert and Mr Gibbons, surgeons.[34]

Record linkage work reveals that Richard Lambert was Master of the Barber Surgeons at the time of the Murder Act. According to Henry Bourne, a local historian, he oversaw a striking building: 'It is very beautiful and not a little sumptuous; it stands upon a tall Piazza, under which is a very spacious Walk'.[35] There was at the front 'a fine Square, divided into four Areas or Grass-Plates, surrounded with Gravel Walks, each of which is adorned with a Statue' of a leading medical historical figure.

There had been a wall at the front to keep out the crowd but it was removed and railings erected instead so people accompanying bodies could see what was happening. Nevertheless Bourne thought the premises where located in 'such a dirty part of the Town'. This explains why Richard Lambert as a leading Newcastle physician had taken the initiative to campaign for the alleviation of abject poverty in the city-centre. He worked alongside leading surgeon-apothecaries to set up Newcastle Infirmary in 1752. Such was the combined anatomical expertise of the penal surgeons who undertook Dorothy Catinby's dissection that three pupils present in the theatre that day later came to national prominence. They represent the diverse and distinctive dissection work for which Newcastle became renowned by 1800.

William Ingham from Whitby was trained by Richard Lambert the physician-surgeon. Ingham became well-known for 'his knowledge of anatomy combined with great manual dexterity' in the North of England and Scotland. His fellow pupil was William Harvey an artist from Newcastle. Harvey had a reputation for doing such remarkable life-drawings of criminal dissections that he was employed by Charles Bell the anatomist in London who trusted him to 'dissect a subject himself and to draw all the muscles as large as life'.[36] It was though William Hexham from Hexham in Northumberland that became one of Newcastle's most famous pupils. In the 1750s Hexham resided with John Hunter and often attended William Hunter's lectures in London. Later he ran well-attended private anatomy lectures at Craven Street in the capital.[37] In this, Hexham took his lead from his mentor, Richard Lambert, who had perfected the art of treating aneurisms in the heart from his anatomical and surgical work, and for this he was much praised everywhere.[38] Newcastle's Surgeon's Hall might have been smaller than its counterpart in London, but its close proximity to Edinburgh meant that it could rival and often outdo its metropolitan ally for medical innovation.

Engaging with the rise of the provincial medical scene involves however taking a chronological leap forward to trace what happened in places like Newcastle by the early nineteenth-century. The sources reveal that standards of dissection in the provinces were surpassing those in the capital fifty years after the Murder Act. By way of example Newcastle doctors like Thomas Giordani Wright wrote that they were very keen to now attend local criminal dissections in the North of England, not for their sensational elements but educational endeavours. After for instance the execution of a Newcastle woman for infanticide on 7 March 1829 he recorded in his diary:

> The poor woman was hung this morning at the old place of execution near the barracks. The procession passed along the street and within sight of my window but I had not the curiosity to join the assembled thousands who crowded the last scene of her existence. The body will I suppose be exposed to the public gaze for a few days when she will be anatomized by Mr Fife.[39]

The doors to 'Old Surgeon's Hall' were left open so that the crowds accompanying the body could satisfy themselves that a child killer had reached medical death. Four days later on 11 March 1829, Wright wrote that only then had the surgeon on duty proceeded to dissection of the criminal brain:

> Mr John Fife on Monday noon gave a very good demonstration on the brain of the criminal who suffered on Saturday. There is to be a course of lecture on this subject free to surgeons and so on, but open to the public on payment of a fee.[40]

There were in total forty-five company members that could gain access to the criminal dissection, with their apprentices required to serve five years. In this Newcastle copied their London counterparts. Yet by doing extensive anatomy of the brain they were also outshining their competitors in the capital: this neurological work is a theme developed in Chapter 6. John Fife who conducted the criminal dissection was typical of many provincial medical men keen to return to their family roots. Born in Newcastle-upon-Tyne, he had trained at the London College of Surgeons, but came back home in 1815 to serve at Newcastle Surgeon's Hall because it was an exciting and innovative place to be by the early nineteenth-century. Soon he could boast that he had become a respected specialist in diseases of the eye and his growing credentials enabled him to help found the Newcastle Medical School by 1834. As a registered member of the London Company he was typical of those that paid a fine to avoid the rotating duty of serving as Master of Anatomy or one of the associated support staff in the capital. The fine was expensive, '£21 0s 0d', but it was a financial cost worth paying to stay *in situ*.[41] To do really cutting-edge work in London would have entailed working for anatomical entrepreneurs like the Hunter brothers who paid privately for exhumed bodies from local graveyards to match medical student demand in the capital. Fife seems therefore to have sensibly decided to return to where he had better business connections, could make his mark rather than being overshadowed by another surgeon, and benefit from exclusive access to bodies from Newcastle's gallows. Others across the South of England followed suit: the Patch family of surgeons based in Devon are illustrative of similar provincial considerations.

On 13 March 1758, Thomas Smale was convicted of murder in Exeter. John Patch (1723–87) a leading surgeon based in the city-centre was designated to do the criminal dissection. Richard Stephens the under-sheriff arranged for 'John senior' (as he was known locally) to sign for safe receipt of the body on 20 March 1758.[42] He also supplied bodies to another family member 'Robert Patch surgeon of Exeter' (he was a nephew of John Patch senior).[43] Despite their medical prominence, the Patch family could ill-afford to be absent from home to work at Surgeon's Hall in London. On the death of John Patch senior in 1787, his widow discovered that his book-keeping had been erratic and very bad for business. She informed the local press that 'her husband had made no entry for professional attendance and therefore she trusted in the honour of his patients for payment'.[44] Her nephew Robert had to help his aunt sort out the debt-collecting crisis. This meant that John Patch junior, having trained in anatomy at Edinburgh, was obliged to maintain the unpaid position of honorary surgeon to the Devon and Exeter Infirmary from 1741 which he could ill-afford. At just eighteen years old he had to try to solicit private fee-paying patients across the county to cement his family credentials once more.[45] It was in fact the next generation, John senior's grandsons—James, John, and various Patch cousins—that eventually secured the family practice by providing a hands-on, personal service, to patients in and around Exeter.[46] Meanwhile other surgeons were making in-roads into Exeter's competitive body-supply market. Thus on 6 August 1808 when George Godbear (alias George Tapp) was convicted of murder, Samuel Luscombe, from another renowned surgical family in the locality, was appointed to do criminal dissections. On this occasion the sheriff of Devon, Pitman Jones, timed very carefully the circumstances of his medical death (brain death having now overtaken a heart-lung diagnosis) before the body was handed over:

> He [Godbear] was hanged by the neck for the space of one hour from 22 minutes past twelve till 22 minutes past one in the afternoon on Tuesday the 16th day of August 1808 until he was dead in the presence of myself, Mr county clerk and diverse others & then delivered to Mr Luscombe.[47]

Opportunities to do original research will be explored in greater depth in Chapters 5 and 6. Meantime, the balance of evidence in this new archival material suggests that most provincial surgeons needed for financial reasons to avoid the official London scene at Surgeon's Hall, and increasingly they were not necessarily disadvantaged by doing so. Staying at home there were plenty of opportunities to publish new findings on criminal dissections in a flourishing medical press. Avoiding relocation expenses meant saving on the cost of a locum to stand in for an absent surgeon, essential if a family practice was to continue to better manage consumer medical fashions. If an inattentive surgeon neglected to manage their casebook for a concerted period then patient loyalty could evaporate: being preoccupied in the capital could undermine years of hard work. Travelling into the Midlands, we encounter an equivalent situation.

The Bolding family of surgeons from Ampthill took a generation to become established at the Bedford Assizes. John Bolding senior treated the sick poor, often free of charge to get started in business. To build his medical reputation and draw in regular fee-paying clients, he agreed to help a local coroner (typically he was legally, rather than medically qualified). In 1723 he likewise became an unpaid local churchwarden to raise his profile in community life. Until his death in 1795, his life's work was to ensure his surgical business was better-networked to cement his family's financial future.[48] His son was thus able by March 1788 to secure bodies from the gallows, like that of: 'The body of convicted murdered John Cooke ordered to be delivered to Edward Jackson of Bedford and John Bolding the younger of Ampthill in Beds for dissection'.[49] The surgeons signed to say that they shared the body: a tradition that kept them on stand-by in the vicinity of local executions. This matches what was happening in East Anglia too. Record-keeping at Lincoln reveals the sorts of young male criminals that became available and for very good reasons kept penal surgeons occupied at home. They sometimes got to dissect without disruption such was the depth of local feeling against a condemned murderer.

On 26 May 1775, *The Gentleman's Magazine* reported that 'William Farmery, of Sleaford, in Lincolnshire' had been arrested for the 'murder of his mother' and incarcerated at 'Lincoln gaol'.[50] The reporter explained why there was such a public appetite to see his body dissected and that surgeons were able to do extensive post-mortem 'harm':

> Having some words with his mother in the morning, he went out and whetted his knife very sharp, and then coming into his room, where his mother was making his bed, he struck her in the throat, as a butcher does a sheep, and then left her weltering in her blood.[51]

At the Lincoln summer Assizes the physical evidence presented in court about the brutality of the killing method and the uncaring attitude of the prisoner regarding his dead mother shocked local people. It was reported that he 'had been determined to murder her for three years', disliked how much 'she corrected him when he was a little boy' and on the morning of the murder resented her 'having words with him'.[52] The case soon attracted widespread publicity and it was this public reaction that local surgeons benefitted from. The *Hibernian Magazine* explained that: 'As this crime was of an extraordinary nature, it drew together great crowds of people' to see justice being done for such cruelty. The reporter took soundings from amongst those assembled and the general gossip was damning. It was written up in an emotive editorial line: 'He was a most stupid, melancholy and gloomy wretch, a great reader of books before and after he was in prison, averse to all manner of labour, prone to taciturnity, disagreeable and unsociable'.[53] Elsewhere it was said that the prisoner was 'twenty-one years of age' and apprenticed to a 'shoe-maker'; this additional information sets in context why his criminal dissection excited so much 'public curiosity'.[54] Farmery was young, fit, and muscular. There was a considerable public appetite for *lex talionis*—the English common law of retaliation authorized by criminal law, in which the punishment corresponds in kind and degree to the injury. Here was an opportunity to do a full-scale dissection; a 'monster' must become 'a demonstration of that monstrosity'.[55]

Available court evidence in many regional parts of England points then to the need to delve deeper under the 'surface anatomy' of everyday life. It is essential to rediscover how exactly penal surgeons worked from 'the outside to inside' on criminal bodies. It is important to keep in mind that regardless of the county location, cutting the corpse could be a complicated legal penalty to carry out and importantly it went on being so under the Murder Act.

OUTSIDE-INSIDE: CUTTING THE CORPSE

Standing over a criminal corpse and making basic decisions about how much to cut the criminal corpse could be a personal challenge for some penal surgeons under the Murder Act. William Hey (1736–1819) of Leeds for instance who conducted most criminals dissections in Yorkshire admitted in a private letter to his son training in medicine that: 'I have often had very solemn reflections in the dissection room; and have when the company was gone, kneeled down in prayer in the midst of these silent preachers of our guilt and misery'.[56] Those that retained a spiritual belief found the work thought-provoking. This understandable human reaction sets in context why there were agreed anatomical methods which many penal surgeons valued. There were a total of four basic ways to go 'outside-inside' a corpse when the Murder Act came into force. These then developed into seven standard methods attached to criminal dissection by the early nineteenth-century: discussed later. Crime histories that therefore give the mistaken impression that post-execution 'harm' can be summarised in a few short sentences in fact misconstrue what was often involved from a medico-legal standpoint. To better appreciate the options available to penal surgeons Table 4.1 (overleaf) sets out the broad working-definitions that developed and their basic equipment in 1752. Importantly although each penalty was in a surgeon's personal gift, he also had to act broadly with decorum before the post-execution crowd.

The leading anatomist-surgeon could first opt to do just an autopsy. This generally meant making a small 'first incision' to look at the body just inside the chest cavity. Then the skin, tissue and muscle with skilled surgical fingers was teased aside, internally going deeper with either a Y-shaped or T-shaped cut known often as a 'simple surface' slice. There then were four general 'deep-seated' anatomical ways to cut up the interior of the condemned in a methodical manner. These ranged from just a nick of the lancet to full-scale incision work. If the crime committed had been heinous then it was accepted practice to undertake a full-scale dismemberment of the corpse, as was the case for William Farmery of Lincoln discussed in Section 2. Generally this full-scale option was known as '*of the extremities and to the extremities*' and it meant more than two thirds disintegration of the human material. There would be very little to bury at the end, an outcome that the capital code permitted. Today in a modern dissection room not more than one third is used as a teaching aid to maintain human

Illustration 4.4   ©Wellcome Trust Image Collection, Slide Number L0022244, '*A Man Thought to be Dead arising from a table in a laboratory and frightening the proprietor*', eighteenth–century drawing, published London, 1790s, details unknown; Creative Commons Attribution-NonCommercial-ShareAlike 4.0 International License (CC BY-NC-SA 4.0)

dignity, but in the case of convicted murderers this was not a consideration in the past.

To learn how to conduct a criminal dissection it was necessary for surgeons from the provinces to attend a wide cross-section of post-mortems. Often the medical press gave advice about how best to proceed. In 1829, by way of example, a contributor to the *Medico-Surgical Review and Journal of*

Table 4.1   Basic ways to cut the criminal corpse in England under the Murder Act, circa 1752

Medical Term	Practical Definition/Method	Cuts to the Body
Autopsy	From the Ancient Greek term '*to see for one's self*—it meant opening the body with the razor or lancet to *observe the cause of death* in the *major organs* of the body—opened in two incisions—one vertical, the other horizontal—to look inside the body for the purposes of general observation of the living state of the human being.	**First Incision**: Cut down the length of the torso & extend to the pubic bone (a deviation to the left side of the navel). **Second Incision**: 1 of 3 options: (**1**) **a large and deep Y-shaped** starting at the top of each shoulder & running down the front of the chest, meeting at the lower point of the sternum (**2**) **a T-shaped** made from the tips of both shoulder, in a horizontal line, across the collar bone region, to meet at the sternum (breastbone) in the middle (**3**) **a single cut** is made from the middle of the neck (at the 'Adam's apple') on males in the torso by transverse section
Anatomy	From the Ancient Greek term '*cut open—I cut on, or upon*' **3 basic branches:** *Animal anatomy* (zootomy) *Plant anatomy* (phytotomy) *Human anatomy, branch of morphology* (developed by Johan W. van Goethe (1790) & Karl F. Burdach (1800)— *Morphology* comes from the Ancient Greek term '*to study form—to research*' In *anatomy* it is the study of the form and structure of the *internal* features of an organism—gives rise to '*the new science of the body*'	5 types of *Human Anatomy*: (1) *Superficial anatomy*—looks at the contours or surface of the body; no cuts (2) *Creationist anatomy*—at an autopsy examines the body's major organs in microcosm, reflects God's sacred creation (3) *Higher or Transcendental anatomy*—opens the body at an autopsy to look at a chain of being and form, to reveal the operation of Natural Laws (4) *Speculative anatomy*—postulates about the physiology and philosophical disposition of living beings, in functioning heart, lungs and brain (5) *Morbid anatomy*—comparative morphology and pathology of diseased organs and tissues, crucially at the anatomical stage seen just with the naked eye, without a microscope

(*continued*)

Table 4.1   (continued)

Medical Term	Practical Definition/Method	Cuts to the Body
Dissection	First requires an anatomical examination and cannot begin until putrefaction of the flesh is visibly observed, ensuring the person is '*truly dead*'. Then disassembles & dismembers a human being or animal form over 3–4 days after medical death is declared, provided a qualified surgeon has confirmed the corpse is in a physical state of '*absolute death*'; normally done on those hanged.	**2 Basic Options:** (1) cuts quickly & somewhat crudely down to the bones with a razor; severs head, limbs and torso; studies muscles & tissue hanging loose by microscope (2) takes an incision (see anatomical options) more carefully and with dexterity using a lancet, extracts flesh, muscles, and tissue, before cutting off limbs, severing head, and quartering the torso; bones are boiled & sent off to make a skeleton for display; body parts put in preservation jars for comparative study of morbid anatomy & pathology
Dismember	*Cuts to the extremities* a dissected body of a criminal corpse or person whose body has been resurrected and/or stolen from a graveyard in the 18th century for the purposes of medical education and/or research. Little will be left for burial, less than one third of the original corpse. Remaining flesh and bones sewn together with a large surgical needle then wrapped in a woollen shroud used as a winding-sheet and buried in a common grave, normally no less than six deep. Lime thrown on each body to accelerate decomposition. No visible sign of burial above ground-level; social death.	**Basic equipment includes:** <u>Scalpel</u> or Lancet (sharpened) <u>Scissors</u> (<u>dissecting scissors</u>) Thumb **forceps** or fine point splinter **Mall probe and seeker** Surgical **spatula** <u>Magnifying glass</u> Needle to test eye reaction in pupils <u>Surgical chain and hooks</u> <u>Razor</u> (used in crude dissections) Rope or cord Surgical blow pipe <u>Surgical prong</u> Syringe of hot water to test heartbeat & brain function <u>Teasing needles</u> Trumpet to blow in ear, auditory test **Pipette** or medicine dropper <u>Ruler</u> or <u>calliper</u> T-<u>pins</u> Dissecting pan/basin/bucket Brush to sweep up fleshy material

Practical Medicine reflected on what he termed 'modern medical ethics' and certain 'state maxims in medicine' in the dissection room. The penal surgeon advised new recruits anxious to obtain bodies from the gallows to follow the crowd to the dead houses of newly built infirmaries. There, they ought to however keep in mind:

> Hospital and Infirmaries: There is now discrepancy of opinion respecting the policy of connecting yourself to a public institution [post Murder Act]... You must by all means, make a collection of disease structures by begging all morbid parts your friends may meet with....In short, there is no part of the body in which a fertile imagination and a good modicum of effrontery may not easily make out traces of disease for the purpose in question...If a further dissection is insisted upon, and more morbid anatomy turns up [at a post-mortem or criminal dissection] you are to ridicule the latter as having anything to do with the disease. All other morbid appearances than those which suit your purpose are to be voted occurrences in the agonies of death.[57]

This lengthy quote is written in a cynical tone and therefore should be viewed with caution. Given however that the article was published in a leading medical journal there must have been some medical substance to the general advice to make it into print. The author claims to be experienced in the artful disguise of diagnosis and dissection. He states that conducting successful post-mortems was all about the way a surgeon performed agreed guidelines to match public expectations. If for example someone died of a brain inflammation and the surgeon had given that diagnosis to the patient's family before death, then at a subsequent dissection it was vital for reputational reasons to stick to what originally had been said. It could damage a medical man's standing to contradict his own fatal diagnosis when the body was actually cut open. The surgeon's practical advice was: 'When the skull cap is removed, you are to knead the brain with your fingers, in the same way that a baker kneads dough in a trough—under the pretence that you are feeling for abscesses'. In any case, in practical terms, the advice continued, 'you will find some portion of the brain softened by the above process.' Technically the dissector was not lying by voicing what people expected to see. It was more professional to do so, rather than revealing what he had just handled inside the head. To make sure however that those present believed in this medical reality: 'These [the softened brain tissue] you are to scrape off your scalpel and triumphantly shew them round as portions of the *suppurated* brain'. The article concluded that although this might seem misleading to your audience, it did not follow that being economical with the material facts was unethical. After all, in traumatic death inflammation of the vessels ceased once circulation of the blood stopped in the head. Only the dissector really knew the finer details when getting down to a deep-seated dissection. Most penal surgeons in this way used their discretion to fuse genuine anatomy with storytelling. To evaluate whether this subterfuge was commonplace and the chosen source material reliable testimony, it is necessary to examine criminal dissections done by penal surgeons who won widespread respect through their virtuosity. Sir William Blizard's work was widely admired everywhere and we therefore return to his record-keeping of criminal dissections which were first introduced in Chapter 2.

Sir William Blizard (1743–1835) was arguably one of the most respected penal surgeons who dissected criminals under the Murder Act.[58] On a regular basis he acquired bodies from Tyburn and Newgate and applied himself to criminal dissections. Blizard started out as an apprentice to a surgeon in Mortlake. He then became a pupil-student of Sir Percival Pott based at St Bartholomew's Hospital across the road from the Old Bailey and Surgeon's Hall. After which, he studied with the famous John Hunter at the London Hospital. There, Blizard was appointed as a qualified surgeon in 1780. Like many he also diversified his business interests. In 1785 he co-founded the McLaurin Medical School. This was a private anatomical venture that he set up at his own expense with a medical partner. To combat his cash-flow problems caused by late-paying, indebted, medical students, he held medical consultations at Batson's Coffee House in Cornhill near the Strand. Here other physicians and surgeon-apothecaries brought their troubling medical histories from difficult fee-paying patients for a second opinion. In 1787 he was elected as a Fellow of the Royal Society, and then appointed as a lecturer of anatomy and surgeon at the Royal College of Surgeons (he was President in 1814 and 1822). For the next twenty years, Blizard won fame and a Royal appointment as surgeon to HRH the Duke and Duchess of Gloucester. He was said to have had a natural gift for teaching, attracting fee-paying students who wanted a hands-on anatomical experience. Small wonder perhaps that he was the founder and first president of the Hunterian Society (1819–22) after being knighted for services to the anatomical sciences in 1803. In short, his anatomical work provides historical insights into the career trajectory of an ambitious man determined to establish his reputation in medical circles by undertaking gallows work so that he could stand centre-stage in the best dissection theatres of London.

The survival of Blizard's case notes and diary detailing dissections of criminal corpses provides therefore an important opportunity to look over his shoulder at the executed body being opened up. Indeed, those who recalled working with him, such as John Abernethy, were fond of quoting his working ethos:

> Let your search after truth be eager and constant. Be wary of admitting propositions as facts before you have submitted them to the strictest examination. If after this you believe them to be true, never disregard or forget one of them. Should you perceive truths to be important; make them the motive of action, let them serve as springs to your conduct.[59]

In May 1815, Blizard took his own advice writing down the details of carefully opening up the body of John Bellingham. He was famously condemned to dissection for the murder of Spencer Percival, the Prime Minister. On this occasion, acting as the leading anatomist-surgeon Blizard checked for medical death at the dissection room of St Bartholomew's Hospital: 'The Right Auricle of the heart moved at irregular intervals, without the application of any Stimulus during the period for nearly four hours from the time of Execution, and did for an hour longer upon being touched with a Scalpel'.[60] These observations were disturbing since the executioner had been very careful to ensure that the prisoner died on the gallows. Blizard thus felt compelled to write down his thoughts based on his accumulated experiences of criminal dissections:

> This Motion is not strictly a contraction, diminishing in any sensible degree in the cavity of the Auricle; it was undulary [sic] and weak, sometimes beginning at the right extremity of the Auricle, and moving to the left; at other times commencing and proceeding in the contrary direction. Not the least motion was observable in the left Auricle; or in either of the Ventricles.[61]

He concluded that these life-signs were in this case organic processes shutting-down, not signs of a life that could be resuscitated or revived. Nonetheless as a precaution he examined the controversial issue of brain death too: 'On the next day the Brain was examined. It was firm and sound throughout. The vessels of the Pia mater were distended with Air. Not a drop of fluid was found in the Ventricles.' The science of the brain being starved of oxygen was known in broad terms but still it was in the end the lack of blood circulating in and out of the brain that signalled 'absolute Death in mind and body'. Blizard was throughout meticulous and he checked the organs of reproduction too, since they were known to show life-signs sometimes when all other physical indications looked fatal: 'The left testis appeared to be reduced in size and loose in texture. The veins on the spermatick [sic] cord were vicarious. The Penis seemed to be in a state of semi-erection'. It was common to see this in gallows bodies and he concluded Bellingham, lacking a sexual reaction, was '*dead*'. Only then did he proceed to take out the 'Stomach and the left Testis' which 'were sent to the College [of Surgeons] Museum'.

Here then we see how a body was opened by delving beneath the 'surface anatomy'. Throughout all the anatomical options described in Table 4.1 were kept open. Later dissection would involve dismemberment and the removal of specimen tissue and organs. Evidence like this reiterates that there were different sorts of post-mortem rites and whilst their timing was variable in the hands of anatomists, there were also agreed methods of cutting the body. There was a medico-legal choreography that was subject to local autonomy and also grounded in a public performance. Sometimes procedures could start and stop, and then restart, or finish earlier than expected. One reason for this trend was that whilst penal surgeons were unconcerned about punishing notorious murderers and did so by matching punishment to cultural norms, they sometimes also had a great deal of sympathy for those executed under the Bloody Code. Blizard took this stance in 1785 in a pamphlet styled, '*Desultory reflections on Police with an Essay on the means of preventing crimes and Amending Criminals*'.[62] His critical attitude took issue with the misleading picture of unfeeling and ill-informed penal surgeons standing with disdain over criminal corpses. Not everyone was simply eager to cut up a body. Often such men were troubled by the harshness of the Bloody Code and its dehumanising side.[63] Yet, at the same time, a recurring theme in the dissection notes of some prominent penal surgeons is that the former social status of a convicted murderer sometimes had a powerful effect on the dissector on duty. One headline case was to fundamentally change how the criminal corpse was cut up by surgeons across early modern England.

The 4th Earl Ferrers (1720–1760) of Staunton Harrold in Leicestershire was found guilty of murder at Westminster Hall in April 1760. The high-profile prosecution led by the Attorney General Charles Pratt on behalf of his fellow peers established that Ferrers was a drunk with a violent temper. He admitted in a candid court statement that he was not a lunatic, to the despair of his defence counsel. Ferrers boasted that revenge was justified by men of good breeding and he thought himself above the law by virtue of his aristocratic birth. The forensic facts were that Ferrers late one night shot his estate steward. He believed him to be too sympathetic to his divorced wife's claims for child maintenance from a family trust fund. The steward named Johnson had called at the mansion to try to resolve a rent discrepancy with Countess Mary Ferrers on behalf of the trustees. She had feared for her life and obtained a separation from Earl Ferrers for cruelty. On arrival, Ferrers drew a gun on the steward, and shot him, but he did not die. Ferrers came to his senses and sent a message for Dr Thomas Kirkland of nearby Ashby-de-la-Zouche to attend the loyal manservant (Kirkland dissected criminal corpses and made his medical reputation in the Midlands, see Chapter 2). In the interim, the steward fearful for his life escaped and somehow stumbled home. By the time Kirkland found him in a chair he was dying from the fatal gunshot wound. Kirkland gave damning evidence in court that Ferrers was cold-bloodied, had committed pre-meditated murder, and was very dangerous. The jury brought in a unanimous guilty verdict. At the conclusion of the case on 5 May 1760, *The Public Ledger*, a popular London journal, thus reported that: 'On Saturday last a great number of eminent surgeons in London had a meeting at Surgeon's Hall in the Old Bailey, in order to consult on proper means for the reception of the corpse of Earl Ferrers after his execution'.[64] At issue, was whether the Murder Act applied to aristocrats or not. By tradition for treasonable offences peers of the realm were beheaded and disembowelled at the Tower of London. This would be the first time an aristocrat had been hanged for a capital conviction and then punished post-mortem by the medical profession.

The *London Evening Post* of 6 May 1760 reported on the precise handling of Earl Ferrers' medical demise by anatomists. Beforehand considerable dispute arose about the cuts to be made with the lancet and how extensive they must be. Journal editors felt that a full-scale dissection was not warranted. Other popular newspapers argued that it upheld law and order to treat men and women with equal retribution in murder cases. One reporter thus wrote:

> The corpse of the late Earl Ferrers was exposed to Publick View on Tuesday evening to a great Number of Spectators in a Room up one pair of stairs at Surgeon's Hall in the Old Bailey – A large Incision or Wound, is made from the Neck to the bottom of the Thorax or Breast, and another quite across the Throat: The Abdomen or lower part of the Belly, is laid open and the Bowels are taken away.[65]

It was noted that Lord Ferrers had been brought to Surgeon's Hall in his satin-lined coffin. Fully clothed at first his body was laid out to view with his 'hat and halter' placed at his feet. Witnesses remarked that his hands on the gallows had turned 'remarkably black'. He had it seemed been hanged for 'one hour five minutes' as a precaution that he was 'truly dead'.[66] There was though considerable dispute in the press about whether these medical statements about his being cut open with a 'crucial incision' to check medical death were accurate or a cover-up. So much so, that newspaper editors on the 9 May 1760 reported that they had been given special dispensation to once more view the corpse before it was buried in a 'leaden coffin'.

> Yesterday people were again permitted to view the corpse of Lord Ferrers from the hours of nine in the morning till one in the afternoon: about five o'clock his lordship's body was so[l]dered up in a leaden coffin, and afterwards enclosed in another covered with velvet and late in the evening set out in a hearse and six to be conveyed to Staunton Harrold in Leicestershire and there interred in the burial place of the ancient and noble family. Lord Ferrers entrails were remarkably sound, the surgeons who opened him having declared that in the whole of their practice, they never observed in any subject that come under their inspection, so great signs of longevity.[67]

This wording was deliberate and measured. The body was essentially anatomized 'under their inspection', and the bowel entrails were removed by tradition because of his social status, but his corpse was not dissected to its extremities. There was no dismemberment of the limbs or cast taken of his brain; nor were any resuscitation experiments carried out on his remarkable muscular physiology. His flesh was not removed and the bones were not scrapped to be made into a skeleton. No study was made of why his hands were 'peculiarly blackened' by the execution rope. In other words, the historical prism of this criminal corpse exposes that an anatomization took place but it fell-short of a standard criminal dissection. Class had shaped the penal punishment rites because penal surgeons had a high degree of discretionary justice in their gift under the Murder Act. Indeed the *London Chronicle* of 29 April 1760 in again well-chosen words confirmed that: 'the unhappy Earl Ferrers is to be Executed and to be anatomized'. No mention was made of dissection *per se* even though it was legally stipulated on the death warrant issued by the Old Bailey judge.

Unsurprisingly perhaps there was a public outcry in the press about this liberal treatment. On 28 May 1760 for instance it was noted that the corpse in its coffin was still lying in 'St Pancras church' because 'for fear of popular resentment' its transportation had been 'deferr'd until a proper opportunity' to get it undamaged out of London. There was a lot of concern that having not been dissected it would be attacked by the mob in the death coach. It thus remained 'deposited with silence and secrecy, directly under the belfry, immediately after removal from Surgeon's Hall'.[68] Later histories revealed that the 'body was re-buried under the tower of the church, and in 1782 was removed to Staunton Harold'.[69] So strong were public sensibilities that it would take twenty-two years before an application was made by the family to remove the corpse from St Pancras to rebury it in Leicestershire.

Earl Ferrers post-mortem rites soon inspired contemporary debates about the ways in which penal surgeons were interpreting the Murder Act at dissection venues. The problem of legal rhetoric versus medical reality seemed to expose obvious short-comings in the surgical application of the capital legislation. Again, by way of example, coinciding with the trial on 8 May 1760 an informative letter was published in the press from a penal surgeon connected to Surgeon's Hall. It began by saying: 'Sir, As there has been great disputes about *dissection anatomizing*, please to give the following a place in your *Public Ledger*.' The penal surgeon conceded that Lord Ferrers' case had highlighted a lot of confusion about the meaning of *anatomization*, as opposed to *dissection*. In the past, he explained, the term *'anatomize'* came 'from its Greek derivation *ana-temno'* and so it was 'originally confined to the interior parts of the human body'. This meant that in the ancient world *'dissection* and *anato-mizing* [sic]' were once 'synonymous terms', but no longer. Moreover the correspondent set out that:

> Anatomy, after several periods of time, was divided into seven parts, viz., *osteology, sarcology, myology, splanchnology, angeiology, neurology,* and *adenology*. I am inclined to think if a human subject is ordered to be *dissected* and *anatomized* the surgeons have a liberty of appointing one of these branches of the art; and this is sufficient, since it is impossible for any surgeon to enter into every one of the branches of art on one single subject [criminal corpse]; therefore, now the law [Murder Act, 1752] is fulfilled in all its intentions, by the operations on the body now before them, as to *anatomizing*: for the branch of this art appointed was *splanchnology*, by which the contents of the lower belly have been displayed, and the intestines and viscera examined; nay, they have gone further, by opening the thorax or the chest, and laying open the heart and lungs in full view; and many have made remarks of some parts diseased, as the kidnies [sic], which numbers have been admitted to view. – This we hope is sufficient to satisfy the laws wisely consulted for good government.[70]

This source suggests that under the Murder Act there was a conventional way to anatomize and this differed from a complete dissection. The terms were not elided together in the way they have been in the standard historical literature. Anatomization by 1760 was commonly interpreted as *splanchnology* (see, Table 4.2). Exposing the body in this way checked for medical death. It was, the newspaper correspondent explained, seen as satisfying the spirit of the capital code and placated the crowds that came to view the criminal corpse at the dissection venue whether at Surgeon's Hall or elsewhere. If a penal surgeon decided to carry out a full-scale criminal dissection then that would involve *all seven* anatomical methods being done *to completion*. That was a hard task given that putrefaction and decaying flesh could only be preserved for a short time, and the dissector would be exhausted by the three-day process.

This medical correspondent was just one of a number of letter writers to London newspapers who pointed out that the general-public were seldom appraised with how old anatomical methods had changed in Georgian England between 1752 and 1760. Up to the 1750s it had been common to remove the flesh quickly to get down to the bare bones and skeleton; whereas surgeons by 1760 were keen to examine the flesh, viscera, organs, brain, glands and so on, which formerly they would have discarded or worked on superficially because the flesh decayed so fast. Later, by the 1790s the desire to carry out original research meant that usually two to

Table 4.2   The seven anatomical methods of a *complete dissection* of the criminal corpse, circa 1760 to 1832

Seven Anatomical Methods	Actual Method on the Criminal Body
Osteology (priority up to 1750s)	study of bones
Sarcology (priority after 1760s)	study of the soft or fleshy parts of the body
Myology (priority after 1760s)	study of the muscular system structure, functions and diseases of muscles
Splanchnology (interpreted as legal method of *anatomization* under the Murder Act 1752)	study of viscera and its vital organs situated in the thoracic, abdominal and pelvic cavities of the body, primarily heart and lungs, but also intestines and kidneys
Angeiology [sic] (original research priority by 1800)	study of the circulatory system and the lymphatic system, including arteries, veins and lymphatic vases
Neurology (original research priority by 1800)	study of the brain and nervous system
Adenology (original research priority by 1800)	study of the glands and hormonal system

three key anatomical methods might be deployed at what would be termed a general dissection. If it was a *complete dissection* then all seven methods were demonstrated. When therefore the condemned was delivered for post-mortem 'harm' the conduct of each criminal dissection was always determined by the actual surgical cuts of the capital code. These were continually in transition in the intellectual research climate of the Enlightenment.

CONCLUSION

Surgeon's Hall was a renowned punishment venue creating material after-lives but it was also a medico-legal space in continual transition. From the 1730s there was a great deal of ongoing discussion about standard methods of punishing the body, even amongst medical elites supposed to be unified on the inside of the criminal justice system. In reality they were considerably divided about the future scientific basis of their unpalatable penal role. The Murder Act coincided with a formative time in the history of anatomy when the 'old study of creation' was giving way to the 'new early modern science' of morbid pathology. Criminal dissections staged these complex cultural exchanges. There was intellectual convergence but also biological divergence, especially about philosophical and religious questions of spirituality and vitality. Forensic standards of heart-lung versus brain death exemplified how medico-legal modifications were being remade, and radically so. Everything was fluid—ideas, methods, and corporate identities—so much so, that there a surprising range of discretionary options that the penal surgeon had in his hands when he raised the lancet to cut the criminal corpse.

Eventually controversial high-profile cases like those of Earl Ferrers compelled anatomists to admit to the press what was well-known behind closed doors. Strictly speaking *anatomization* was defined as *splanchnology* when the capital statute was applied in punishment venues. This outcome reflected closely the controversy surrounding medical death before dissection took place. There were altogether seven anatomical methods linked to a better scientific understanding of biological functions, but in an era when preservation techniques were cruder not even the most skilled anatomist could hope to do more than about one third of the anatomical options before putrefaction. To do all seven was a physical endurance test, even for the most dedicated and fast-working. This meant that generally penal surgeons proceeded to 'dissection to the extremities' once medical death was established, unless they took the opportunity to do original anatomical research in ways that will be elaborated later in this book (see Chapters 5 and 6). In which case they adopted a pick and mix approach to suit their personal preferences. The key skill was to dissect the maximum amount as the biological clock ticked. This meant that many aimed to get down to the skeleton of the bare bones that would be sent for cleaning and wiring. Others wanted to pay more attention to the morbid anatomy of skin, flesh, tissue, and so on. The most conservative decided that there was no need to be meticulous or careful about preserving certain body parts in the process of decay, leaving behind at best one-third of an identifiable criminal body.

In all of this autonomous decision-making there remained theatricality; an inherent sense of drama was a powerful subtext by virtue of the organic instability of the criminal corpse. This was essentially what attracted diarists and early modern crowds to criminal dissections. It meant that encountering the executed was about getting a chance to do something illicit, entering the '*Under-door*'. Going into an anatomical space was where normative assumptions about the history of the body were over-turned, seen upside-down, and looked at from the outside-inside: what Lewis Caroll would later see as an *Alice in Wonderland* perspective of the Natural order of things. The range of emotional intensities this stimulated was influenced by a synaesthesia that smelled foul, looked rotten, felt strange. Audience members like Silas Neville and Richard Hodgkinson joined in a shocking, thrilling, repugnant, blood-stained, enticing experience, even when it became commonplace, routine, and just plain mundane by the 1790s. In Chapter 5 we travel to provincial anatomical scenes to rediscover a body supply network because this was an experience for every-body in Georgian times.

NOTES

1. Fay Bound Alberti (2009), 'Bodies, Hearts and Minds: Why Emotions Matter to Historians of Science and Medicine', *Isis*, Vol. 100, Dec. Issue, No. 4, 798–810, quote at p. 799.
2. D. Cozens-Hardy ed. (1908), *The Diary of Silas Neville, 1767–1788* (Oxford: Oxford University Press), 15 September 1773 entry, p. 205.
3. See, Susan C Lawrence (1996), *Charitable Knowledge: Hospital Pupils and Practitioners in Eighteenth-Century London* (Cambridge: Cambridge University Press), p. 145, endnote 108, for a synopsis of his medical career.

4. Refer, notably, Walter Thornbury (1878), *Old and New London: Volume 1*, (London, Cassell Ltd), chapter XXII, 'St. Paul's Churchyard', pp. 262–74 and Henry C Shelley (2010), *Inns and Taverns of Old London, Part II: Coffee-Houses of London*, (Bremen, Germany: Europaeischer Hochschulverlag GmbH & co) pp. 1–21.

5. Cozens-Hardy *Diary of Silas Neville*, diary entries for 16 September–17 October 1767, at pp. 24–5.

6. Precise anatomical definitions are elaborated in Sections 2 and 3.

7. Cozens-Hardy *Diary of Silas Neville*, diary entries for 16 September–17 October 1767, at p. 25.

8. A. Cunningham (1997), *The Anatomical Renaissance: The Resurrection of the Anatomical Projects of the Ancients* (Aldershot, Hants: Scolar Press), p. 67.

9. Refer, W. S. Blunt (1909), '*John Baker's Horsham Diary* edited by Wilfred Scawen Blunt', *Sussex Archaeological Collections*, Vol. 52, 38–83, quote at p. 60, 31 August 1773.

10. Thomas Pennant (1790), *Account of London*, (London: Robert Faulder printers), pp. 216–7.

11. Dr Fordyce was a member of the Royal Society and Fellow of the Royal College of Physicians. In 1794 his public lectures on brain death attracted the theatre-going crowd to Surgeon's Hall.

12. Figure cited by Simon Chaplin (2012),'The divine touch, or touching divines: John Hunter, David Hume and the Bishop of Durham's rectum' in Mary Terrall and Helen Deutsch eds. (2012), *Vital Matters: Eighteenth Century Views of Conception, Life and Death* (Toronto, Canada: University of Toronto Press) chapter 10, pp. 222–46, quote at p. 225, endnote 13, and compiled from: *The Account Books of the Company of Surgeons 1745–1800*, Library of the Royal College of Surgeons; J. F. South (1866), *Memorials of the Craft of Surgery in England*, (London, Cassell & co), pp. 274–96; C. Wall (1937), *The History of the Surgeon's Company 1745–1800* (London: Hutchinson), pp. 91–109.

13. Sydney Young ed. (1890), *Annals of the Barber Surgeons of London complied from their records and other sources* (London: Blades, East and Blades publishers Ltd), 'The Beadle', pp. 299–306.

14. Signed receipts styled 'executioner' in the Royal College of Surgeons Collection.

15. All original entries in this detailed discussion are taken from Royal College of Surgeons [hereafter RCS], COS/3/1, Financial Records, *The Company of Surgeons, 1745–1778, Volume 1* (six volumes in all). They are listed by date.

16. Illustration 4.1 ©Wellcome Trust Image Collection, Slide Number M0015855, '*Edward Stanton at the Saw and Crown in Lombard Street*

London' (1754–61): 'lancet-maker: maketh and selleth all sorts of surgeons instruments likewise razors scissors penknives knives & forks… note: lancets and other instruments carefully ground and sett', business card; creative commons license, authorised for academic purposes.

17. The National Archives [hereafter TNA], PCC, Abstract of Wills: Wills Proved at the Prerogative Court of Canterbury, 1 April 1761, Edward Stanton, London's instrument Maker, buried Saint Mary Woolnoth, City of London.

18. Again, all original entries are taken from RCS, COS/3/1, Financial Records, *The Company of Surgeons, 1745–1778, Volume 1* (six volumes in all). They are listed by date.

19. Ibid., again, cited entries are by date.

20. All original entries are once more taken from RCS, COS/3/1, Financial Records, *The Company of Surgeons, 1745–1778, Volume 1* (six volumes in all). They are listed by date. The horse-shoe design of the central table is referred to in RCS Library, White, after [Daniel?] Dodd, "The Body of a Murderer exposed in the Theatre of the Surgeon's Hall, Old Bailey", c.1779, published in "Malefactors Register, or, a Tyburn and Newgate Calendar" (London: printed for Alexander Hogg. [1779]), vol. 4, facing p. 168, is an engraving. The image can be viewed online at: Bridgeman Art Library Collection, Reference XJF140216: The Body of a Murderer Exposed in the Theatre of Old Surgeons' Hall London dated 1760–90. The costs of licensing a reproduction for use in this open-access, not-for-profit book, have been prohibitive and so it cannot be illustrated here.

21. Ibid., cited by date and in this case also referenced, as p. 72.

22. S. Chaplin, (2009), 'John Hunter and the Museum Oeconomy, 1750–1800' (unpublished PhD, King's College London), image of the dissection table, p. 52.

23. Medical correspondent (1819), 'Consummate Depravity—Disinterestedness—Anecdotes', *The Imperial Magazine*, Vol. 3, March, Issue 1, p. 266.

24. The deal on offer suggests he had not been convicted for homicide since post-mortem punishment for murder was generally automatic. It is likely he was a notorious housebreaker or highwayman but no substantive details were reported.

25. *London Lives online, Fire Insurance Registers: 1777–86*, policy number 380040, John Hopper the carpenter, near the George Drury Lane London.

26. *London Lives online, Middlesex Coroners Court Records*, CO/IC, 1747–1803, lists Thomas Pacey, John Mansell and John Hooper or Hopper acting as jury men on 'the View of a Body of a Boy unknown then and there lying dead' who was discovered floating downstream in the River Thames.

Bruised and battered as he passed each bridge until fished out with a hook, his flesh much bashed: verdict 'found drowned'—This was one of the most common ways to account for an unexplained death that was probably homicide but unprovable in court.

27. This source has been cited recently in Andrew Cunningham (2010), *Anatomy Anatomis'd: an Experimental Discipline in Enlightenment Europe* (Farnham, Surrey: Ashgate), and dated to 1734 but it exists in several earlier draft forms written for private consumption at Leicestershire Record Office, Henry Halford MS and medical papers, DE107/261/1-10 1734.

28. Ibid.

29. *London Lives online, Middlesex Sessions, Justices Working Papers,* LMSMPS502480076, 23 November 1727 a signed deposition by Abraham Chovet the surgeon about a dangerous assault he witnessed on his doorstep.

30. See, Anon, (1736), *A catalogue and particular description of the human anatomy in wax-work, and several other preparations; to be seen at the Royal-Exchange,* (London: T White) for example cited in A. W. Bates (2008), "Indecent and Demoralising Representations": Public Anatomy Museums in mid-Victorian England', *Medial History,* Vol. 52, January, Issue 1, 1–22, endnote 13.

31. Old Bailey trials online, *The Proceedings of the Old Bailey, 1674–1913,* http://www.oldbaileyonline.org/, cited by trial date reference.

32. See, RCS, COS/3/1, Company of Surgeons, 1745–1778, Volume 1, Financial Accounts, for payments in date order.

33. Daniel Defoe (1723), *Curious and Diverting Journeys through the whole Island of Great Britain* (London: G. Parker, 1723), p. 106.

34. Moses Aaron Richardson (1841-6), *The Local Historian's Table Book of Remarkable Occurrences Connected with the Counties of Newcastle-Upon-Tyne, Northumberland and Durham, Historical Division, Volume 1,* (Newcastle Upon Tyne: Richardson Booksellers), p. 56.

35. Henry Bourne (1736), *The History of Newcastle upon Tyne or the Ancient and Present State of that Town by the Curate of All-Hallows in Newcastle* (Newcastle: John White Publishers), p. 138.

36. Eneas Mackensie (1827), *A Descriptive and Historical Account of the Town of Newcastle-upon-Tyne including the town of Gateshead, Volume I,* (Newcastle: Mackensie and Dent publishers), quotes at p. 506, & 587.

37. Alexander Chalmers (1814), *The General Biographical Dictionary Containing An Historical and Critical Account of Imminent Persons Volume XVII,* (London: J. Nichols and Sons), p. 436.

38. See for example his work on the vitality of the heart praised by Glasgow surgeons in, *The Critical Review or Annals of Literature,* Vol. 55, (1783), p. 183.

39. Alastair Johnson ed. (2001), *The Diary of Thomas Giordani Wright Newcastle Doctor 1826–1829* (Woodbridge, Suffolk: Boydell), pp. 292–3, entries 5 March 1828, 7 March 1829 and 11 March 1829.

40. Ibid.

41. See, fines paid by 5 company members totalling £105 between 14 July 1762 and November 3 1762, RCS, COS/3/1, Company of Surgeons 1745–1778, Volume 1, Financial Accounts.

42. TNA, Sheriff Cravings for Devon, E389/242/227 (Assize Calendar), 20 March 1758.

43. See for instance, Devon Record Office, Mortgage Bond, John Patch, Surgeon of Exeter, 59/7/4/10/1 13 March, 1740/41 and transfer of mortgage, 1142 B/T22/118-119, 1769; TNA, Prerogative Court of Canterbury, Will of Robert Patch, Surgeon of Exeter, PROB/11/1599, 17 December 1817.

44. *Trewman's Flying Post*, 4 January 1787, death notice for John Patch senior, surgeon, Exeter.

45. See, http://www.exetercivicsociety.org.uk/plaques/485/.

46. See, F. G. B. Watson (1939–40), 'Thomas Patch (1725–1782): Notes on his Life, Together with a Catalogue of His Known Works', *The Volume of the Walpole Society*, Vol. 28, 15–50. Thomas was the middle son of John Patch junior and he became an artist.

47. TNA, Sheriff Cravings for Devon, E389/252/217, Assize Calendar, 6 Aug 1808.

48. TNA, Prerogative Court of Canterbury, PROB11/1265, Will of John Bolding, Surgeon of Amphthill, Bedfordshire, 3 September 1795.

49. TNA, Sheriff's Cravings, 389/248/196, Assize calendar, Bedfordshire, body of John Cooke, dated March 1788.

50. *Gentleman's Magazine and Historical Chronicle*, Vol. 45, (26 May 1775), p. 299.

51. Ibid.

52. Anon (1775, 2nd edition), 'Summer Assizes report, Lincoln 1775', reported in *The Annual Register or a View of the History, Politics, and Literature for the Year 1775*, (London: J. Dodley of Pall Mall), pp. 154–5.

53. *The Hibernian Magazine, Or, Compendium of Entertaining Knowledge*, Vol. 5, (September 1775), report of 'hanging of William Farmery at Lincoln on 4 August 1775', p. 562.

54. Referred to in a report of the facts attending the case, in, Edmund Burke (1775), *Dodsley's Annual Register*, (London: Printed for R and J Dodsley in Pall Mall), September issue, p. 155.

55. Lambeth Palace Library, Medical Licences issued by the Archbishop of Canterbury for Lincoln, 1535–1775, 4.19, pp. 36–7, lists the medical men in the vicinity as 'CRITCHLOE (Thomas), of Grantham, EVERITT

(Richard), of Horncastle, FIELDHOUSE (Geoffrey), of Lincoln,, FLEMING (John), of Billingborough, HATFELD (E) (John), fellow of the College of Bonhommes, Ashridge, MORLEY (Henry), of Lincoln, ROGERS (John), of Stamford, TAYLOR (Nicholas), of Gainsborough, VINCENT (Brian), B.D., of Stamford, Lincs., WILDREN (Thomas), clerk, of Horncastle, WRAY (John), of Brant Broughton; the material fact of being sent for dissection was recorded at Lincoln Record Office, Summer Assizes, 4/8/1775, and reported in *Leicester and Nottingham Journal*, 12/8/1775 with no pardon given the brutal nature of the murder.

56. Leeds University Special Collections, MS 504/1/3, Manuscript of Mr. William Hey, Senior Surgeon to the Leeds General Infirmary.

57. Medical Correspondent (1829), 'Modern Medical Ethics; or State Maxims in Medicine by Philoethicus & C', *The Medico-Chirurgical Review and Journal of Practical Medicine*, October issue, 145–9.

58. Rita R Auden FRCS (1978), 'A Hunterian pupil: Sir William Blizard and the London Hospital', *Annals of the Royal College of Surgeons of England*, Vol. 60, 345–9, summarises his career.

59. Ibid., p. 347.

60. RCS, William Clift Collection MS007/1/6/1/1, 'Record of the Bodies of Murderers delivered to the College for Dissection', circa 1800–1814.

61. Ibid.

62. William Cooke (1835), *A Brief Memoir of Sir William Blizard, read before the Hunterian Society October 7th 1835, with additional particulars of his Life and Writings* (London: Longman, Rees, Orme, Brown and Co), p. 36.

63. Auden, 'Sir William Blizard', p. 349. Although Blizard's remarks were cited in his medical biography written by a former colleague, and could be dismissed as hagiography, his philanthropic work, particularly as a leading member of the Samaritan Society, points to a charitable character with some considerable human feeling. He abhorred what he called 'evil persons' but he equally thought that 'outright sin' was rare in society: there was more good than bad in people was his attitude of mind.

64. *The Public Ledger* or *The Daily Register of Commerce and Intelligence* (London), Monday 5 May 1760, Issue 98, page 2, column 1.

65. *The London Evening Post*, 6–8 May 1760, Issue 5072, Page 1, column 2.

66. *The Public Ledger* or *The Daily Register of Commerce and Intelligence* (London),Tuesday 6 May 1760, Issue 99, page 1, column 2.

67. *The Public Ledger* or *The Daily Register of Commerce and Intelligence* (London), 9 May 1760, Issue 102, page 1, column 3.

68. *The Public Ledger* or *The Daily Register of Commerce and Intelligence* (London), 28 May 1760, Issue 118, page 2, column 2; *The London Evening Post*, 8 May 1760, Issue, 5073, page 1, column 2, reporter followed the coffin coach from Surgeon's Hall but wrote that he could not trace its destination because everything was being conducted with such

'great secrecy'; whereas the *London Chronicle* 8 May 1760, Issue 526, page 1, column 2, claimed that the body was 'privately carried from Surgeon's Hall and interred at St. Pancras'.

69. British History Online, http://www.british-history.ac.uk/report.aspx? compid=65576, *Survey of London, volume 24, The parish of St Pancras, part 4, Kings Cross neighbourhood*, (London, 1952), pp. 147–51.

70. *The Public Ledger or the Daily Register of Commerce and Intelligence* (London, England), Thursday 8 May 1760, Issue 101, page 1, column 4.

CHAPTER 5

Mapping Punishment: Provincial Places to Dissect

INTRODUCTION

In 1723, Messenger Monsey started practicing as a physician in Bury St. Edmunds. He advertised his services in both physic and surgery to try to attract new clients to a town considered to be a medical backwater in East Anglia. Even so, for a man born in Norfolk, setting up business in an area he thought he knew well proved to be a major career challenge. Neither his degree in classics from Pembroke College in Cambridge, nor his apprenticeship to a physician at Norwich, prepared him for the financial cut and thrust of provincial doctoring. He soon found out why small provincial towns like Bury did not attract ambitious medical men in the eighteenth-century. Monsey thought, incorrectly, that in medical business he would be advantaged by a lack of qualified doctors in Suffolk. He presumed that patients would pay his fees promptly to secure reliable medical services. It was worrying when he calculated that his gross medical profit was about '£300 per year' and it was not sufficient to keep him solvent.[1] He worked unceasingly because it was a constant financial strain to pay the rent on his business premises, the livery fees for several fast horses and a sturdy carriage, and manage the slow cash-flow of indebted clients. Monsey was then very relieved when he won the local patronage of 2nd Earl Godolphin, son of Queen Anne's Lord Treasurer and grandson of the first Duke of Marlborough. The young peer had suffered an 'apoplectic compliant' [a minor stroke] one night returning to his family seat at

© The Author(s) 2016
E.T. Hurren, *Dissecting the Criminal Corpse*, Palgrave
Historical Studies in the Criminal Corpse and its Afterlife,
DOI 10.1057/978-1-137-58249-2_5

Newmarket. His manservant redirected the carriage to the nearest town, at Bury, and called on Monsey who saved Godolphin's life. In return, he offered to sponsor Monsey's application as a physician to Chelsea Hospital in London. Monsey's obituary-writer recounted that few could have predicted this successful outcome for someone so unseasoned in medical fashions:

> He began business at Bury [St Edmunds], where he experienced the common fate of country practice – constant fatigue, long journeys and short fees; and in rusty wig, dirty boots, and leather breeches, might have degenerated into a hum-drum provincial doctor, his merits not diffused beyond a county chronicle and his medical errors concealed in the country churchyard.[2]

Other diarists wrote about the difficulties of making inroads into a local medical market-place regardless of how well-connected qualified surgeons had become since the Murder Act. Richard Hodgkinson (encountered in Chapter 4 on a visit from the country to Surgeon's Hall) corresponded on 6 March 1828 about the history of business strife in the North of England:

> Mr Bedford (whom I think you may remember) was considered an eminent Surgeon, but tho' he married in Bolton, and had good connections, he was obliged to leave the Town some Years ago and settle in Liverpool. A Mr Moore who has long resided in Bolton has the leasing Practice as a general Surgeon and Apothecary and has the best Families but there are those who vie with him and some of inferior Rank. A Dr Black some years ago settled in Bolton as a Physician, he married but died young and was succeeded by another Dr Black (no Relation I believe) who, on his coming, was so violently attacked by the whole body of the medical men in Bolton that he was obliged to turn on his Assailants and fairly write them down in Public prints. The common Surgery business is all engrossed by a man named Taylor, a Relative of the Oldfield Lane Doctor whose Practice he imitates.[3]

Getting established as a surgeon required tact, patience, fortitude, determination, patronage, good local connections, and sharp elbows. If these did not work then the surgeon could be run out of town. Unless, that is, he was prepared to take some sort of evasive action by harnessing the power of the press, publishing original research in respected medical journals, or taking on post-mortem work. Above all, it was essential to watch one's back because competition was fierce and medical rivalries all too common. This meant that few could afford to ignore the chance that criminal dissections afforded. Measuring business success was though dependent on three elusive factors: first, maintaining a sense of local belonging especially if an outsider coming into the community; second, developing a reputation for medical innovation; and thirdly, displaying a steely determination to succeed in the neighbourhood. It was increasingly necessary not just to be able to identify each patient's troubling symptoms, but to verify their diagnosis too. Medical consumers needed convincing that doctoring was not just the art of storytelling but incorporated genuine medical improvement. This required surgeons to produce new opinions based on a working knowledge of dissection. Hence, this chapter is all about where exactly post-mortems could potentially be improved by criminal dissection work in England. It examines execution places, their actual supply networks, and the distribution of corpses on a regional basis. This empirical picture is also concerned with the spatial architecture of dissection spaces and their alignment in communities, as well as their economies of scale. In the historical literature the number of bodies, their delivery and dispersal, developed at a provincial pace that remains very poorly understood. In a theatre of punishment, the capital looked centre-stage but it lost its starring role by the early nineteenth-century. Historians of culture, crime and medicine need therefore to rediscover medico-legal realities outside the metropolis, and to do so over the long duration. It is insufficient to rely either on a basic reading of the legal rhetoric of the capital legislation when it was first passed or to maintain a blind faith in accepted theories about the condemned body in a history of ideas without testing them in the archives. Instead there needs to be a spatial mapping of actual punishment provision beyond London. Research reveals that dissecting was not a theatre of make-believe but a compelling material showcase in even the remotest parts of England.

Dissection cases known to have been generated by the Murder Act have therefore been mapped in Section 1 of this fifth chapter. Data is presented for the first time on actual supply mechanisms. A quantitative analysis of Sheriff's Cravings found in the National Archives and court records retained by country record offices together form the bedrock of this chapter's empirical findings. Financial expenses claimed back by executioners from central government have been used to reconstruct just how many murder cases had a secure conviction and dissection verdict actioned on a county-by-county basis. These figures provide accurate information about the chain of supply and its regional profile from 1752 to 1832. A significant discovery is the provincial spectrum of large, medium and small-scale punishment venues for post-execution rites. In the North of England and Midlands many penal surgeons used small medical dispensaries;

others worked in the lobby of Shire Halls. These were very public places to dissect in the community before voluntary hospitals were built. In the Southern counties and West Country, the condemned were dissected either in the dead-houses of newly constructed infirmaries or on the business premises of penal surgeons who sometimes made use of a prison room for post-mortem work too. What all the venues had in common was their symbolic architectural alignment near local courtrooms. There, crowds of people gathered to satisfy their 'public curiosity' and stayed behind to get involved in the post-execution encore.

Travelling to these punishment sites involves selecting places that represent their region and then studying them in-depth in Section 2. A sight-seeing tour of dissection rooms thus begins in major 'hanging-towns' like Lancaster in the North West. These are compared to equivalent punishment sites in growing industrial areas across the Pennines in cities such as Leeds and other major towns in Yorkshire. Moving then down to Derby, a semi-industrial location rich in archive sources, illuminates how leading surgeons set medico-legal standards across the Midlands. It is feasible to explore who got criminal corpses, what they did with them, and where exactly punishments happened. The evidence highlights some of the logistics of operating a local economy of supply and those findings are then compared to smaller towns in East Anglia akin to Bury St Edmunds where Monsey Messenger worked. The aim being to find out a lot more about what dissection days were really like, who organised them, and which penal surgeons staffed the sessions. To achieve this, it is also necessary to travel into Devon, Dorset and down to Cornwall. Our sight-seeing in this way takes a circular route using Map 5.1 (see, Section 1), starting at the top of England and ending at its equivalent West Country nexus. Exploring these forgotten punishment spaces is all about revisiting their symbolic meaning on location and walking figuratively with the condemned ready-made for criminal dissection.

FACTS AND FIGURES: SUCCESSFUL CONVICTIONS AND CRIMINAL DISSECTIONS

In eighteenth-century homicide cases presiding judges delegated the capital sentence to a local Sheriff. He had the legal right to reclaim from central government medico-legal expenses incurred in order to carry out the death sentence and punish the condemned post-execution. This included the cost of transportation, scaffold-building, rope manufacture, paying the hangman, and any surgical fees. These financial returns are known as Sheriffs Cravings and they were reimbursed annually by the Exchequer of the Treasury in London. For this book, they have been cross-matched to available Assizes records held in county record offices. Together extensive record linkage work has produced quantitative data that shows in Figure 5.1 how a total of 1, 150 murderers were successfully convicted and sentenced to dissection under the Murder Act in England between 1752 and 1832.[4] These were the official number of bodies that were made available to designated penal surgeons. Since only eight per cent (93 people) were pardoned before execution and thirteen per cent (147 bodies) were hung in chains on the gibbet, rather than dissected, the vast majority of seventy nine per cent (908 criminals) corpses entered the medical economy of supply.[5]

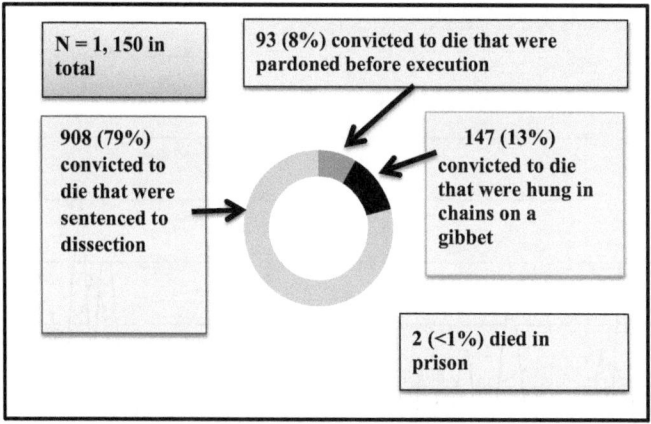

Figure 5.1 Condemned bodies sentenced to death with post-mortem punishment under the Murder Act, circa 1752 to 1832.

Figure 5.2 then plots the actual rates of supply per annum, with conviction rates (the solid black line) displayed against a 5-year moving average (the dotted line). The data shows four distinctive periods of dissection supply. From 1752 an average of 17 bodies was supplied each year until 1800. In a second phase around 1801 there were just 4 in an annual cycle, but this recovered to 15 by 1809. Supply figures during a third phase were more erratic. They reached 18 a year by 1812, rose to 25 by 1813, but fell back to 15 in 1815, before doubling to a sharp peak of 30 by 1816. In a fourth phase, from 1817 supply fell to 10 bodies per year until just before the Anatomy Act was passed in 1832. The symbolic importance of these supply levels for law and order should not be under-estimated and needs therefore to be carefully set in its historical context. Yet, it is self-evident that the official body business was inadequate for the needs of an expanding medical fraternity. It was debatable that future professional recognition could be so dependent on the expansion of human anatomy teaching when there was such a significant shortfall in supply from the gallows. The medical irony was that not enough criminals were committing homicide. To appreciate the differences between a London education and one in the provinces it is essential to utilise the data-set to compare and contrast supply levels in the capital with a regional picture for the time period too.

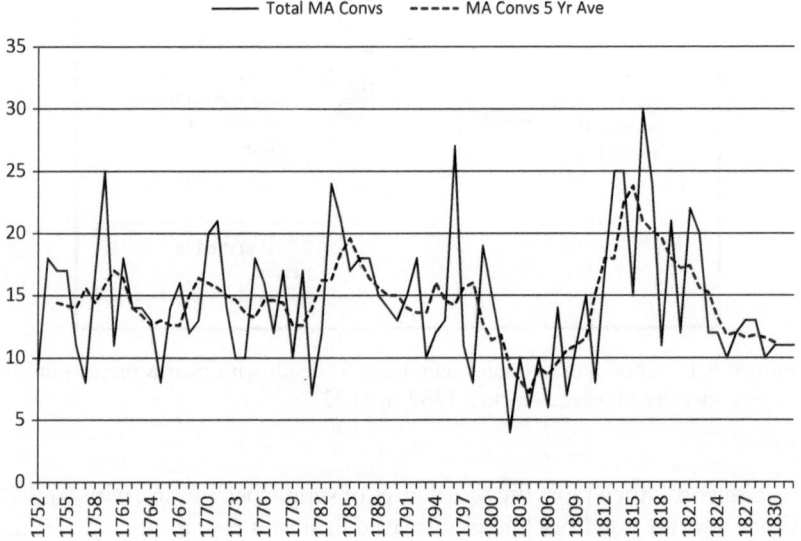

Figure 5.2   Convictions under the Murder Act, 1752 to 1832 (including Admiralty cases).

Simon Chaplin has estimated that '80' bodies were acquired by Surgeon's Hall between 1752 and 1832.[6] If correct, this means that of the 908 available in the entire chain of supply only 8.81 per cent came into the main dissection venue that has been so dominant in the historical literature for London. Using more accurate figures compiled for this study from the Sheriff's Cravings and cross-matching these to Old Bailey records means we can begin to engage with change over time, regional variation, and reassess the supply trends in the capital in a more sophisticated way. Inside London, 170 bodies were recorded as sentenced to dissection for murder and, of these, 148 entered the chain of supply (ten were pardoned & twelve committed suicide on the eve of execution). This equates to some 16.29 per cent (almost double Chaplin's original estimate) of the overall total of 908 bodies. Figure 5.3 then shows that London body-supply networks had primacy, but this dominance only happened until 1800.

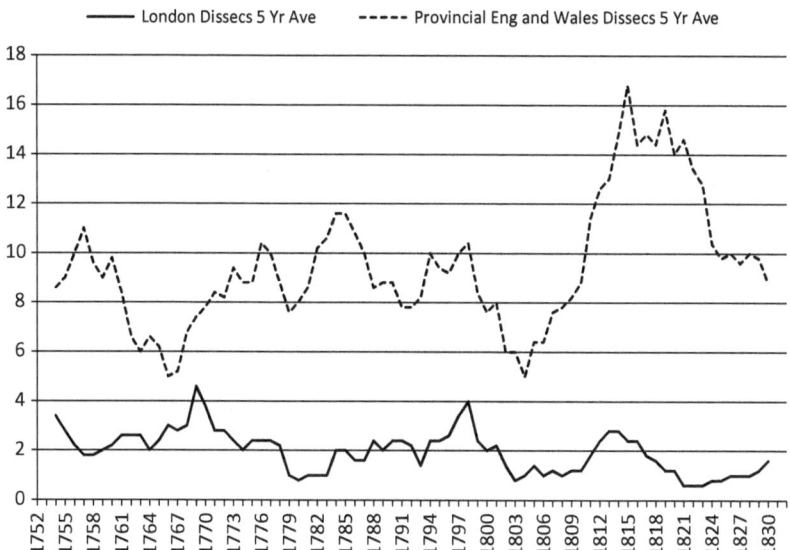

Figure 5.3 Corpses made available to surgeons under the Murder Act, in London compared to provincial England, circa 1752 to 1832.

At a time when Surgeon's Hall started to be seen as lacklustre, its supply lines were on a downward trend. This sets in context why the Hunter brothers for instance by acquiring fresh corpses exhumed from churchyards relied on an illegal supply-mechanism to dominate private medical education in London. It is then a noteworthy finding that after 1804 a penal surgeon had a much better chance of dissecting on a regular basis from legal sources that became available in the provinces, rather than the capital. If he wanted to attract fee-paying pupils to boast his income streams it was prudent to be networked into medico-legal circles in a county setting.

It follows that between 1812 and 1824 when the death of the brain at criminal dissections was regarded as a medical frontier, an ambitious penal surgeon relied on the local Assizes outside of the capital to supply his body needs for original research and teaching. There were of course areas of the country that were better to be located in than others to take advantage of official supply mechanisms. Before we map these and their punishment places on a regional basis (see, Map 5.1, Section 1 below), it is important to appreciate that what mattered to medical men was whether or not when a criminal was sentenced to dissection that body actually entered a chain of supply in each county. The death sentence could after all be changed to hanging in chains, or the judge might be merciful and hand back the body to the relatives. They also had the discretion to pardon or lessen the sentence before leaving town. Of those that were made available in an area, some bodies might be moved to another location out of the county in which they were generated. This often happened in 'hanging-towns' like Lancaster and across Yorkshire where convicted criminals were executed together and so there was sometimes a supply surplus. It made sense to send a selection of corpses to another area before they decayed too fast to be dissected. Later the theme of body-redistribution across county boundaries is elaborated in Section 2. In the meantime, some basic observations about bodies actually made available in English counties can be made from Figures 5.4, 5.5 and 5.6 that appear sequentially below, on the next few pages.

Figure 5.4 shows that there were ten key locations in England where the condemned were obtained for dissection on a regular basis, excluding London.[7] In the North of England, the main body-supply places were Lancashire (a total of 35 corpses) from an area extending down into Greater Manchester, accompanied by the main 'hanging-towns' across Yorkshire (with 53 bodies overall) like Leeds, making up a total chain of 88 supply-bodies. In the Midlands, it was Warwick that was the most active supplier totalling 33 criminal corpses. Across the West Country, the counties of Devonshire (53), Gloucestershire (31), Somerset (33), and Wiltshire (33)

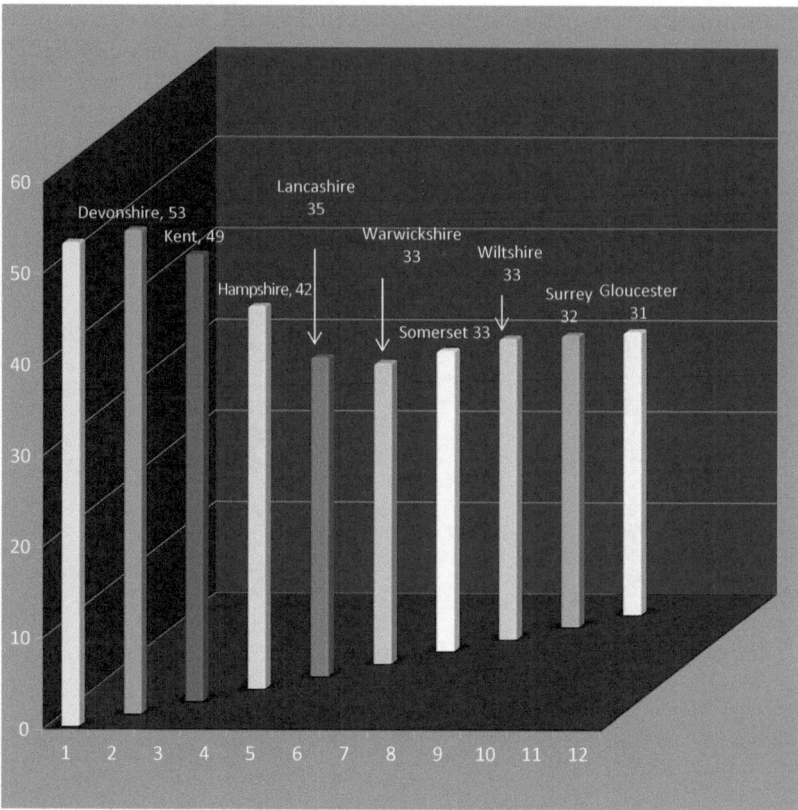

Figure 5.4   The first-rank of ten leading English counties that sentenced the condemned to dissection and punished them post-mortem circa 1752 to 1832.

together delivered 130 bodies for dissection. That meant that in the South of England, Kent (49) and Hampshire (42) were the two leading counties supplying together 123 condemned bodies. It is notable that all of these key suppliers shared a skilled hangman with those in the second rank of county suppliers (cross refer, Figure 5.5). He kept up an efficient supply in his two main hanging towns of Lancaster and Warwick, and did lots of related work for Leicester and Gloucester too. Unsurprisingly penal surgeons made sure they got to know him personally. On location they were thus by 1800 remarkably well placed to exploit county towns linked in the supply chain by their chosen hangman. Their London counterparts lacked an equivalent geographic coverage. In summary then 394 bodies were supplied to penal surgeons from the top ten shires that hanged and dissected on a regular basis under the Murder Act in England.

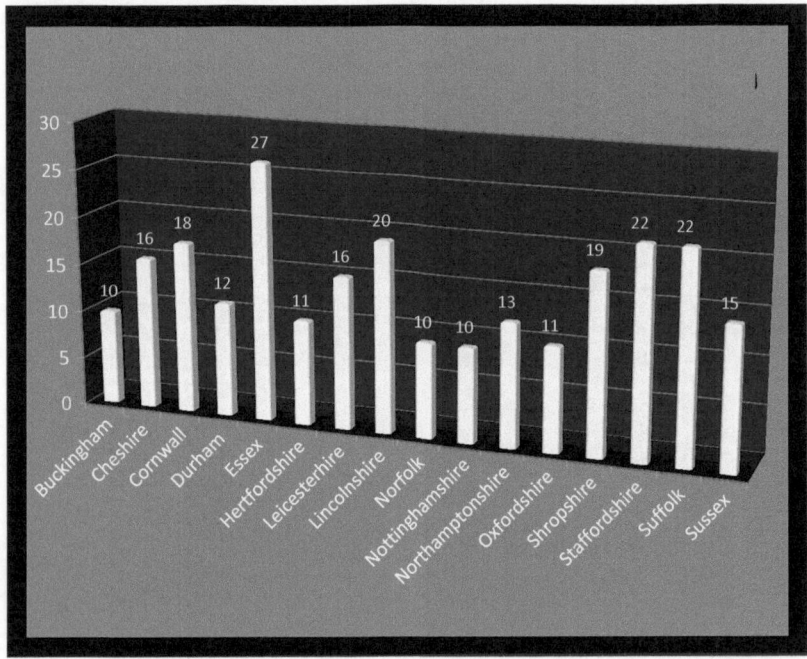

Figure 5.5 The second-rank of body-suppliers in English counties where criminal corpses were made available for dissection, circa 1752 to 1832.

Figure 5.5 reveals those places that fell into a second-rank chain of supply amongst English counties. None of these places identified ever supplied more than thirty bodies in total over the timeframe but the symbolic importance of their delivery schemes should not be under-estimated. In the North West the palantine of Durham with 12 bodies and its equivalent at Cheshire with 16 bodies predominated. In an area covering the Midlands heartland and East Anglia there were 113 corpses in the chain of supply. Stretching down into the South Eastern counties 38 of the condemned were dissected.

In the Southern belt around Oxfordshire there were 21 cadavers, and as far down as Sussex 15 gallows bodies were utilised by local surgeons. This left Cornwall with 18 corpses, the most distant outpost from London. Later we will encounter the different types of anatomical venues in these vicinities and what made the nature of the dissection work undertaken there so distinctive. Meantime the collected data shows that all of these mid-range suppliers delivered 252 bodies in total for criminal dissection.

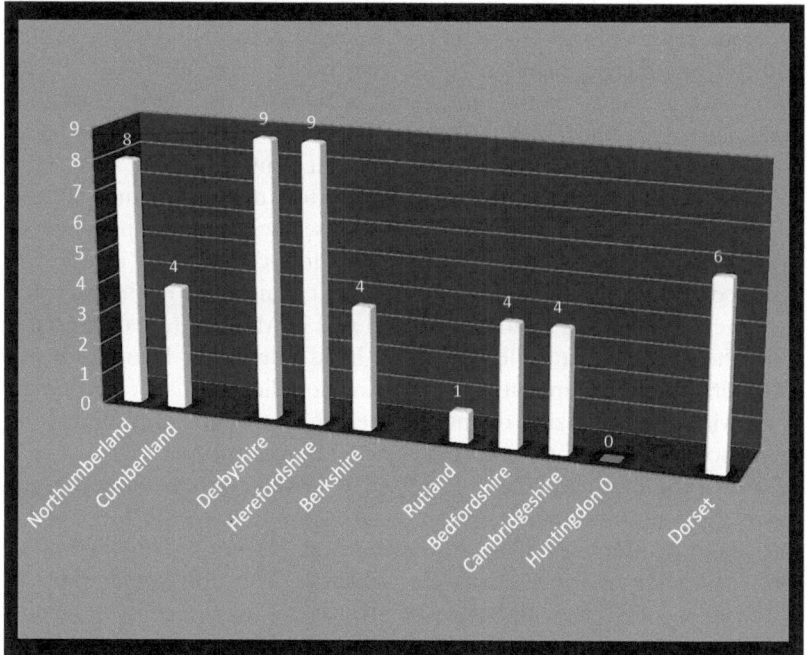

Figure 5.6 The third-rank of body-suppliers in English counties where criminal corpses were made available for dissection, circa 1752 to 1832.

Figure 5.6 concludes this geographical picture of English counties by identifying the third-rank of body-supply locations. These were places that never supplied more than 10 criminal cases in total: although again their smaller numbers must not be understated symbolically. The public reception of a criminal dissection was usually more enhanced where it was a rarity. It really depended on the circumstances of the murder conviction, the scientific credentials of the penal surgeons doing the post-mortem punishment, and where exactly the corpse was dissected. Derby for instance supplied a total of 10 bodies in this time period. That supply figure looks insignificant but each criminal dissection, as we shall see later in Section 2, created widespread publicity. It also involved some sort of original research into the death of the criminal brain. Such findings stress why it is important to avoid broad generalisations about the insignificance of provincial rites. Often cultural studies neglect to appreciate that some locations were sometimes chosen for strategic reasons, so much so that bodies from Nottingham (in the second-rank) were often shared with places such as Derby (in the third-rank) because the dissection work done in the latter was considered prestigious. It thus sometimes could enhance the deterrence value of punishment in violent criminal cases.

To bring all of these locations together into a national picture of post-mortem 'harm' it is then necessary to analyse their body-supply network by mapping them together. This book has found that these can be categorized broadly in one of four punishment zones that relate to typical post-execution venues on location illustrated in Map 5.1 (see overleaf).

In the North of England to the left-hand side of the Pennines it was common to dissect criminal corpses in small public dispensaries, especially in growing industrializing towns like Preston. Cities like Manchester and Liverpool were the exception to this rule because they already had constructed a voluntary hospital in 1752 and 1749 respectively, with a morgue in which to do criminal dissections around the time of the Murder Act (see, <u>Band A</u>, Map 5.1). The same pattern can be seen to the North East too. Small medical dispensaries were utilised as dissection venues in places like Halifax and Wakefield. Meanwhile Newcastle developed its Surgeon's Hall (as we saw in Chapter 4) at a time when York boasted from 1740 that it had a fully-equipped dead-house at its voluntary hospital to house post-mortem rites. Eventually however it was Leeds from 1767 that established itself as *the* place to dissect in Yorkshire. It competed nevertheless with small medical dispensaries for supply for longer than many historians of the period have appreciated.

Moving then down the country into a Midlands heartland it was the local Shire Hall that became the main criminal dissection venue. It was usually located conveniently in the centre of a semi-industrial town like Derby, where the Assizes was held, and so could be used for criminal dissections. This arrangement continued until bodies were moved to local

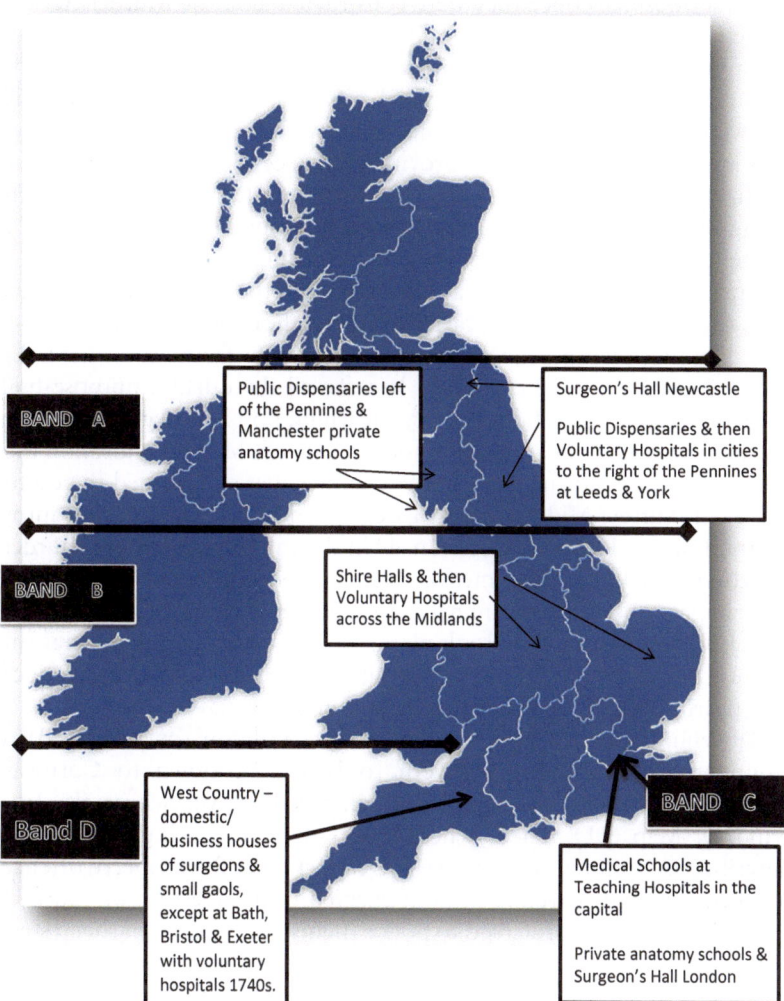

Map 5.1   Geography of punishment zones (Bands A, B, C and D) and their corresponding dissection venues in England, c. 1752 to 1832.

voluntary hospitals once they were constructed after the Murder Act (see Table 5.1, below). It took time for local elites to raise money to establish a hospital venue from charitable funding. When they eventually did each provided a dedicated anatomical space for local surgeons and facilities were often shared with coroners to save money. The Shire Hall was thus a pivotal medical space in provincial life (see, <u>Band B</u>, Map 5.1) until gradually at Leicester (1771) and Nottingham (1782), for instance, the laying of a cornerstone of a new voluntary hospital marked a change of venue for the post-mortem journey of the criminal corpse.

Travelling south on the Great North Road to the capital meant encountering a different scale of medical market-place the closer one got to London (see, <u>Band C</u>, Map 5.1). The capital was different from everywhere else, except Newcastle and Edinburgh, because all three had a purpose-built Surgeon's Hall. Inside London, large teaching hospitals began to develop and alongside them anatomical schools relocated. Medical education thus gradually became formalized. In the counties that surrounded the capital like Kent, Surrey, Middlesex and Essex, often bodies were moved into a central London location. Occasionally in the case of very violent murders the condemned was sometimes dissected on location because of the symbolic impact of staging a criminal dissection for its deterrence value in the actual place that a murder took place. Yet, this was not the general rule because of the reorganisation of institutional structures of medical education nearby and the availability of dead-houses.

Travelling then by express coach out of London towards the West Country, across the River Severn to Bristol or down to Cornwall, something different was happening compared to everywhere else (see, <u>Band D</u>, Map 5.1). These where places that Peter King has recently described as remoter areas where the capital code for property offences was resented and therefore seldom enacted in the Western outreaches.[8] Instead local people preferred to police themselves in cases of sheep stealing, housebreaking and highway robbery. Fewer 'extras' were available because criminal justice was so localized. Homicide however remained a different order of criminal offence. It was punished severely but in anatomical spaces that tended to be more domestic and small-scale before the 1790s. The premises where individual surgeons lived and worked were often used to conduct punishments, and this made sense because local autopsies on behalf of coroners were done in these sorts of places too. Individual surgeons tended to bid for criminal corpses at the local Assizes by making a personal application to the courthouse. They might also co-opt a room at the prison for reasons of convenience. Once however

Table 5.1   Establishment of English provincial voluntary hospitals where post-mortem punishment is known to have taken place after opening, circa 1730 to 1810

Date opened	Location	Date opened	Location
1735	Bristol	1752—**Murder Act**	Manchester
1736	Winchester	1755	Gloucester
1740	York	1755	Chester
1742	Bath	1766	Cambridge
1743	Devon and Exeter	1767	Leeds
1744	Northampton	1767	Salisbury
1745	Worcester	1769	Stafford
1747	Shrewsbury	1770	Oxford
1749	Liverpool	1771	Leicester
1751	Newcastle	1772	Norfolk and Norwich
		1782	Nottingham
		1797	Sheffield
		1810	Derby

Source: Ernest Reginald (1988), *The Life and Times of the Royal Infirmary at Leicester: The Making of a Teaching Hospital 1766–1800* (Leicester: Leicester Medical Society), p. 535

local voluntary hospitals were established this is where bodies were then sent by the 1820s too. In Bath, Bristol, and Exeter, a voluntary hospital had been established in the 1740s and having coincided with the Murder Act there was a public morgue to dissect in. But in outlying areas of Dorset and Launceston in Cornwall, bodies were given to individual surgeons to dissect at home or sent back to a local gaol surgeon to depose of as he saw fit. Over our entire chronological focus the majority of criminal dissections happened up to the 1790s in either a Shire Hall, medical dispensary, the domestic premises of a surgeon, or a local gaol. After 1800, law and justice was more formulaic taking place inside a dedicated dead-house or morgue of a voluntary hospital.

This data taken in its entirety reveals why it was that local people increasingly saw criminal dissections as a medical event they could and should attend because they were happening in their community by the early nineteenth-century. We therefore need to visit actual dissections days staged in rural and urban England. The aim in the next section is to embark on a medical sight-seeing tour of Map 5.1. In so doing, we will be looking in-depth at archetypal places of punishment, those congregated at the scene, and how exactly they were staffed by medical men and others associated with capital rites. The representative examples have each been recently rediscovered in provincial archives.

DISSECTION DAYS: 'HALF SUFFOCATED AND SQUEEZED TO A JELLY'

In all the data illustrated on Map 5.1, we saw that **Band A** to the left of the Pennines had a main dissection venue at Lancaster. In the eighteenth-century it was known as 'The Hanging Town' in the North of England. This was because no other Assizes outside of London hanged more criminals in the long eighteenth-century. It has been estimated that between 1782 and 1812 some 71 convicts were hanged for capital offences, and of these 71 some 35 (fifty per cent) became available for dissection. Lancaster was situated in the Duchy Palantine of Lancaster which since 1351 had special judicial powers adjudicated to try all capital offences for the North West, as far down as Manchester and up to Liverpool. By legal custom Lancaster was the only Assize court in the vicinity of a large sweep of manufacturing towns and rural villages until 1835.[9] Eighteenth-century paintings suggest that prisoners were normally tried together for various capital offences accompanied by huge crowds that congregated outside Lancaster Castle. Architectural drawings of the Castle of 1822 likewise establish this medico-legal scenery.[10]

Before 1800 all those convicted of homicide were hanged on Gallow's Hill on the Lancaster moors, close to an area known today as Williamson Park. By tradition the condemned walked in street-procession taking a last drink at the Golden Lion public house in the town. Coffins were carried in a cart and beside it the condemned walked to their fate accompanied by the crowd. Local accounts stress that people came from across the North West to witness 'The Hanging Day'. After 1800 however executions were moved to the outside of the precincts of Lancaster Castle, near the churchyard. The hangman now erected a gallows at Hanging Corner between the Tower and wall on the east side of the terrace of the Castle. This meant that in 1752 the medical men in the vicinity had to negotiate out on the moors to obtain the body for post-mortem punishment amidst crowds of up to ten thousand. By 1800 however they could make a more discrete body deal just outside the Castle Walls where average crowds of five thousand were encouraged to accompany the dead body down the hill into the town. The actual post-execution penalty was thus choreographed in three ways. First, the sheriff, his deputy, and the constables, together with the hangman and a designated surgeon, checked for medical death when the body was cut down at the Castle Walls. This was always a preliminary examination to make sure the criminal was not what local people called '*the half-hanged*'. It was then carried aloft by the constables to a nearby medical venue where social justice could be seen to be done over the next three days in the presence of the townsfolk (Illustrations 5.1 and 5.2).

Illustration 5.1 Hanging Corner at Lancaster Castle (photographic image, supplied by the author, 2013). Note: *Just outside Lancaster Prison was used as a site of execution for capital offences post-1800 until the Drop Room was opened in 1865. There was originally a drain for all the detritus at the Hanging Corner, now filled in.*

Illustration 5.2 Lancaster Dispensary, Castle Hill, Lancaster (photographic image, supplied by the author, 2013).

In Lancaster, and towns like it across the North West of England, the primary venue for dissection was a local medical dispensary. At the bottom of Castle Hill some two hundred yards from the gates of Lancaster Castle was a medical space in which by custom criminal corpses were dissected to complete the capital sentence of the Murder Act. When by way of example Ashton Worrall aged twenty-five was convicted of the murder of Sarah McLellon in 1831, his body was executed at Hanging Corner. After basic resuscitation procedures were tried and failed, his corpse by nightfall was taken down the hill from the Castle by the constables to Lancaster medical dispensary. Overnight the medical men monitored his life-signs to double-check medical death at the anatomization stage of the punishment process, and then he was made ready for dissection the next day. John Pickstone explains why this venue came to prominence:

> The Lancaster Dispensary was begun by Dr David Campbell, an Edinburgh graduate ... In 1795 both Lancaster and Kendal were prosperous confident towns, not greatly depressed by wars. Beside their dispensaries, each had a relatively new workhouse (Kendal 1769, Lancaster 1787) and these were well kept. Friendly societies, including some for women, had developed since mid-century, and especially during the 1780s. Lancaster had eighteen...and since the male membership of Lancaster societies was over 1, 100, and the number of families in the borough about 2, 000, a high proportion of those families must have been insured against the sickness of their wage earners....By the end of the Napoleonic Wars, Lancaster's trade had stagnated, but in the early 1820s the Dispensary continued to treat between 1, 000 and 1, 500 cases per year...By contemporary standards Lancaster was well-equipped with medical institutions and these enjoyed general support.[11]

Medical dispensaries tended to be built to promote religious duty and a sense of civic loyalty through positive healthcare initiatives. Funded by major manufacturers, amongst townsfolk they promoted social stability. Their charitable rules included free treatments for general illness, epidemics, and industrial accidents because these were not good for business profits, trade cycles or robust public health. The dispensary venue had important medico-legal purposes too. It could be used for a local coroner's court instead of a public house, or act as the designated legal space where post-mortem punishment took place after execution. Dispensaries were generally seen as ideal places to send criminal corpses in the North West because they were very popular and so they became an accessible public relations vehicle for the anatomical sciences. Those at Lancaster, Bolton, Halifax, Kendal, Preston, Wigan, and Wakefield, all provided non-domiciliary medical care and were known locally as the 'people's friend'.[12] If the medical profession in the provinces sought to really gain acceptance of dissection then it made sense to use dispensary facilities to gain popular acceptance of post-mortem 'harm' before the establishment of local voluntary hospitals. The medico-legal fraternity promoted this forum in which local people already felt comfortable, trusted the services on offer, and were encouraged to actively seek a medical note given by a subscriber. They did so aware that their participation in dissections was symbolically about endorsing their deterrence value too, in return for medical help. If elites wanted to get local people talking about feeling a sense of belonging for retribution, then this location was ideally placed to start that official process of crime and punishment.

At Lancaster the dispensary was located in a small town house, a walk downhill of no more than five minutes from the gallows at the Castle. This made the public dissection day an accessible experience in close proximity. The crowd had an opportunity to engage first-hand with the criminal corpse. Most could literally lean over a finger, hand, leg, foot, chest, head, inspecting the cold pallor, smelling the scent of decay, noting the hair colour, length of nails, height and weight, as they walked around the condemned lying dead. Normally they were laid out on a dissection table placed just inside the dispensary front door. Dissecting in a dispensary space like this thus tried to set up a much more interactive, more positive public relations relationship with the local community in terms of post-execution legitimacy. It was also the case that the Old Poor Law discouraged people from becoming claimants around Lancaster fearing that sickness was a long term drain on the parish rates. Local people likewise often resented being sent to, constrained by, or indeed judged within, an expanding voluntary hospital system. In dispensaries, they felt no sense of shame and could participate in the range of services on offer without fear of censure. In other words, dispensaries were seen as more democratic spaces of social justice and increasingly they became conduits for medical services in ways that have been neglected in standard crime histories.

Record linkage work introduces us to those medical men present at the dispensary on the night that Ashton Worall's corpse was carried down from Castle Hill to the Lancaster dispensary after execution on 14 March 1831. Dr. Christopher Johnson (1782–1866) was the honorary surgeon waiting to receive the executed body from the Castle constables. He was an ideal choice to undertake the post-mortem work. Johnson was fascinated by the new science of forensic medicine, having already published the well-received *The Signs of Murder in New Born Children* (1813). George Howson explains that like many provincial self-made surgeons, Johnson was largely self-educated, from a humble background (an orphan child), but ambitious. He concentrated on climbing the career ladder in the vicinity where he could gain a strong financial foothold.[13] Seeking further anatomical education abroad in Paris or Leyden, or in London, was not economically viable. Instead he had to rely on local connections to improve his reputation and research credentials in the immediate medical market-place. Apprenticed to a Preston surgeon-apothecary, Johnson earned enough to study for his medical degree at Edinburgh before returning to Lancaster to establish his reputation in the North West. His honorary surgical position with the Lancaster dispensary was necessary to attract clients from across the social spectrum. If he did not diversify his income-streams, then he might not survive the cut-throat medical business. In addition to his dispensary work he thus acted as honorary surgeon to the part-time Lonsdale Local Militia and by 1815 held the same position with the newly established House of Recovery in the town. A prominent member of the Lancaster Medical Book Club (1823) and Leeds Mechanic Institute (1824) he was well-networked by the time he raised the lancet over Ashton Worrall.

Standing beside Christopher Johnson in the dissection room was Lawson Whalley (1782–1841).[14] He was a Quaker, and honourable Physician to the Lancaster dispensary. Whalley's role was to oversee the dissection in 1831. He was tasked with watching over the knife skills of his surgical colleagues. He too went to Edinburgh to study medicine but coming from a wealthy Quaker family could afford to train for seven years as a physician, qualifying in 1804. Nevertheless, he found it necessary to attract loyal fee-paying appointments because of the topsy-turvy nature of medical consumers with cash-flow problems. So he secured the position of medical officer to the Eagle Insurance Company, as well as his honorary dispensary position. Whalley soon established himself as the pre-eminent medical man in Lancaster, seeking election to office, helping to set up the medical services of the local asylum, becoming a magistrate in 1836, and joining the same medical networks as Christopher Johnson. John Pickstone has found that there were often professional tensions between physicians and surgeons in local dispensaries located across the North West connected to Manchester.[15] Yet, this detailed study has not uncovered such rivalries in the case of Lancaster. Perhaps because it was

'The Hanging Town' and more bodies became available from the gallows there seems to have been less to dispute about access to, and medical authority over, criminal dissections in the town.

A third medical man present at the dissection of Ashton Worrall in Lancaster dispensary was Dr Jonathan Binns. He was described in the *Annual Medical Register* as 'Extra-Licensate of the Royal College of Physicians for Lancaster'.[16] Binns had strong medical connections to Liverpool, being listed as a branch member of the Society for the Abolition of the African Slave Trade in 1788. He also worked for a time with Dr Currie at the Liverpool dispensary before taking up an appointment as Physician superintendent to Ackworth Quaker School in 1795. At this renowned Yorkshire educational establishment he wrote *An Introduction to English Grammar* in his leisure hours. By 1807 he had moved jobs to Lancaster Infirmary as an honorary Physician where he became an expert on the treatment of scarletina and appears to have developed a keen interest in child medicine.[17] Taken together then, these three men had medico-legal jurisdiction over Ashton Worrall's corpse.

Other towns in the North West had been anxious to get hold of Ashton Worrall's criminal corpse, but their requests were rejected. Penal surgeons did not however give up hope of being supplied by Lancaster gallows on the night that Worrall was executed. Lancaster had a policy of doing double and triple hangings, and this meant that there were two other bodies executed alongside Worrall. Those in Preston thus pressed hard to get their fair share. William Worrall aged thirty-eight (the older sibling of Ashton), was an accessory to the same murder of Sarah McLellon of Failsworth near Oldham in Greater Manchester. Although Ashton and William had been hanged together, the Sheriff used his discretionary powers to redistribute the two brothers in death. William Worrall had offended the hangman because he refused to walk quietly to the gallows outside Lancaster Castle. Local newspapers reported that he would not stand still. He was very agitated and aggressively kicked off his shoes, hurling them at the executioner. The Judge had sent for Samuel Haywood the skilled hangman from Ashby Magna in Leicestershire who normally covered the East Midlands area (refer Chapter 2). Even so, local histories stressed that it was no easy matter to do a fast execution of William Worrall. Eventually, with the help of the constables the hangman tied the rope round his neck and he was hanged for over an hour. To avoid any further disruption, and keen to assist other surgeons nearby, William Worrall's corpse was sent to Preston medical dispensary. This outcome reveals how the economy

Table 5.2   Preston physicians, surgeons and apothecaries, circa 1831

Preston Physicians	Preston Surgeons	Preston Apothecaries
Dr Thomas Cuncliffe	Dr William Alexander	John Fallowfield Snr
Dr R Watson Robinson	Dr Edmund Armistead	John Fallowfield Jnr
Dr William St Clare snr	Dr Edward Briggs	Thomas Fallowfield
Dr William St Clare jnr	Dr William Farrar	William Gilbert
	Dr William Gilberston	James Mounsey
	Dr Thomas More	John Taylor
	Dr James Swift	John Thompson
	Dr S. Sherlock	
	Drs Walton and Lodge	
	Dr James Wood	

of supply operated through body-distribution schemes managed by the Sheriff and hangman who consulted medical men across the North West.

Conveniently the Preston dispensary had a discrete entrance in Surgeon's Court just off Lune Lane in the town, and visitor numbers could be controlled.[18] Yet this was also a public space and one valued by the poor who needed cheap medicine dispensed from its charitable funds. William Worrall's dissection was not a large-scale entertainment but instead it took place amidst ordinary people who congregated into a small room at the front of the building to see, smell, brush past, gaze at, and talk about the bruised condition of the criminal corpse. This type of dissection venue was again all about being in close proximity. It was expected that there would be an accessible viewing of the abnormal criminal reduced to an anatomical normality of bone, skull, and body shell. In this socio-medical space, the labouring poor, industrial workers, people who lived on the threshold of relative to absolute poverty, participated in a spectacle that expressed community and belonging, criminal deviance and what it meant to be the ultimate social outcast.

There were a lot of medical men working in Preston who could potentially attend the dissection of William Worrall in March 1831. Table 5.2 above lists those trading in town. In addition to those listed, a Mr Greenwood acted as dispensary apothecary, Mr Richard Oldfield was his dispensing assistant, and the work of everyone on site was overseen by Dr William St Clare junior, an honorary Physician, and a renowned medical figure in the locality. As Preston was the economic heartland of the district, rapidly becoming one of the richest trading and most expansive towns in industrial Britain, it is perhaps unsurprising to find such a crowded medical market-place in operation. The four prominent physicians, ten surgeons and seven apothecaries, together served most of the county of Lancashire. There was thus a competitive atmosphere and high demand for the corpse of William Worrall in Preston. If everyone was permitted to attend the enclosed space—and there was some local debate about this in many localities—there often was not much elbow-room.

Opened on the 25 October 1809, Preston dispensary had treated '12, 239 patients' by 22 October 1817. It was a busy, crowded medical space, described in a *History of the Borough of Preston* some ten years before William Worrall's body was being dissected inside, as:

> This noble edifice has a fine polished stone front, well lighted by eight elliptic and square windows, fronting Fisher-gate, and is palisaded, with two flights of steps up to the main entrance from the street, ornamented with an elevated lamp, for the purpose of giving light, by gas, in the winter time, which considerably embellishes the front, The inside is well planned for the purposes it was built for, consisting of a room for the medicines, with a room on the ground floor [consulting space] and a kitchen below for the matron, together with drawing and other rooms [for dissection], so useful and necessary.[19]

Early histories of the Preston dispensary stress however that until the 1830s surgeons were excluded from practicing from the premises. It was physicians who dominated the medical charity to the chagrin of the surgeons.[20] This explains how it was that Dr William St Clare junior became a respected physician and leading political commentator in the town. He described election days and execution times as raucous affairs. When working from the dispensary he said he often felt: 'half suffocated and squeezed to a jelly' by the assembled crowds.[21]

John Pickstone observed that from 1821 several surgeons tried to challenge the dominance of physicians like St Clare at the Preston dispensary but their chief motivation for doing so were never made public.[22] This study's recent finding is that the ten local surgeons (see Table 5.2, Section 2) pressed to get official appointments at the Preston dispensary by 1831 because for a decade they had wanted to benefit from the body-delivery scheme controlled by Lancaster Castle. Only it facilitated medical research opportunities and yet they were excluded from local justice that should have come within their official purview. Preston surgeons were not alone in their desire for better professional recognition through improved access to criminal dissections in the North West. Manchester surgeons were equally concerned to get their fair share from the gallows.

Illustration 5.3   © Wellcome Trust Image Collection, Slide Number L0011830, Samuel Austin (1831), '*The Manchester Infirmary, dispensary and lunatic asylum*', line engraving; Creative Commons Attribution-NonCommercial-ShareAlike 4.0 International License (CC BY-NC-SA 4.0)

There was a third body that was distributed the night that the Worrall brothers were executed, called Moses Fernley. Fernley had been found guilty of a separate homicide, the killing of his five year old stepson at Hulme a town suburb in Greater Manchester. The *Manchester Guardian* thus reported that although Ashton Worrall died almost instantaneously and William struggled a lot' but in the end 'appeared to suffer very little pain', Fernley's death 'was long and violently convulsed, before life was extinct'.[23] The timing of medical death differed in all three executions and it was troublesome to the very experienced hangman and penal surgeon on duty at Lancaster. Since Fernley had taken much longer to die, and his crime of homicide had been originally committed in Hulme, Samuel Haywood the executioner agreed to send his body back to Manchester. He did so on the basis that post-mortem 'harm' belonged by customary rights to the community in the vicinity of the original murder. Fernley's body was despatched by the coaching express to the Manchester Infirmary, Dispensary and Lunatic Asylum (see Illustration 5.3 above) where the leading anatomist-surgeon Thomas Turner presided over the criminal dissection.[24] The chosen medical venue was not a coincidence. The records of the Lancaster and Chester Antiquarian Society explain that the infirmary had been established in 1752 and moved to new buildings in 1755 along Piccadilly in central Manchester. Here the majority of the sick poor were alleviated on a daily basis through a medical dispensary system.[25] These arrangements coincided with the Murder Act and so it was logical to use the premises for dissection work too. When Moses Fernley's body became available in 1831 it was returned to a vibrant research and teaching space. Here Manchester anatomists were anxious to prove their credentials to their counterparts at the Royal College of Surgeons.

In infamous murder cases a lot more information has survived about the sorts of dissection venues used and therefore issue of representativeness need to be evaluated. The value of such detailed record-keeping is that it is a useful starting point for retracing body-supply schemes in leading areas like the North West. These can then enable historians to start to build upon a wide range of recent scholarship on healthcare and welfare too. It is feasible for instance to envisage how post-mortem 'harm' fitted more broadly into a local economy of makeshifts. This was organized regionally and looked distinctive as you travelled up or down the country. Steve King's work on 'regional states of welfare' shows that the more West and North you travelled to the Left of the Pennines the harsher the Old Poor Law welfare system became in practice from 1750 to 1850.[26] Martin Gorsky has equally found that in these same areas there tended to be a greater number of friendly societies because poorer people had to generate alternative ways to save for a rainy day in times of sickness or to bury their dead.[27] Medical dispensaries (funded from charitable resources) thus became intrinsic to the mixed-economy of welfare in the North West. This backdrop stresses the importance of dissection spaces in ordinary people's lives and why they took on such a symbolic importance in the vicinity.

Before leaving the North of England it is worth exploring briefly the sorts of dissection venues to the right of the Pennines in the North East. In prominent medical places like York and Leeds, the Assizes was held on a regular basis covering the East and West Ridings of Yorkshire. There were similar medical dispensary buildings being used for criminal dissections as those in Lancashire. In manufacturing towns like Halifax, Sheffield and Wakefield, provincial physicians and surgeons were anxious to dissect in community spaces like those in use across the North West. Although therefore Newcastle had its own Surgeon's Hall, it resembled the spatial architecture of Edinburgh and London venues, and was thus atypical of communal premises designated for use across the North East. One representative example provides useful context about the setting of an average criminal dissection in northern towns and cities. On 7 February 1817 Michael Pickles and John Greenwood were charged with the robbery and murder of Samuel Sutcliffe. He was killed at a well-known beauty spot called Hardcastle Crags, near Hebden Bridge, not far from Fountains Abbey estate near Ripon.[28] In court verifiable evidence was presented which established beyond reasonable doubt that whilst Greenwood had robbed the victim, it had been Pickles who had strangled him with his bare hands in a vicious attack. The Judge decided to make an example of Pickles and ordered that he be hanged at York. He was then taken down, anatomized to check his medical death, and sent by express coach to Halifax dispensary to be dissected in a medical venue not far away from the murder scene. Justice had to be seen to be done locally to make it a community affair. Such new findings reiterate that popularizing criminal dissections was intrinsic to the post-mortem 'harm' of the criminal corpse. It took place in a wide variety of manufacturing towns with busy medical dispensaries in the North of England from 1752 to 1832.

Moving then down Map 5.1 into **Band B** (see, Section 1) covering the Midlands reveals what was happening across the central belt of England. In for instance eighteenth-century Derbyshire there were a number of medical gentlemen with provincial standing who were to rise to national importance after the Murder Act. Aspiring men, like Erasmus Darwin the physician who had qualified in medicine at Cambridge, cultivated a large circle of up-and-coming Natural Scientists. Together they founded the famous Lunar Circle based in Birmingham.[29] Known today as the Lunar Society, historians have seen it as a vehicle for ambitious men who exuded Enlightenment ideals.[30] Surgeons connected to the criminal justice system were keen to join from across the area.

Membership of this tight-knit group featured dynamic characters like Joseph Wright of Derby. He was a renowned painter whose enquiring mind extended to scientific endeavour. Wright was fascinated by links between engineering design and resuscitation techniques. These were famously depicted in his paintings of *The Orrery* (1766), *Experiment on a Bird in an Air Pump* (1768) and *The Alchemist in Search of the Philosopher's Stone* (1795). From the 1760s he painted those pursing scientific eclecticism like his near-relation Richard Wright the surgeon

(1730–1814) of Derby (see Illustration 5.4) who operated in the provinces under the capital code. These were the sort of characters that also purchased phrenology heads (see Illustration 5.5) made in porcelain in the nearby Staffordshire potteries (see, also Chapter 6).

Illustration 5.4   © Wellcome Trust Image Collection, Slide Number L0013434, Joseph Wright of Derby portrait of '*Richard Wright (1730–1814) surgeon of Derby*', oil painting; Creative Commons Attribution-NonCommercial-ShareAlike 4.0 International License (CC BY-NC-SA 4.0)

When therefore it came to establishing the Derby Infirmary and designing its new dissection room, the team of assembled experts was dominated by men with leading medical and scientific interests. All were concerned to improve health and safety for patients, practitioners, and penal surgeons. By the 1820s the 'new sciences' of electricity had really taken hold in Derby.

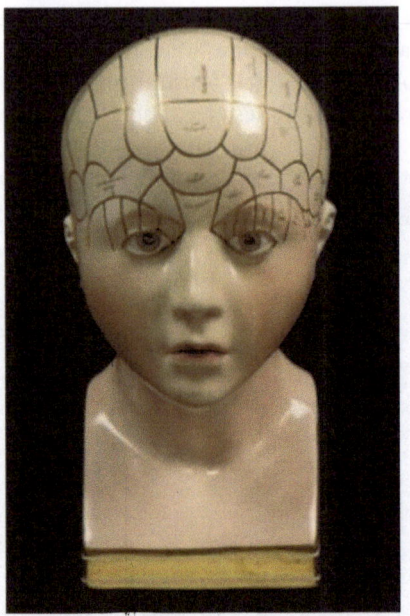

Illustration 5.5   © Wellcome Trust Image Collection, Slide Number L0058695, *'Porcelain phrenological bust, tinted skin colour, divisions labels and numbers marked in gilt, probably in Derby'*, made at the Staffordshire Potteries early nineteenth century, object held in the Science Museum, A642806, clay cast; Creative Commons Attribution-NonCommercial-ShareAlike 4.0 International License (CC BY-NC-SA 4.0)

It was these sorts of enquiring subjects and minds that also joined the newly formed Derby Mechanics Institute. Access to fresh cadavers was thus one, amongst a number, of medical channels to improve a growing and impressive body of knowledge. In the archives the historical prism of Derby illuminates the sorts of 'natural curiosity' that criminal corpses engendered in the Midlands that was influential elsewhere in England.

Traditionally in Derby town there were three sites of execution for general capital offences—at Nun's Green up to 1807—then at Friar Gate from 1812 to 1828—and at Derby Jail from 1833 until 1907. Data collected for this chapter confirms that at Friar Gate some 58 prisoners were hanged between 1756 and 1825 covering most of the time period of this book. The majority of murderers were sentenced to death at the Shire Hall where the county Assizes was held four times a year. They were then taken outside to the front of the building into a courtyard and hanged. Once the jerking ceased, the condemned body was brought back inside the front door of Shire Hall. Here it was dissected in full public view of the assembled crowd: a pattern of local justice that was repeated across the Midlands. The judiciary decreed that retribution had to be transparent and conveniently located in the vicinity of original criminal trials. The application of the Murder Act essentially expressed local sentiments that there must be a combination of legal, professional, religious, and scientific eyes gathered together in one place for completing the capital punishment rites. One example stands in for many at the time.

Mary Dilkes was convicted of the murder of her bastard child on 29 March 1754. She was conveyed in a cart from Derby jailhouse to the Assizes courtroom at Derby Shire Hall. Found guilty of murder, on pronouncement of the death sentence, she was executed on a new gallows that had been erected outside in the main courtyard of the law courts. In the vicinity, she was one of the first murderers to be hanged like this after the Murder Act. Her case thus set a number of medico-legal precedents about the accepted choreography of punishment rites. The Judge sentenced her to be 'dissected and anatomized' (he refused a pardon). This was done after she was cut down by the hangman and taken just inside the entrance to the left door of Shire Hall (see, images below).[31] By tradition the designated penal surgeon occupied a house at 44 Friar Gate in the town. Handily, he lived close by to be called upon by the hangman on execution days. Spatially the execution and its post-mortem rituals were therefore aligned carefully into Derby's urban design. Derby Shire Hall, or 'Court of Justice' as it was became known, was thus described as 'long the pride of the Midland Circuit; longer the dread of the criminal and the client; but the delight of the lawyer'.[32]

After 1752 building modifications to the main courtyard reflected closer medico-legal ties. From the street side the gallows was protected by a five foot high wall over which the assembled crowd glimpsed executions (see, Illustration 5.6); later this was replaced by high black railings (seen in Illustration 5.7). The latter was erected for better for crowd control. The railings were strong enough to hold back a mob pressing forward but at the same time afforded a better view of what was happening for those standing at the rear. It was hoped that better visibility would stop the crowd rioting. The 'Derby Dissection Door' of the Shire Hall building (see, Illustration 5.8) was regarded as the equivalent of the '*Under-door*' at London's Surgeon's Hall, but locally it was left open to everyone assembled to see the condemned being carried inside. Post-execution, the crowd were permitted to flood through the iron-gates and walk around the criminal corpse. Surviving contemporary engravings confirm that this was a busy social space. The courtyard-design outside was used for concerts, plays and gatherings, as well as the Assizes. When Derby's New Assembly Rooms were built in 1714 the building became an exclusive Crown Court. This was when an inn, wine vaults and stables were added to the left side in 1795 and judges' lodgings to the right flank around 1811, eventually creating a u-shaped enclosure. The entire Bloody Code could be self-contained inside the ready-made facilities.

Illustration 5.6   © http://www.capitalpunishmentuk.org/derbygaol.html, '*Derby Gallows*', woodcut, late-eighteenth century [*cross-reference Sketch 5.1*] showing walled off area, academic fair use made of open access image; Creative Commons Attribution-NonCommercial-ShareAlike 4.0 International License (CC BY-NC-SA 4.0)

Illustration 5.7 © Picture the Past Digital Images Collection, Derbyshire County Council, Derbyshire Record Office, IMAG 300050, from a painting by S. H. Parkins C. 1800, image can also be viewed at http://www.capitalpunishmentuk. org/derbygaol.html, 'Derby Shire Hall and Assizes Court', woodcut, late-eighteenth century [cross-reference Sketch 5.2]; Creative Commons Attribution-NonCommercial-ShareAlike 4.0 International License (CC BY-NC-SA 4.0)

Illustration 5.8 Derby Magistrates Court, located at Old Shire Hall, Derby Town, Courts of Justice, (photographic image supplied by the author, 2013). The door is to the left

EXECUTION OUTSIDE NOTTM SHIRE HALL. ~   ~

PRINTED FROM WOOD BLOCK CUT BY E.WILDD, del·sc, ORIGINAL BLOCK IN
POSSESSION OF A.C.VICE, PRINTER, NOTTM ~.   ~   ~

Illustration 5.9   © Picture the Past Digital Images Collection, Nottingham City Council, Nottingham Record Office, NTGM 015347, image can also be viewed online as '*Nottingham Assizes, Shire Hall, Gallows*' at http://www.nottshistory. org.uk/articles/shirehall.htm, woodcut sketch of public hanging; Creative Commons Attribution-NonCommercial-ShareAlike 4.0 International License (CC BY-NC-SA 4.0)

Since Derby did not have an infirmary or public dispensary until 1810, the chosen medico-legal setting at Shire Hall was a convenient and commodious place.[33] This was similar to Nottingham where a gallows was constructed in front of the Shire Hall for the same ends of local public justice (Illustration 5.8 depicts the Derby location today, compared to a similar contemporary Illustration 5.9 at Nottingham above).

Turning then to the actual execution day, new source material reveals that Derby anatomists were keen to establish that heart-lung failure was not a reliable indicator of medical death on the gallows around 1810. Thus after a typical execution a hinged table was brought into Shire Hall and placed just inside the left-hand door. It was crude in design but could be carried and the rough wood washed down easily. The top was a basic front door-design that detached from its pedestal. At one end was a head-shaped hole for a bucket to catch any bodily fluids. In the recess, brain

Illustration 5.10 © Science Museum, Science & Society Picture Library, '*eighteenth to nineteenth century dissection table*', Image 10572151, circa 1750–1870; used in early modern provincial anatomy schools, dispensaries and voluntary hospitals. Note: image also used at Museum of London exhibition, 2013; Creative Commons Attribution-NonCommercial-ShareAlike 4.0 International License (CC BY-NC-SA 4.0)

work was done once the skull top was sawn off. Nail marks down the table side indicated where pins were put into the arms to keep them in the right position or rope was used to tie down the limbs, especially once rigor mortis set in. Anatomists generally preferred to elevate the lower limbs. They did this by placing them at right angles to the body in a prone position (face up in this instance) usually with the feet placed flat, again nailed or tied down on the bottom of a portable table, or in detachable stirrups. Muscles of the thighs and calves were exposed forward in a large triangle for dissection. The table design was thus basic but simple to handle if the 'mob' decided to riot (Illustration 5.10). The cross bar underneath could be flicked up with a shoe, folded vertical, and carried or transported by hand or cart. It was an ideal length to fit in most anatomical spaces—about seven feet in length—but narrow being just two planks width—not more than three feet wide—often held under the shared ownership of coroners and penal surgeons. It generally stood by the banks of a river in drowning cases, was propped open under suicidal hangings, and yet accessible to crowds that demanded to be included in the post-execution spectacle. Onto this innocuous piece of anatomical furniture some of the most notorious criminals were brought to justice in Derby and provincial places like it. Comparing and contrasting 'hanging-towns' in the Midlands with those in East Anglia is instructive about the basic environment and equipment used everywhere.

In the central Midlands belt, the medical fraternity in places like Derby, Nottingham and Leicester, were competing with colleagues in Cambridge, Bury St. Edmunds, and Norwich, for their share of an expanding medical market-place. Their chosen dissection venues were Shire Halls in which Assizes courts were held too. Medico-legal matters were however then transferred to newly-built infirmaries. Some had been constructed around the time of the Murder Act, with many more becoming established in the 1810s and 1820s. These new venues afforded a more medically-focused space (still open to the general public) in which to view the criminal corpse across provincial settings in **Band B** on Map 5.1. That did not however lessen the symbolic importance of opening up bodies at Shire Halls or elsewhere across East Anglia. If anything it made them of greater symbolic importance, especially once more people could squeeze inside purpose-built but smaller-scale voluntary hospital morgues known as dead-houses. In by way of example the famous case of William Corder hanged for the Red Barn Murder and taken to be anatomized at the Shire Hall in Bury St Edmunds a lot of spatial anatomical detail was reported about the change-over of venues in the local press. This was done to reassure the general public that such a notorious murderer had received the full legal penalty. The *Morning Post* stated for instance on 14 August 1828 that:

> About half an hour after the execution the body was removed to a private room in the Shire Hall where Mr Creed, the county surgeon, assisted by Mr Smith and Mr Dalton, made a longitudinal incision along the chest, as far as the abdominal parts, and deprived it of its skin, so as to exhibit the muscles of the chest...They were going to move him to the hospital but this was objected to by Mr Foxton (the completer of the law) until he had first stripped him of his trousers and stockings.[34]

The newspaper reporter explained that the first anatomical duty was to try to accommodate '5, 000' people determined to accompany the body on its post-execution journey. Their actions reflected a strong public reaction to the infamous murder. The reporter continued:

The anxiety to see Corder's body was as great as at his execution: crowds flocked around the doors. He was put in the Nisi Prius Court, and the people entered at one door and departed at another; he was placed on the table, and, with the exception of his trousers and stockings, he was naked; there was not much change of countenance, but it appeared there was a great affusion [sic] of blood about his throat. Such was the anxiety to see him that we heard several females boasting that they had been in to see him five times after his head was shaved![35]

These two excerpts confirm key findings already presented in this book. There was a lot of 'natural curiosity' amongst the crowd participating in a renowned murder case after the body was cut down from the gallows. The accounts also suggest that local people were determined to accompany the corpse from the gallows. In terms of the choreography of the capital sentence, one third (legal death) took place on the hangman's rope but two thirds of the indictment, anatomization (medical death) and dissection (post-mortem punishment) happened elsewhere. If the execution was then a spectacle, what happened next was a thrilling encore for the crowds assembled at Shire Halls everywhere.

In East Anglia the procession to dissection venues was a very important but still understudied aspect of the Bloody Code. The extracts cited in the case of Bury St Edmunds confirm that again there were practical differences between the first anatomical cuts and a full-scale dissection. And, once more, we can see this in the transitional use of Shire Hills to voluntary hospitals in which to do post-mortem punishments. Local newspaper reports thus contained a candid admission that there was a 'completer of the law' and this happened post-execution. In the Red Barn case the hangman Mr Foxton had a customary right to take the prisoner's trousers and stockings but only once a preliminary check had been made for life-signs by Mr Creed the county penal surgeon and his pupil assistants. Both officials were responsible for checking medical death had occurred because sometimes it did not. Of note too is the physical attraction that a shaven corpse had for a female audience. Corder was a dangerous alpha male lying dead on a table in the courtroom causing quite a stir of mixed emotions for the assembled crowd. This seems to explain why the location of the initial anatomization took place in a room of Shire Hall before the body was moved to the county infirmary for dissection the next day. Of necessity, there were two different medico-legal checking mechanisms to ensure a public death: one confirmed death, the other harmed the condemned in death. Each was staged separately to emphasis their separate medico-legal functions. In the meantime, the Shire Hall door was left open to visitors until 6 pm when the doors were closed. A local reporter thus remarked that:

> We heard that his mother intends to apply for his body after being dissected; but this we trust she will not do, as it must only pain her already too much of agonized feelings to meet a refusal. It is the intention to preserve his skeleton. He is the first body that ever was dissected at the Infirmary.[36]

The reporter was determined to get the full post-mortem story and so he followed a select group of medical men to the Suffolk General Hospital[37] in the town the following morning, the crowd having satisfied themselves that Corder was 'truly dead'. The next day, Tuesday at 12 noon, there was:

> A great concourse of medical gentlemen, with a crowd of students, assembled at the county hospital to witness the dissection; among whom were all the practitioners round the neighbourhood and even some from Norwich & Cambridge & c...
>
> Mr Creed junior assisted by Mr Smith and Mr Dalton, commenced the operations; they firstly minutely dissected the muscles of the chest, and having elevated the sternum, and examined the lungs, they took out all the intestines, all of which appeared in a most healthy state. From the formation of the chest, it did not appear that Corder would have been a likely subject for pulmonary affection. The Medical Students heard demonstrations about the respective parts... There were some Italian artists there who took two or three casts of his head and also a celebrated Craniologist who informed us that the organs of 'Destructiveness and Secretiveness' were strongly developed.[38]

There is one final detail worth noting about William Corder's post-execution rites. His case caused such a sensation that canny booksellers brought out remarkably detailed accounts of his material demise. These commentators explained that there had recently been a minor but important legal change that impacted on the body-supply of corpses to the surgeons. The passing of *A Bill entitled an act for the consolidating and amending the statutes in England relative to offences against the person 1828 (255) [9 Geo. IV.],* changed how judges issued legal warrants for criminal corpses. The anatomical fate of bodies to be dissected was now declared on an official warrant when the death penalty was pronounced in court in cases of convicted homicide. The corpse could not be moved until this was done especially in cases when an offensive weapon, like pistols in the case of William Corder, had resulted in murder. The warrant had to state explicitly the designated dissection venue. This official piece of paper, akin to a post-mortem passport, had to accompany the body until it reached the grave. William Maginn's popular book thus explains:

> The warrant for Corder's execution differed slightly from the form in which former warrants for the execution of murderers had been drawn up. The alteration was made in consequence of Lord Landsdown's late Act for Malicious Injury to the Person. The old form of the warrant merely ordered the body to be given to the surgeons to be anatomized and dissected; the present form appoints the hospital at which the dissection will take place.[39]

Though it seemed some time off, this pivotal procedural change was part of a much bigger trend to create a body supply system that would expand when the poorest became staple subjects of the dissection table after the Anatomy Act.

Meantime in small country villages residents were equally very interested in criminal dissections across the Midlands and East Anglia. In an example that is typical of what tended to happen in remoter country areas, when Elizabeth Morton was executed for murder on 6 April 1763 the judge decreed that her body had to be punished at the actual site of the homicide she had committed. This was in a village called Calverton about seven miles north east of Nottingham on a small tributary of the Dover Beck. There Morton was laid out for inspection, anatomized on day one, and then dissected to the extremities for two days in full public view. The official report said that 'her body was dissected by a surgeon [unnamed] at Calverton and the public curiosity awakened all the curiosity of the surrounding villages who flocked in crowds to the back premises of the surgeon's house'.[40] It was also reported by those present that 'her features were rather attractive than repulsive: she was strongly made, and tall, considering her age...18'. What was left, less than two-thirds, was buried in a common grave at Sutton-in-Ashfield. We see the same trend in the West Country.

In Map 5.1, **Band C**, when John Anderson, a constable, apprehended Elizabeth, known as Betty Marsh, aged 14 on 21 January 1794 for the murder of her grandfather there was great excitement in the Dorset countryside.[41] Betty thumped John Nevill of Mordern over the head with a blunt instrument whilst he slept in bed. He died that night from his fatal injuries. Betty was charged with homicide, tried and convicted. She thus became the first person to be hanged at Dorchester County jail on 17

March 1794. The question was where could her body be 'dissected and anatomized'? Bath city was a renowned medical market-place awash with surgeons of all descriptions, so too was Bristol, a port city. Yet both were too far away. The corpse would be stinking by the time it got there. It was spring, St Patrick's Day, the turn-pike roads were very muddy and in a cart it would take two good horses a 2–3 day journey. On the other hand, this was a young girl and therefore a valuable anatomical specimen. John and Philip Coombs, surgeons of Dorchester, were on standby to benefit from a rare, but exciting opportunity to take home the criminal corpse of Betty Marsh for their personal research at their domesticated business premises.

Moving down into Cornwall meantime we see further examples of this pattern of punishment. In 1814, Williams Burns aged twenty-one, an Irish army recruit of the Royal artillery, murdered a sailor named John Allen after a drunken night in a public house in Penzance. Burns was committed to Bodmin gaol, tried and found guilty of homicide at the Assizes for the Western Circuit at Launceston.[42] A surviving bill in the Sheriff's cravings reveals the expenses connected to his punishment:

Gaoler's Bill:	£	s	d
Paid Deacon his Bill on executing William Burns	4	14	6
J Chapple sending a waggon to Bodmin for the Drop	2	2	0
Turnpike	0	6	0
4 Guardsmen 1 day @ 7/- each	1	8	0
J Chapple and Horse 1 day	1	5	0
Attendance of William Burn's Executioner	0	10	0
Total	**10**	**5**	**6**[43]

Quite often in Cornwall criminal corpses were handed back to gaol surgeons to punish post-mortem because of the rate that bodies rotted at. This sets in context why a gaoler's bill was being reclaimed on this occasion. If Burns' body had been sent back to Penzance where he committed the murder then the only fast means of transport was according to the *Cornwall Visitor's Guide* of 1814 to catch 'the great mail coach from London-to Falmouth, Penzance &c'.[44] It advised, 'Quitting Bodmin the road proceeds to Truro, twenty-two and a half miles, passing over the gorse-moors (eight miles in length)…The road to Penzance branches off between Truro and Penrhyn' through lots of small villages. But this meant covering a total distance of almost forty-eight miles between execution at noon and nightfall. On those main routes resurfaced across Cornwall in the 1790s the mail-coach travelled on time, but on minor roads the carriage wheel often got stuck in the mud and pot-holes. When bodies were handed over to penal surgeons they thus tended to be dissected by prominent figures in town politics like Dr Coryndon Rowe, senior alderman and magistrate, who was living in Launceston with a business premises close to the Assizes execution site.[45] William Burns however was dissected by Joseph Hamley, surgeon and coroner for the Eastern district of Cornwall since a decision was taken by the sentencing judge for reasons of convenience to dissect him at Bodmin gaol where he had been remanded pending trial.[46] In out-laying rural areas involving provincial anatomical studies after the Murder Act, location mattered. This was why the economy of supply in criminal corpses in the West Country and remoter counties looked different compared to elsewhere.

CONCLUSION

English hangmen are not celebrated for their medicinal abilities in histories of crime and justice under the Murder Act: a general observation recently substantiated by Owen Davies.[47] Many home-grown executioners were inefficient and lacked a basic knowledge of death's infinite variations compared to their European counterparts. All manner of anatomical features were confusing about the executed—'age, body size, recurring disease patterns, ambient body temperatures, air movement', and apparent manner of death.[48] Yet, the medical fraternity increasingly relied on the co-operation of those confused by death's dominion. They handled a chain of supply needed to carry out official criminal dissections in English counties. The geographic reach of these medico-legal networks and their basic punishment provisions have not been documented on location, until now. This chapter has provided for the first time a model of supply mechanism under the Murder Act. It has correspondingly shown an historical appreciation of the spatial architectural setting of dissections and their actual placement in communities. That research has revealed that whilst the state increasingly sought to limit the crowd's interaction with medico-legal officials at the execution site itself, they did the reverse at local dissection venues. This outcome better explains why Simon Devereaux for instance has found that scaffolds were being built much higher by the 1790s.[49] This physically distanced the crowd from the hangman and increased the theatrics of the punishment event. But that change of practice did not happen in medical isolation. In terms of crowd control, the forces of law and order could afford to be more distant because a change of execution rituals was accompanied by a subtle shift in the legal performance of a medical choreography *in situ* by 1800. Crowds still came to see, be seen, and talk about the punishment spectacle at the hanging-tree. They accepted however more detachment from the punished criminal in return for greater access to the corpse about to be opened up to public inspection in a dissection space in their community centres. The execution event thus became a case of less is more, as Devereaux observes, but this was because what came afterwards was made more accessible and high profile. It was now in the purview of crowds assembled to see and be part of a process of post-mortem 'harm' in the vicinity. Over the long duration then the number of murderers sentenced to death declined historically, but this meant that criminal dissections had to have a more symbolic placement in provincial English society to be as effective. The four punishment zones identified in this chapter set in context where this happened, why a particular location was medically chosen, and thus we see anew the different layers of spatial meaning created.

In the Midland's heartland voluntary hospitals were being built at a rapid pace in England. They were displacing medical dispensaries by offering more specialized surgical services in expanding market towns by the nineteenth-century. The journey of the criminal corpse from court-room to dissection venue reflected this local medical reality. This book's central finding is that post-mortem 'harm' was always located in public spaces in which it would gain greater acceptance by a wide cross-section of the community between 1752 and 1832. In the North of England medical dispensaries were generally used, whereas Shire Halls were occupied in the Midlands and across East Anglia up to 1800. Elsewhere either a purpose built anatomy theatre in Edinburgh, Newcastle or London was refurbished. Meanwhile across large swathes of Devon, Somerset and Cornwall criminal dissections were more variable, taking place in Exeter at a voluntary hospital because it was one of the earliest built in the country, and likewise at Bath and Bristol. By contrast in the rural hinterland of Dorchester and Launceston we find the domesticated business houses of local surgeons and a local gaol being commandeered for use. Once voluntary hospitals were built everywhere by the 1820s the majority of criminal dissection work was transferred there because that made sense as teaching facilities expanded in London and large provincial cities like Birmingham and Manchester. Changes to judges' warrants, which ordered the criminal body to be moved to a specified local hospital for dissection, reflected a gradual bureaucratic restructuring and systemization of the post-execution ritual. Although it was not foreseen or intended at the time, medico-legal officialdom was making advance preparations to accommodate what was to become a national system of body supply. It was one connected intimately to healthcare and welfare provision in the provinces, and this was one of the Murder Act's more subtle medical legacies that historians of the early modern era have missed. The spatial alignment of post-execution rites and their physical placement was accommodated by an Old Poor Law infrastructure of small medical dispensaries and public infirmaries. These then expanded their service provision; so much so, that the poorest in society needing basic healthcare, but dying from common diseases of poverty and financial privation, could be made to supply the dissection table by the time of the New Poor Law and its Anatomy Act in the 1830s.[50] The Murder Act facilitated then a lot of medical enterprise. It enabled surgeons to find ways to overcome their supply issues in practical ways. They aligned with health and welfare facilities, peopled by a wide cross-section of the community; some later came to regret their 'natural curiosity' for medical enlightenment that would in turn exploit their impoverishment.

There is no doubt that Surgeon's Hall in London was an iconic venue for criminal dissections under the Murder Act. Its infamy and longevity has however been overstated. The data-set on body supply trends shows that only 16.29 per cent or 147 bodies of the total number of 908 criminal corpses supplied officially from the gallows actually ended up being dissected there. This was double Simon Chaplin's original estimates; nevertheless the majority of condemned bodies were dissected before 1800. This trend sets again in context the lacklustre reputation of the London Company from the 1790s. By the early nineteenth-century the dominance of body supply from the provincial gallows was a meaningful occurrence in the majority of English counties. Ranking suppliers on three levels has revealed 'the hanging-towns' that predominated and the locations where a surgeon could aspire to do original criminal dissection work outside the metropolis. Wherever those punishments took place they were always carefully orchestrated in early modern England. Many provincial post-mortems were marked by distinctive dissection work. Inside the system ambitious medical men congregated together in towns like Lancaster, Preston and Wakefield. Fewer bodies came to Derby but when they did crowds took a great deal of interest in and were active participants, fascinated by heart-lung-brain research It is time then to get closer to this hidden anatomical world to engage with the disintegration of the criminal corpse, its career-making opportunities, and the subsequent remaking of the material afterlives of the condemned throughout England.

NOTES

1. Henry Wilson (1822), *Wonderful Characters, Comprising Memoirs and Anecdotes of the Most Eccentric Characters, Volume 3,* (Boston, Lincolnshire: N. H. Whitaker), pp. 132–3.

2. Royal College of Surgeons Collections, MS0396, Correspondence of Messenger Monsey, c. 1732–1788, quote by his obituary writer 'Mr Wadd'.

3. Florence and Kenneth Wood eds. (1992), *A Lancashire Gentleman: The Letter and Journals of Richard Hodgkinson 1763–1847* (Stroud: Allan Sutton Press), entry 6 March 1828, p. 338.

4. Sherriff Cravings, circa 1752–1832, E389-243 to E389-254; 90–147 to 90–169, primarily used to compile database of corpses sent for dissection (individual cases, cited in endnotes).

5. Of the 908, 15 bodies came from the admiralty and were executed by naval authorities at execution dock.

6. See, Appendix, for list of bodies acquired, in, Simon Chaplin, (2009), 'John Hunter and the Museum Oeconomy [sic], 1750–1800' (unpublished PhD, King's College London).

7. London is the only place with a 150+ body-count category, by virtue of its size, and court system.

8. Refer, Peter King (2010), 'The Impact of Urbanization on Murder Rates and on the Geography of Homicide in England and Wales 1780–1850', *Historical Journal,* Vol. 53, 1–28.

9. There were further legal changes to Assize jurisdictions in 1873 and 1971. The Queen remains however Duke of Lancaster. The Duchy's continued importance is recognised by the appointment of a Chancellor of the Duchy of Lancaster by the Prime Minister with a seat in Cabinet today.

10. The architectural design can be viewed online at: http://www.british-history.ac.uk/vch/lancs/vol8/pp4-22 [accessed 28 July 2015], taken from 'The parish of Lancaster (in Lonsdale hundred): General history and castle', in *A History of the County of Lancaster: Volume 8,* edited by William Farrer and J Brownbill (London, 1914), pp. 4–22.

11. J. V. Pickstone (1985), *Medicine and Industrial Society: A History of Hospital Development in Manchester and its region, 1752–1946,* (Manchester: Manchester University Press), pp. 65–6.

12. See, Hilary Marland (1987), *Medicine and Society in Wakefield and Huddersfield 1780–1870,* (Cambridge: Cambridge University Press).

13. George Howson (2002), 'The Lancaster Doctors: Three Case Studies' in Sue Wilson and Jenny Loveridge eds. (2002), *Aspects of Lancaster: Discovering Local History,* (Barnsley: Wharncliffe Books), chapter 5, pp. 53–63.

14. Refer for career highlights, Susan Wilson (2002), *Aspects of Lancaster: Discovering Local History*, (Barnsley, South Yorkshire: Wharncliffe Books), pp. 59–61.
15. See, general context in, Pickstone, *Medicine and Industrial Society*.
16. See, Anon (1811), *The Royal Kalender or Complete and Correct Annual Register for England, Ireland and America for the Year 1811, and so on*, (London: J. Stockdale), p. 304.
17. See for example, Jonathan Binns (1807), 'Letter to Dr James Hamilton, Senior Physician to the Edinburgh Royal Infirmary on the Cure of Scarlatina, from Physician Lancaster', *Edinburgh Medical and Surgical Journal* Vol. 3, (April), 135–44. He wrote against purgative medicine for children with scarletina.
18. http://lanternimages.lancashire.gov.uk/index.php?a=advanced&s=item-&key=XYToyOntzOjY6IkNPTE9VUiI7czoxMDoiTW9ub2Nocm9tZSI7czo2OiJNRURJVU0iO3M6MTg6IlBob3RvRvZ3JhcGhpYyBwcmludCI7fQ==&pg=70 [accessed 28th July 2015], Digital Collections, Lancashire Lantern, Lancashire County Council, Record Number 615, '*Old Dispensary, Surgeon's Court, Lune Street, Preston*', image taken circa 1937 of the remaining building. See also, Lancashire Archives, Preston Dispensary Records (1809–1866), Accession 5512, HRPD/1-9, NRA 49305. It cannot be reproduced in this book as its copyright is uncertain.
19. Marmaduke Tulket (1821) *A History of the Borough of Preston*, (Preston and London: P. Whittle). p. 80.
20. Pickstone, *Medicine and Industrial Society*, pp. 65–6.
21. Lancaster Record Office, MSS, DDWh/4/99, Whittaker of Simonstone.
22. Pickstone, *Medicine and Industrial Society*, p. 70.
23. See, wide coverage in, *The Manchester Guardian* 19 March 1831 and *The Manchester Times and Gazette*, 19 March 1831 and the *Lancaster Gazette and General Advertiser* 19 March 1831.
24. E. T. Hurren (2011) *Dying for Victorian Medicine: English Anatomy and its Trade in the Dead Poor, 1832–1929* (Basingstoke: Palgrave), chapter 7, recounts his dissection work. See also, W. J. Ellwod, A. F. Tuxford eds (1984), *Some Manchester Doctors: A Biographical Collection to Mark the 150th anniversary of the Manchester Medical School 1834–1984* (Surrey and Manchester: Unwin Brothers), pp. 75–6. His career path represents the eventual ascendancy of provincial medicine over its metropolitan counterpart. Turner (1793–1873) was a much travelled and well educated man. He was considered to be a leading medical figure in Manchester, renowned for his commitment to improve medical education standards in the North of England. By the time that Moses Fernley's body had become available in 1831, he had established himself as one of the most engaging surgeons in the North West.

25. W. Farrer and J. Brownbill (1911), 'The city and parish of Manchester: Introduction', in *A History of the County of Lancaster: Volume 4*, (London: Victoria County History Series), pp. 174–187—http://www.british-history. ac.uk/vch/lancs/vol4/pp,174-187 [accessed 23 January 2015] cites that: 'The Infirmary was first established in Garden Street, Shude Hill, in 1752, and removed to new buildings in Piccadilly (then called Lever's Row) in 1755. In front of it were the old Daubholes, afterwards transformed into a piece of ornamental water, with a fountain; this was removed in 1857. A lunatic asylum was added in 1765, public baths in 1781, and a dispensary in 1792. The building was refaced with stone about 1835. The lunatic asylum was removed to Stockport Etchells in 1854. Lever's Row was so named from the estate and town house of the Levers of Alkrington': information taken from *Lancaster and Chester Antiquarian Society Records*, Vol. 20, p. 238.

26. S. A. King (2000), *Poverty and Welfare in England 1750–1850* (Manchester: Manchester University Press).

27. Martin Gorsky and Sally Sheard (2006), *Financing Medicine: The British Experience since 1750* (London: Routledge).

28. The case has been reconstructed from two key sources: Anon (1830), *Yorkshire oddities, incidents and strange events, volume 1* (York; private publication), p. 266 and 'Trials: Lent Assizes York', *The York Herald and General Advertiser*, Issue 1386, March 22nd 1817, pp. 1–2. A contemporary history of the Halifax Dispensary can be found in John Crabtree (1836), *A Concise History of the parish and vicarage of Halifax, in the county of York*, (Halifax: Hartley and Walker publishers), p. 346.

29. Jenny Uglow (2002), *The Lunar Men: Five Friends whose Curiosity Changed the World* (London: Faber and Faber).

30. Peter M Jones (2009), *Industrial Enlightenment: Science, Technology and Culture in Birmingham and the West Midlands, 1750–1820* (Manchester: Manchester University Press).

31. Stephen Glover (1829) *The history of the county of Derby, drawn up from actual observations and the best authorities*, (Derby: Henry Morley and son) p. 615.

32. William Hutton (1817), *The History of Derby: From the Remote Ages of Antiquity* (London and Derby: Nicloas, Son, and Bentley), p. 32.

33. Desmond King-Hele ed (2007), *The Collected Letters of Erasmus Darwin* (Cambridge: Cambridge University Press), p. 575, note 1, explains that William Strutt ran the infirmary and was a close friend of the Darwin family.

34. *The Morning Post* 14 August 1828 carried extensive reports of the execution and post-execution events and the reaction of the early modern crowd over several days.

35. Ibid.

36. Refer, endnote 34 above.
37. Suffolk General Hospital was founded on 4 Jan. 1826; after 1832 it was called the West Suffolk Hospital.
38. Refer endnote 34 above.
39. William Maginn (1831 edition), *The Red Barn a Tale Founded on Fact*, (London: John Bennett Publishers), p. 690.
40. Nottingham County Record Office, BB34.8, 'Gallows Rememberancer for 1763', p. 25.
41. Dorset Record Office, NG/PR1, Prison Admission and Discharge Registers, 1782–1901,' Case of John Anderson Constable arrest of Elizabeth, otherwise known as Betty Marsh'; D/SEN/3/7/8, 22 June 1792 confirms a bond between the two main surgeons of Dorchester, John Coombs and Philip Coombs; D/SEN/3/7/10-11, 24–25 February 1794 likewise indicates a conveyance of property where John lived at Sturminster Newton and Philip at Shillingstone after their respective marriages, although they worked from medical premises in Dorchester.
42. Editorial (1814), 'Murders in Cornwall', *Journal of The Universal Magazine of Knowledge and Pleasure, Provincial Occurrences*, December issue, pp. 517–8, where it was alleged that William Burns may have killed a man in Dublin too; see also, Rita Margaret Barton (1970), *Life in Cornwall in the early nineteenth-century: being extracts from the West Briton newspaper in the quarter century from 1810–1835* (Cornwall: D. Bradford Barton Ltd).
43. The National Archives, (hereafter TNA), T90/170, Burns (Cornwall, 1814), ref. 7306891, expenses claimed back by Joseph Edwards, Undersheriff.
44. Rev. Daniel and Mr Samuel Lysons (1814), *Magna Britannica Volume 3, Cornwall*, (London: Cadell and Davies in the Strand), p. cxiii.
45. TNA, PROB 11/2066/422, Will of Dr Coryndon Rowe, Doctor in Medicine of Dockacre Launceston, Cornwall. His portrait is on display today at Launceston Museum, where his anti-slavery stance is celebrated in the county history. The Sheriff Cravings confirm he dissected bodies in 1793, 1796, 1812, 1815, 1821, & 1821.
46. See, TNA, Sheriff's Cravings, E389/253/355; E389/253/359, 1814/03/31, Burns/Hamley dissection. and is listed as doing dissections for 1820, 1821 and 1828; *Lancet*, Volume 2 (1836), p. 685, calls him a penal surgeon and coroner who gave evidence on the Medical Witnesses Bill. It debated the cost of an autopsy and who should open the body. Hamley was unusual for being medically not legally qualified to serve as coroner.
47. I am grateful for an advance copy of Owen Davies and Francesca Matteoni (2015), '*A virtue beyond all medicine*: The hanged man's hand, gallows tradition and healing in eighteenth and nineteenth-century England', *Social History of Medicine*, (May, 2015), Issue 2, in draft copy pp. 1–37, point made at p. 30.

48. Jessica Synder Sachs (2002), *Time of Death: The True Story of the Search for Death's Stopwatch* (New York: QPD for William Heinemann at the Random House Group Ltd), p. 6.
49. Simon Devereaux (2008), 'Recasting the Theatre of Execution: the Abolition of the Tyburn Ritual', *Past and Present*, Vol. 202, 127–74.
50. For a complete history of body trafficking after 1832, see, Hurren, *Dying for Victorian Medicine*.

CHAPTER 6

The Disappearing Body: Dissection to the Extremities

INTRODUCTION

In eighteenth-century England no matter where the punishment of the criminal corpse took place, the essential humanity of the condemned was about to be eroded. In a strong oral culture it has however been historically difficult to substantiate with documentary evidence what it was like for the crowd assembled, containing medical men, the middling sort and labouring poor, to together be confronted with physical despoliation. This sixth chapter takes a new approach to this historical problem. Staging the punitive rites shaped the archaeology of emotions of those present. These were triggered by a 'natural curiosity' that everyone was capable of experiencing, even though many crime histories omit it. The presence of ordinary people in such large numbers indicates intention by the majority to act in some personal capacity, and to repeatedly do so, as so often featured in contemporary newspapers of the period. This found expression in a public performance of a narrative of belief, emotion, participation, exclusion and of sentiments, attached to an ever-present synaesthesia. The act of being there and expressing curiosity about the criminal corpse being opened up to public scrutiny can consequently be read as a story in itself. This fresh approach is necessary if historians are to relocate amongst the crowd the experiential history of punishment rites, in the way that anthropologists and ethnographers have done in death studies. This chapter thus investigates the cultural stimulus of 'natural curiosity' since it

© The Author(s) 2016
E.T. Hurren, *Dissecting the Criminal Corpse*, Palgrave
Historical Studies in the Criminal Corpse and its Afterlife,
DOI 10.1057/978-1-137-58249-2_6

provides a completely different sense of how ordinary people experienced the drama of anatomy and its spectacular denouement.

Anticipating that there would be sustained interest in criminal dissections, the *Universal Magazine* in August 1770 soon declared:

On the Dissection of a Body

Observe this wonderful machine
View its connection with each part,
Thus furnish'd by the hand unseen
How far surpassing human art!...
See how the motion of each part,
Upon some other still depends
When all the material aid impart
Conducing to their various ends...
These tubes convey'd the purple juice
Which with new strength suppl'd the whole;
And here branch'd forth the nerves, whose use
Was to keep converse the Soul.[1]

Yet, many anatomists acknowledged that being present could also 'harrow up the feelings'.[2] In material terms the criminal corpse became a 'disappearing, dirty body' by the time it was being cut 'on the extremities and to the extremities'. People naturally reacted in different ways when entering the dissection space. Some stood abreast in a tight crowded room which had the advantage that if anyone fainted they would not fall down. Many others walked in slow procession past the criminal body enabling them to say they had participated. At any time they retained the option of exiting if the spectacle was too disturbing to sensibilities. They could recoil and then come back for a second look having acclimatised to the awful sight on display. Even privileged audience members kept changing seats, staying for the removal of the heart but reserving the right to leave before the skull craniotomy. Everyone faced an anatomical storyline with an emotional sub-text of some sort; the legal aim being to make it impossible to be impassive to verify standards of justice. Here was a novel opportunity to experience new ways of seeing and believing in medicine, religion and science. This chapter is hence all about the dramatic re-enactment of the contemporary notion of post-mortem 'harm' in the popular imagination of Georgian England.

Early modern histories have yet to convey the repopulation of people—coming and going—looking and standing over—handling and letting go—exiting and returning—part of a human wave of eyewitnesses at criminal dissections. Many of those that turned up suspected the corpse was on the boundaries of life/death. It was still possible to 'harm' the criminal at this stage of the punishment process and this painful subtext ignited 'natural curiosity'. The overall aim in Chapter 6 is therefore to envisage criminal dissections as a form of immersive theatre. Audiences got closer to the action because there was something for everyone involved. The criminal body should have lost its embodied identity as it became a human being despoiled, but whether it ever did when a material afterlife was created was an emotive issue, and remains so in museum culture. Section 1 thus begins by looking at what happened once the criminal corpse was cut deeply and specifically what it meant to 'harrow up the feelings' when watching the actual dissection taking place. We will see that as the external and internal appearance of the criminal changed shape, it became less than human to the lay-public. Section 2 then asks what it was about the brain and nervous system that enthralled many of those present and how exactly did those speculations start to change punishment rites. One outcome was a medical desire to dismiss the crowd before deep dissection got underway. Yet this was complicated by the agency of ordinary people too. Many however feared contagion by the criminal corpse. There was always the push and pull of public/private consumption being played out at dissection venues. Section 3 expands on this theme by exploring what it was like for medical students to engage with dissection and what a newly qualified doctor did with that experience to get on the career ladder. The aim is to rediscover some of the opportunity costs of doing criminal dissections and if ambitious surgeons relocated to areas where executed bodies were more plentiful. Finally, Section 4 engages with material afterlives and the ways that the embodied identities of criminal corpses were changing in a museum culture by 1832.

The Disappearing, Dirty Body: 'To Harrow up the Feelings'

At an eighteenth-century criminal dissection over two-thirds of the human material was generally disposed of during the punishment rites. Today (it is worth reiterating that) in modern dissection rooms no more than one third of a body is used for anatomical teaching.[3] That limit preserves human dignity. To dissect over two thirds in the past meant there was very little left to bury at the end.[4] Most of the human waste was collected in buckets. It was then either washed away with disinfectant chemicals or put out with the rubbish because of the dangers of contagion. In a typical punishment scenario, body parts survived a lot longer than a whole body shell. Thus William Hey senior a leading Leeds surgeon in an enlightening letter on this subject wrote to Walter Spencer Stanhope on 21 March 1785. He explained why 'the extremities' were so valuable: 'as the subjects generally die in health, the bodies are sound, and the parts distinct'.[5] There was however no guarantee that the body on delivery would be unblemished. It was well-known that many convicted prisoners facing the death sentence tried to commit suicide in gaol. In a representative case on 31 August 1822 at the Surrey Assizes the condemned: 'had made a most desperate effort to deprive him-self of life by inflicting several dangerous wounds upon him-self while in prison'.[6] A newspaper report explained that in the case of 'D. Thomas sent for execution...After hanging the usual time, his body was cut down and taken to St Guy's Hospital' in central London 'for dissection in pursuance of the death sentence'. Yet, on arrival his was a 'badly damaged' corpse. The parts were worth more than the whole to the dissector on duty. There had to be a great deal of discretionary justice in the hands of the penal surgeon for this practical reason. Perhaps then it is not surprising to discover deep ambivalence in some cases about the appropriate level of post-mortem 'harm' enacted. Dissectors had to consider the nature of the capital offence, the class of the murderer, and public sensitivities. They were also under a lot of pressure to put on a good performance once criminal dissections transferred to newly built hospital infirmaries from 1810 because hospital boards were not always convinced that post-mortem punishment should be permitted on the premises.

The previous chapter explained how around the turn of the nineteenth-century provincial dissection rooms moved location. Most were repositioned inside a courtyard or underground of local hospital infirmaries. This architectural realignment coincided with humanitarian debates about the need for executions to be carried out behind a walled-off area. At Leicester Infirmary by way of example the governing body together decided to allocate 'three rooms in the cellar' where the dissection room had a 'vaulted roof'.[7] They decreed that there should be an ante-room to prepare the body, leading to a space to dissect measuring '15 foot by 12 foot'. There was 'a window, overlooking the...frontal court area' just above ground level to let in enough light to work but keep prying eyes out. New equipment was also ordered. 'Mr Carley' the carpenter was commissioned to 'make a Dead board for the corpses' to a bespoke design to better facilitate brain research for the penal surgeon's private use. Yet these vigilant arrangements still caused two practical problems. Keeping the general public outside could provoke the excluded mob. Indeed the Leicester Infirmary minutes of 1816 reveal how on execution days that it was 'absolutely necessary [sic] to have a strong guard for the protection of the garden and walls...for the prevention of mischief and depredation'.[8] An added difficulty was that the internal move now placed criminal bodies beside an inadequate drainage system which was easily contaminated by blood, tissue and material human waste (a recurring problem in Chapters 4 and 5 at Surgeon's Hall from 1752). Just because original research was being pioneered on the criminal corpse did not mean that dissectors could avoid the logistics of contagion. Not only was this was one the biggest public health nuisances that all provincial anatomists faced after the Murder Act, but ironically it proved to be such a difficult problem to resolve that it led to better professional recognition that dissection should be the preserve of the medical fraternity.[9] Conditions in the Midlands are enlightening and reveal the strong connection there was between practical considerations and the emotions they often stimulated.

Derby Infirmary was proud of its new brain research on criminal corpses but preventing contagion from decomposing bodies was testing. If the local medical fraternity wanted to enhance their reputations for doing original research, it was important that they publicize what steps they were taking to sort out basic health and safety issues. Nobody was going to be convinced by research on criminal corpses that ironically kept killing penal surgeons! Erasmus Darwin throughout his career took a keen interest in these practical problems for personal and professional reasons.[10] His young cousin Charles Darwin aged just 20 when a medical student caught a fatal infection after 'dissecting the brain of child which had died of hydrocephalus' at Edinburgh' in 1778.[11] A careless nick of the lancet was a well-known hazard, causing cross-contamination in the anatomist's hands. Thus when Thomas Alcock RCS, and a leading light in the Medical and Chirurgical Society published his much-praised, *An Essay on the Use of Chlorurets of Oxide of Sodium and of Lime as Powerful Disinfecting Agents and so on* (1827), he highlighted best practice established in dissection rooms in places like the Derbyshire General Infirmary. He advised:

The covering of tables with lead, or any substance incapable of absorbing moisture is advantageous, whilst a simple contrivance, a conducting pipe, for conveying any liquids from the table to the bucket underneath, prevents them from flowing upon the floor. Each table or pair of tables should have a water pipe with a stopcock and a moveable spout, and a constant supply of water, to be used when occasion may require...The material of which the floor is composed, should be incapable of absorbing water – stone or stucco answers this purpose...The most strict attention to cleanliness, and to the removal of useless parts, should be carried into effect.[12]

This room design, basic equipment, and procedures for carefully disposing of body-parts after dissection, were evidently not in place when Shire Halls and small medical dispensaries had been requisitioned for the post-mortem punishment of criminal corpses from the 1750s to the 1810s. The transfer of the condemned body by Judge's warrant to purpose-built infirmaries around 1815 thus heralded a noticeably more 'professional' style of dissection work. Alcock's health and safety reforms also reveal important details about the odour of the room and the way the body looked once the dissectors got to work. He advised: 'The floor should be washed with chlorureted water' [chloride] to disinfect it but this also gave the room a pungent smell of cat urine, a necessary but offensive odour. To prevent putrefaction of the corpse Alcock suggested applying a 'solution of chloruret of lime or of soda over the subject each time before beginning to dissect, removing with a sponge all superfluous moisture'.[13] He reminded his medical audience and general readers that body fluids did go on leaking even two or three days after medical death and that dissectors needed to re-sprinkle the corpse with chloruret each day which although malodorous was better than a fouler stench emanating from a decomposing corpse. It was also best he observed to cover the 'subject with a coarse cloth or cloths moistened' in the 'solution of chloruret ... cloths should be renewed night and morning'.

There were then a number of techniques that could be used to preserve the criminal body to extend its use as a teaching aid and it was these that tended to have an emotional impact once dissection became a preservation race against putrefaction. Of these, Alcock claimed to have perfected injecting 'a saturated solution of pure muriate of soda with a little nitre [saltpeter]...into the arteries without heat: this will considerably retard putrefaction; but does not preserve the florid appearance of the muscles'.[14] If the dissector needed to 'distend the blood vessels' in the case of major arterial vessels he injected 'red lead' so students could see this run through the corpse. If working on minute vessels it was better to do a 'fine injection consisting of spirit varnish and vermillion'. White lead for example mixed with oil on vermillion and thinned to the right consistency with turpentine worked well—it ought to be the 'consistency of treacle and should be gently injected to the arteries'. In the case of the veins it was advisable to use a different colour, 'either rose-pink or a powder-blue' paint base. Many medical men nonetheless still preferred working in wax because it preserved parts for longer. Whatever the chosen preservation method, the visual effect was that the actual corpse changed its physical appearance at an alarming rate in front of the crowd. The greatest challenge Alcock pointed out was working in warmer conditions when 'care must be taken to prevent the contact of flies, lest a breed of maggots should result from negligence'. If this happened chloruret sponged onto the 'mucus or slime that formed on the exposed parts' must be done. The dissector he emphasised must not touch the corpse or risk his own death from cross-contamination. The crowd could not avoid seeing bacteria eat the human flesh: a medical epitaph that was eye-catching to all sensibilities in the room, revealing how exactly 'an untimely or dramatic death began to create, as well as test, the new kind of fame' that notorious murderers started to generate in the English press and print culture of the period.[15]

Alcock elaborated on how best to handle a situation in which ordinary people in the crowd turned up out of 'natural curiosity' to the view of the corpse and did not anticipate just how emotional they might feel about the deceased being dissected. It was necessary under these circumstances to show 'proper attention' to how the body might be perceived by a lay audience. It was unacceptable he remarked and could 'harrow up the feelings' if a body part being examined had been 'strewed upon the table' or treated with 'the carelessness of juniors'.[16] There was no doubt that if the body was not dissected quickly and it thus was 'far advanced in putrefaction' when examined that the 'clothes worn by the operators and assistants' would be 'rendered useless by the intolerable odour' that was retained. Even after being removed and laundered dissecting clothes never lost their foul smell. There was also the well-known problem of what he called 'the unsightliness of the body after examination'. In other words, once dissection was underway the process became distasteful even to those that should have been inured to the defacement of the criminal corpse.[17] The most 'disgusting occupation' of all, in his professional opinion, was when 'making anatomical preparations, particularly of the bones'. It was a standard procedure 'to employ maceration till the soft parts become

putrid and decomposed'. This involved 'the removing of the soft parts
and the cleaning of the bones is ... not unfrequently attended by danger'.
The advantage of using a chloruret solution in the macerating vessel was
that it dampened down the smell of decomposition, cleaned the bones,
and so whitened them, before the bonesetter wired them back up to make
a skeleton for display.

What all of these unpalatable details suggest is that historians can
retrieve an experiential sense of criminal dissections. The archive evidence
again reveals that over time it remained offensive to look at, repugnant to
smell; there was an eerie soundscape of dripping blood and chemical fluids
flowing down temporary water-pipes into buckets. The corpse felt slimy to
handle and nauseating to rub between the fingers once it started decomposing. Hair and nails had to be shaved and cut into piles left on the stone-
floor for sweeping away. The flesh treated with chemicals to disinfect it
smelt of cat urine from the chloride used to slow down putrefaction; above
all, everyone present breathed in a lot of stagnant smells. Standard historical accounts tend to neglect this synaesthesia and produce instead a rather
sterile view. Medical men did seek better professional recognition by securing a private and privileged access to criminal dissections. Yet, the balance
of evidence also reveals that there was another side to this medical history
too. The majority in the crowd were keen to attend the anatomization but
most wanted to depart once dissection went deeper. This was understandable for reasons of personal safety and natural sensibilities. Many commentators remarked that it took a cast-iron stomach to see post-mortem 'harm'
taken to its logical conclusion. Even the audience's clothes had to be discarded if someone spent too long in close proximity to the body. Hands
had to be scrubbed twice a day to get rid of the lingering 'putrid effluvia'.
Ordinary people heeded the medical pamphlets and publicity generated
by experienced dissectors such as Thomas Alcock. General discussion provided timely warnings that 'ill-health is too frequently the consequences of
close application in the dissecting rooms'.[18] In an era when major epidemics like typhus fever or cholera contagion from dead bodies was a genu-
inely frightening prospect, few sensible people wanted to risk catching a
fatal condition to satisfy their idle curiosity by lingering over a socially-dis-
eased corpse about to be opened up to medical enquiry. Medicine therefore did dominate the criminal corpse and had the discretionary justice
to do so. Yet that authority was also very much derived from an agency
in the crowd, their consensus, based on popular curiosity and revulsion,
about what really happened to the corpse dissected to its extremities.

Essentially dangerous criminals in death created a dangerous duty. It was monotonous and mundane, harrowing and hazardous. This was part of the emotional conundrum of seeing social justice done. The biggest medico-legal challenge was how to get the timing right of the crowd's orderly departure. If it was not voluntary then chaos could ensure: several archival cases illustrate the public order logistics.

Two Scottish brothers Alexander (aged 35) and Michael Keands (aged 24) attacked a publican's wife Mrs Blears and murdered her servant Betty Bates at Winton near Worsley in Lancashire in 1826. Even after the guilty verdict was pronounced on 17 August at the Lancaster Assizes, neither man would admit to fatally cutting the victim's throat. The presiding Judge therefore ordered the full punishment rites according to the Murder Act. At Lancaster Castle a 'temporary scaffold covered with black cloth was erected... and two chains were attached to the fatal tree and placed very near each other'.[19] If neither sibling would confess then each would be made to look into the eyes of the other in the hope that one would cry out his innocence and one his guilt at the fatal time. Alexander was the tougher of the two brothers and so the executioner made sure that 'his arms closely [were] pinioned, and the halter about his neck'. He was an audacious character and underneath the scaffold he 'turned his face towards the wall' looking away from the assembled crowd. The executioner pulled a black cap over his head and then did the same for Michael whose countenance was said to have altered as he neared death. This was taken as a sign of guilt for the murder of the servant. At the appointed time, a local reporter noted that Alexander did an odd thing. He 'took Michael's right hand in his left'. As the rope tightened, Alexander 'stopped jerking' within minutes whereas Michael 'struggled violently': emphasising some thought the latter's culpability. There was a lot of excitement in the town about whether the post-mortem punishment rites would reveal the true nature of their sinister characters.

Local accounts stress that 'several applications from individual surgeons were made to the sheriff for the bodies, but he declined granting any of them'. He had already allocated one body to Manchester infirmary—that of Michael—and, the other—of Alexander—was given to Lancaster surgeons. In the case of Alexander, a new hospital infirmary had now been built in Lancaster during 1823. This was where his criminal corpse was taken to (instead of the local medical dispensary at the bottom of Castle Hill, as we saw in the last chapter). Newspaper reporters noted that the chosen medical men had acquired the legal right under judge's warrant to receive the body and do the anatomization and dissection in private. However, in attempting to do so, they badly misjudged the mood of the post-execution crowd:

> The body of Alexander was delivered to the infirmary on Wednesday evening and on Thursday morning, the surgeons commenced dissecting it. So great was the public curiosity to see the body, that a great crowd of persons forced their way into the dissection room, and filled it to such a degree as to put a complete stop to the operation [of the law]. The body was then taken into the yard, and exposed, and for some time, to the view [of the major organs] of all who chose to go and look at it. The dissection was resumed and, we believe, completed yesterday.[20]

The Lancaster medical men had neglected to include the post-execution crowd in the initial anatomization process because for such notorious men medical death had been checked three ways before reaching the dissection venue. Firstly the penal surgeon and the sheriff had done so at the scaffold, the bodies having been hanged for longer than usual (approximately an hour and a half) to make sure they were killed in front of the crowd. Secondly, each body was laid out side-by-side inside the Castle prison and the sheriff arranged for the designated Lancaster surgeons to send their assistants to collect the body from Mr Thomas Higgins keeper of the Castle.[21] Before release, Higgins double-checked that Alexander and Michael were deceased by monitoring their breathing and pulse. Thirdly, both bodies were left at the Manchester and Lancaster infirmaries respectively for twenty-four hours before dissection was due to commence to check that decomposition and putrefaction had started; another fail-safe indication of medical death. In Lancaster these careful procedures however excluded the crowd, and so they bombarded the dissection room door pushing it open to view the body. In so doing, their collective actions give important new historical insights into the processes of emotional engagement in the crowd which William Reddy has called 'emotives'.[22] Rather than simply 'speaking' themselves into an emotional state, changing the nature of a capital punishment experience through the self-realisation of deep-felt emotions being voiced, it is possible to argue in this context that the crowd members were equally capable of seeing, smelling, and feeling themselves into a deeper emotional awareness too. The historical cliché, 'actions speak louder than words', salient on this occasion, was a physical demonstration of the archaeology of emotions people who pushed and pulled were capable of, which in turn powerfully expressed the level of sustained sensations being generated at controversial criminal dissections. As a reporter for the journal *The Age* noted in the Sunday 26 August 1826 special edition, 'it would appear that the townships' around Lancaster 'had sent forth their entire population' to the execution and it was these 'groupes [sic] of people hurrying' to accompany the body that presented a new public order challenge for the dissectors.[23] It was though remarked on in local newspapers that those present whose emotions had been aroused did not seek to touch or interfere with the criminal corpse. The crowd wanted access to the body to make sure it had paid its legal dues on the gallows, but remained disinterested in actually handling dissection *per se*: an important distinction. A second archival case reveals that these acts of human curiosity that triggered strong emotional and physical pressure on the part of the post-execution crowd appear to have been rooted in a collective narrative of belief, emotion, participation, exclusion and of sentiments that merit closer historical scrutiny.

James Brodie was a blind man executed for murder on the 15 July 1779. He was aged 23 and had murdered an eight year old boy. The homicide victim was a poor lad, paid a few pennies to be Brodie's guide. He was beaten to death with a stick on a turnpike road from Mansfield to Farnsfield. Brodie was evidently an angry and frustrated man. He despised his plight and took it out on a defenceless child. Brodie then tried to conceal the boy's body by burying it in a shallow grave, but it was discovered by the lad's distraught mother who came looking for him when he went missing. Brodie's violent temper was revealed in court. He started biting and kicking the constables as the judge sentenced him to death. Yet what happened next arrested the local authorities because the crowd acted with emotional intensity linked to 'the perils of curiosity'. The incidents that followed at the criminal dissection at the Shire Hall, though extreme, did expose the range of shared emotions the mob was capable of, sometimes to the shock of 'polite' society in the Midlands. It was thus reported that James Brodie was in Nottingham

> submitted to the surgeons for dissection and was afterwards exposed at the County [Shire] Hall. Crowds of men, women and children, indulged their morbid curiosity by thronging the scene of this most repulsive spectacle. The mutilated and frightfully excoriated corpse lay extended on a long table with no other covering other than a loose drape thrown over the loins. Words can scarcely give an idea of the coarse and brutal jests, the obscene remarks, and the horrid and blasphemous oaths to which the display gave voice...In the height of the exhibition, when amongst the crowd were a

number of women, a disgruntled fellow treated the body in a manner which most completely violated every feeling of decency and which his conduct excited the disapprobation of a few, and the screams and exclamations of the females, but to the great majority it served to elicit great peals of laughter. Brodie was a native of Dublin and lost his sight by an attack of the smallpox in his infancy.[24]

Issues of representativeness and an emotive language in this newspaper account need to be investigated carefully to assess these contemporary sentiments objectively. Nonetheless from the standpoint of the medico-legal fraternity what took place was unedifying. The crowd's reaction was very emotional, baying for revenge. Females were present but repelled; others seemed determined to express their anger by laughing over the body and seeing it exposed. Feelings of anger and revulsion were expressed in foul language. One man allegedly grabbed a body-part in a sexual way. The implicit suggestion is that he may have held the erect penis in an illicit gesture (executed males often had an erection on the rope). That the murderer was a blind man, marked by smallpox scars on his face from birth, may have made him look ugly, and this could have lessened his humanity for the crowd. Yet the newspaper reporter present attributed the level of public anger to his notoriety as a convicted child-killer. Intriguingly no mention was made of his disease or disability or ethnicity igniting public anger. There was also a striking silence on the part of the surgeons. They did not step in or call for extra support from the sheriff to try to control the public order threat. According to first-hand reports, a chorus of baying seems to have continued throughout the post-mortem 'harm'. This later justified the corpse being 'excoriated' by dissection. The authorities evidently judged that it was better to let people vent their collective anger and utilise that approbation to 'harm' the condemned flesh. If Brodie's execution had been a dramatic tragedy, this was surely its calamitous encore.

An emotional feature of these and many other cases like them relocated in the archives is the early modern concept of 'curiosity'. As Neil Kenny explains about eighteenth-century popular mentalities:

> The point of invoking curiosity was almost always to regulate knowledge or behaviour, to establish who should try to want to know what, and under what circumstances. Talking or writing about curiosity was very often a way of trying to stop some people from wanting to know or do something and a way of trying to allow – or even to force – other people to know or do it. Discourse about curiosity, while not confined to questions of knowledge, played a crucial role in the production, acquisition, control and circulation of knowledge.[25]

It could moreover act as a 'salutatory warning' about the dangers of some types of curiosity—as seems to be the cases above. The post-execution crowd was criticised in contemporary newspapers for having strong emotions trigged by 'public curiosity' and/or 'morbid curiosity'. It was accepted that some type of 'natural curiosity' was necessary because its purpose was 'to legitimatize the Scientific Revolution' in which the 'new anatomy' of the 1790s was to figure so prominently during the Enlightenment. This meant though that those living in early modern society were adopting and adapting to a medico-legal discourse that 'talked about curiosity for conflicting reasons'. These aspects of dissecting the criminal corpse have been lost to posterity because of the historical predicament of oral cultures, setting in context why they are still missing in crime studies. The emotional capacity and character of the crowd in the punishment drama remains misunderstood, unless (as this chapter argues) historians accept that acts do reveal intentions and their underlying curiosities are a storyline to be read as evidence too. As Kenny points out: 'To talk or write about curiosity was usually to enter an arena'—in this case the dissection venue—'within which some of the period's basic anxieties and aspirations about knowledge and behaviour were thrashed out'.[26] In early modern England curiosity came to be seen as 'usually a passion, desire, vice, or virtue in human subjects'. Post-execution crowds had curiosity because they wanted access to a body of knowledge, but this human impulse also depended on the powerful attraction of the focus of their curiosity. In terms then of controlling access and emotional reactions to the criminal corpse, there was by the 1790s a fundamental shift in the cultural meaning of the sort of 'curiosity' attached to post-mortem 'harm'. It was considered to be a vice (generally in spiritual terms if questioning the theology of the afterlife of the soul) and yet a healthy passion (usually from a secular, scientific perspective about the need to engage with the anatomical punishment of the body). To have 'natural curiosity' tended to be regarded as simultaneously 'either defective, or morally neutral, or admirable' and this made it 'immensely controversial'.[27] Under these circumstances it is not difficult to comprehend why it was that criminal dissections became 'even more of a battleground to distinguish good knowledge or behaviour, from bad'. Emotional impulses associated with different types of curiosity at criminal dissections can thus be illustrated in a paradigm covering eighteenth-century England, thereby providing a revisionist perspective (Figure 6.1).

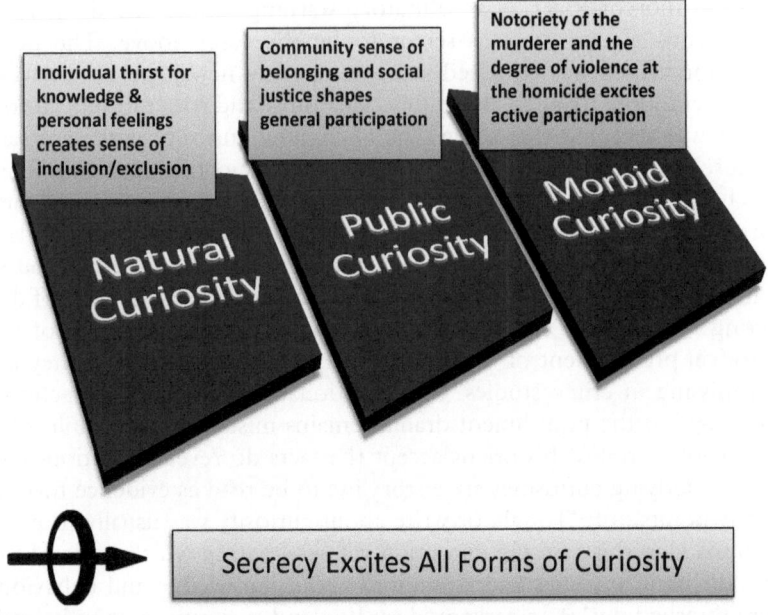

Figure 6.1 A new model of eighteenth- and nineteenth century curiosity at criminal dissections.

Encircling the paradigm is an early modern sense that 'secrecy excites curiosity' because to be curious is to recognise that there will be 'conflicting uses of curiosity made by different groups, often for crucial ends'.[28] State power seeks to suppress, but instead will often stimulate that which it fears, the forces of 'radical curiosity'. Such human reactions set in context why it was that at criminal dissections the actions of the crowd came to express on many occasions an insatiable curiosity that reflected narratives of beliefs, emotion, participation, exclusion and sentiments, commented on in contemporary newspapers. The key logistical issue for the early modern state was that in exciting all this 'popular curiosity' to stimulate support for capital punishment in the transition from a moral to political economy, there was little that legal officials and penal surgeons could do when that 'curiosity took on a life of its own, behaving in ways that no single person or institution could control, however much they tried'.[29] There

Table 6.1 Audience participation activities at criminal dissections, 1752 to 1832

Body, Condition	Penal Punishment	Official Status	Crowd, Agency
Alive, Anti-mortem	Execution	Legal Death	Active, present
Dead-Alive, Liminal	Anatomization	Medical Death	Active, present to check 'truly dead'
Truly Dead, Decay & Putrefaction	Dissection, Dismemberment	Post-Mortem penalty	Active, but recoils & retreats
Curated Skeleton, Skull & Body Parts	Display	Post-Mortem, curated	Active, returns to see the spectacle

was only one practical solution under the circumstances and that was to make criminal dissection very offensive when taken to its logical conclusion. The 'perils of curiosity' had to somehow be deployed for political and professional ends, and a discourse on contagion, real and imagined, served this purpose well. There was thus a point in the dissection itself when the reality of the dirty, decaying corpse became very distasteful. The crowd retreated and, often quite sensibly so, until skeletons, skulls, and body-parts were brought back for display. We have then through the prism of the post-execution crowd's curiosity discovered a fourth step on the punishment journey of the Murder Act (Table 6.1):

Chapter 5 mapped provincial venues into four typologies covering different areas of England. These staged a medico-legal choreography that was fluid and framed by the four procedural steps identified above. Even so, there is one general point to keep in mind. Dissections were not a single event in the life of a community. They were usually medical performances staged on average over three consecutive days. This meant that the audience kept changing its profile. The execution crowd came first, they filed past, vacated the premises, but as there could be up to 25, 000 people walking around the criminal body during day one there was no let-up in the number of those taking part in the initial ritual performance. If a body was hanged at noon, cut down at 1pm, then the anatomization done by 2pm, and the doors left over for the next ten hours to be closed at midnight, then in that timescale there could be as many as 40 people a minute (for a crowd of 25, 000), 20 a minute (for 10, 000), 10 a minute (for 5, 000) and so on, walking past the body. It was a logistical headache to move people efficiently through a space, especially a gaol room or small dispensary, but challenging even in a Shire Hall reception area or dedicated dead-house of a voluntary hospital. If those in charge got the timing of the crowd control wrong then it could be chaos in minutes.

On day two, another audience of educated people from the local area had pre-purchased a ticket to gain entry to the medico-legal space. Again this might be as many as 500 wanting to walk past and then sitting down to see the start of a criminal dissection. As they were paying spectators the onus was on the surgeons to show say the brain and to take more time doing so. It helped that this second audience might not choose to stay all day; it really depended on the research interests of those conducting the dissection and what they elected to present over sessions lasting 2–3 hours typically. Occasionally at night women of the elite and middling sort might be permitted to attend an anatomy demonstration on day two (on say the eye, or face, or heart) but this would have a very different atmosphere from those which females of the labouring sort attended to stare at dangerous men laid out for their delectation. By day three, medical men from the vicinity were admitted exclusively, usually for surgical and scientific purposes, generally accompanied by their apprenticed pupils. Provincial dissections were therefore a product of the curious, educated, or research-driven interests of those pressing to see different daily sessions, and each depended on the age, gender and ethnicity of the criminal. At the end there was an opportunity for receiving back material afterlives to be studied, and once more the crowd entered to gaze. To reach what was soon dubbed the 'dead-end' of a criminal dissection thus looked like that in Figure 6.2 overleaf.

William Hey, a senior surgeon working at Leeds Infirmary carefully scheduled his criminal dissections to the basic timetable illustrated below.[30] His private papers indicate that in the first hours after death he displayed the criminal corpse for crowds of up to twenty thousand people that accompanied a body sent from York to Leeds. Hey charged the audience a nominal price of 'three pence per visitor' to view the disreputable cadaver, but nobody was permitted to touch it, or take souvenirs like clothing or a body part. After the execution of Mary Bateman hanged at York for murder and witchcraft in March 1809, Hey said he raised a total of 'thirty pounds' at '3d per person' from some '24, 000 spectators' at the public viewings. Many people paid the nominal fee but the labouring poor were permitted to enter free-of-charge to prevent a public riot (sure to follow

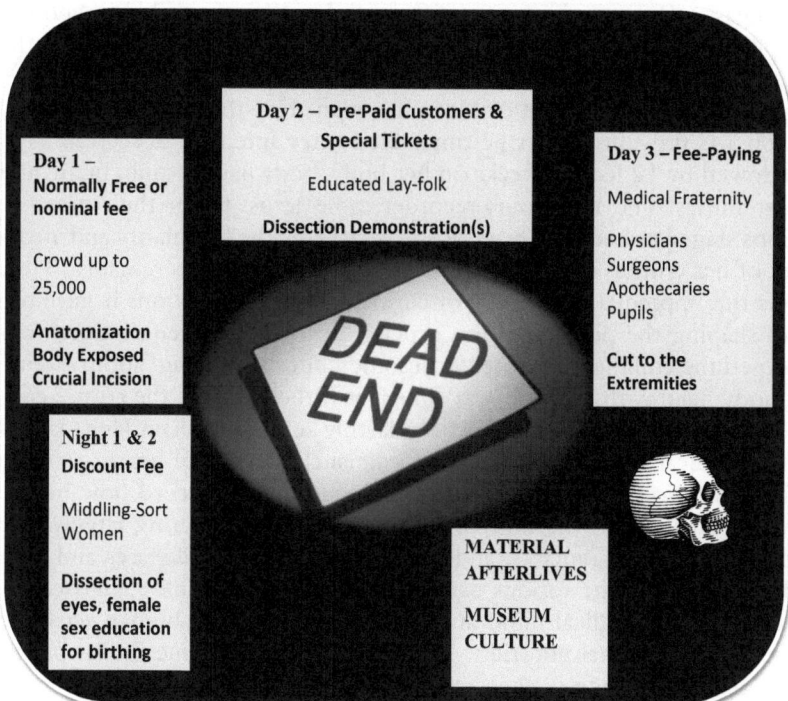

Figure 6.2 The Dead-End of the criminal corpse and its daily audiences.

had they been refused on the grounds of expense). On day one, people filed past the body at 40 per minute such was the level of local interest. They generated charitable monies donated to Leeds Infirmary: a clever ruse to minimise bad publicity.

Once the corpse looked lifeless, the crowd were then encouraged to take a respite and medical students came forward for a fee of '10s 6d each' to see the major organs expire.[31] Hey's pay-as-you-view scheme was choreographed to ensure that ten surgeons and their apprenticed pupils from the vicinity moved to the front of a long queue once original research got underway on day two. Hey also had an eye to making more charitable profits and so he sold 'special tickets' to the professional men of Leeds: 'about 100 tickets were available to gentlemen who paid five guineas' each to see dissection of the muscles. Women could buy another type of discounted ticket for Hey's 'special lectures for ladies upon the eye', the 'windows of the soul', a fitting spectacle for female sensibilities. At the famous dissection of Mary Bateman, Hey thus banked a total of '£80 14s 0d'. Altogether he timetabled three intensive dissection days, followed by 12 lecture weeks on her body, body parts, trunk, brain, and extremities. The York court recorder came across to see the entire sessions staged at Leeds: 'The curiosity excited by the singularity and atrocity of her crimes, extended to the viewing of her lifeless remains'. Here was the unpredictable nature of curiosity and the emotions it gave rise to, shaping the post-execution experience for the audience. There was something chilling, he wrote, 'gruesome', but 'fascinating' about 'seeing a body filled with guilt'. It was almost as if the inside of the corpse contained a latent power of its horrible earthly actions. William Hey insisted that medical students should overcome such feelings of horror to learn from 'a rational amusement'. Yet, his private notes record how he too was fascinated by 'the body of the malefactor ... entering largely into the physiological remarks, and more slightly into the diseases and accidents to which the various parts of the body were liable'.[32] Dissection educated the medical mind and trained the irrational human senses in the face of criminal notoriety. In the provinces meantime similar profit-margins were made in the first 3-days at Cambridge, Manchester, and Preston (places we will be visiting later in this chapter). Alighting at Derby it is feasible to explore the underlying medico-legal and social tensions—between on the one hand the desire to drive forward original research—and on the other hand the need to somehow redress the discomfort many felt about the sorts of opportunity costs associated with being part of a criminal dissection. Although the anatomy of the criminal brain captured the general public's imagination, it also troubled medico-legal officialdom too.

Dissecting the Criminal Brain: *The Nervous Energy of Original Research*

Next the long nerves unite their silver train,
And young Sensation permeates the brain.[33]

Derby Physicians	Derby Surgeons	Assize Legal Officials	Other Interested parties
Richard Forester Forester – Consulting Physician, emeritus	William Bainbridge	Charles King, Sheriff's Officer, grocer and dealer in small wares	Thomas Noble, gent and editor Derby Reporter
Francis Fox senior – retired, & influential	William Bennett	William Webster, Sheriff's Officer & bailiff	Joseph Lee maker of artificial limbs
Thomas Tycross	Charles Burrough	**Thomas Keeling* Police Officer**	William Child stationer and from 1810 secretary to the Derby infirmary
William Baker	Peter Brown		William Stevenson druggist and chemist
Thomas Bent	Samuel Davenport		John Saywell, artist and his associates
George Ferguson	Thomas Eaton		
Francis Fox*junior Leading Physician	**Douglas Fox,* Prison Surgeon**		
	Richard Bennett Godwin* Thomas Haden Henry Hadley Thomas Harwood John Hull junior William Hoare Francis Huggins John Johnson John Jones **Joseph Wright***		

*handled the body

Figure 6.3 William Webster's corpse, dissection day, Derby, 1807 (**Source***:* William Webster's corpse, dissection day, Derby, 1807, reconstruction of those present. Record linkage work compiled at Derbyshire Record Office on 'William Webster's Corpse, Dissection Day, March 1807, Derby Shire Hall').

At Derby, dissections that focused on brain function had become a respected research priority involving the criminal corpse by 1800. Thus in a well-documented and representative case, when William Webster was convicted of murder at Derby Assizes on 20 March 1807 medico-legal officials and lay-people assembled together at his criminal dissection. Webster had 'poisoned to death Thomas Dakin, Elizabeth Dakin and Mary Roe'. The judge decreed that for a triple murder and such a despicable method of homicide—poison being the silent killer—his body must be delivered for extensive dissection. The designated venue was Derby Shire Hall (Derby Infirmary was not opened until 1810). Yet, events subsequently caused considerable controversy in the medical press.[34]

An array of medical men stood awaiting the body of William Webster in Derby town. Record linkage work reveals that there were seven physicians, seventeen surgeons, two legal officials connected to the sheriff's office, and six other interested parties, who had a right of access. All were anxious to have a hands-on experience (see, Figure 6.3, page 235). The actual dissection was done twenty-four hours after the death sentence by a leading physician (Francis Fox junior) and three surgeons (Richard Benet Godwin, Joseph Wright and Douglas Fox); as well as someone not named but described as an 'operator' (likely to have been one of the two sheriff's men present) tasked with shaving, washing the body and preparing the anatomical instruments. What primarily interested all the leading medical men in the town was the nervous energy in Webster's brain. The *Medical and Physical Journal* thus reported:

> Many, medical men of Derby and its vicinity attended the dissection, and some were desirous of seeing the state of the brain in a person destroyed by hanging. It was therefore examined by the gentleman [physician] who conducted the dissection, who was very much accustomed to the task, and by others who were present.[35]

Brain research had been pursued at Derby since the Murder Act. As we saw in the previous chapter, Erasmus Darwin and his scientific circle in the Midlands stimulated a lot of ambitious research on the function of the brain. Their work was being talked about in the town and its vicinity. In William Webster's case, Francis Fox junior, the leading Derby physician in 1807, raised his dissection instruments to cut the top off the skull. Fox had established his reputation in the town by attending Erasmus Darwin in his final illness in 1802.[36] He had also been doing criminal dissections in the Midlands since the 1770s, as we again saw in the last two chapters. A local Derby history published in 1829 thus described Francis Fox as: a 'physician, philosopher and chemist' who by 'superior talent is rising into considerable eminence'.[37] He was ably assisted by Douglas Fox the leading surgeon in the town who 'filled the situation as Surgeon to the County Prison some years with universal satisfaction to the Magistrates of the County and the prisoners under his care'.

By tradition the hangman in Derby completed the death sentence either on or behind the gallows. 'John Crossland the infamous hangman' until 1705 was said to have had responsibility for: 'The bodies of the executed were his prerequisite; signs of life have been known to return after execution; in which case, he prevented the growing existence by violence'.[38] After the Murder Act the responsibility for legal death remained with the Derby hangman in 1752 but the task of medical death was delegated to the prison surgeon who assisted the Sheriff. This meant that over a time-span of five decades once the capital legislation was passed, Francis Fox and Douglas Fox were well-versed in handling condemned men and medical death by 1807.[39] They knew how to act efficiently post-execution. In this, Francis seems to have been influenced by Erasmus Darwin, his private patient, and fellow physician. Darwin had long been fascinated by the 'sensorium of the mind':

> The word sensorium is used to express not only the medullary part of the brain, spinal marrow, nerves, organs of sense and muscles, but also at the same time that living principle, or spirit of animation, which resides throughout the body, without being cognizable to our senses except by its effects.[40]

On the boundaries of life and death, especially *in extremis* at executions, Darwin claimed that medical men must be alert to the physical fact that dying could be prolonged and protracted because contraction of the sensorium in the mind might be slow until a physical state of general diminution was reached.[41] Or, it might only result in partial diminution causing a state of suspended animation; worse still, in a criminal it might go into reverse, looking dead, but actually be a form of extreme cold in which an accumulation of the sensorium was happening as the brain rested to recover its powers. At a time when galvanism, resuscitation, and suspended animation were very fashionable topics, the dissection room was where the boundaries of life/death might be glimpsed together.[42] Against this intellectual backdrop William Webster's dissection was attended with considerable publicity in 1807.

The medical press noted that although everyone present agreed about the need to conduct original research on the dissection of the criminal brain in Derby, there were differences of opinion about how much experimentation figured under the original Murder Act. It was understood that post-mortem 'harm' was intended but the medical limits of potential human vivisection in the criminal brain remained controversial. An onlooker wrote to the *Medical and Physical Journal* on 26 August 1807 admitting in print that he 'had some awkward feelings' which he kept to himself during the dissection of William Webster.[43] He claimed that he did not walk out or intervene with the experimentation because 'he did not conceive himself in a situation to interfere'. This stance was ridiculed by the London press. In the capital surgeons questioned whether provincial practitioners should be permitted to attend criminal dissections if they were so ill-qualified to remark upon them. At Surgeon's Hall leading men like Sir William Blizard had pioneered criminal craniotomy, ran one robust reply from a London correspondent. He went on with bitter satire to challenge the rather conventional attitude of some provincial practitioners who had access to brain dissections at Derby and places like it—'I dare say that he had some awkward feelings of incapacity and inferiority'. Angered by the insinuation that local doctors were second-rate, the same 'Derby practitioner' in a follow-up letter went on the offensive in the medical press: 'I am well able to prove that this dissection, as it respects the brain, was not conducted on the principles of scientific anatomy' but out of consideration to his profession he chose not to elaborate further in print.[44] There was a lot of dispute by 1807 as to whether heart-lung failure, strangulation by the rope, or brain death, marked the end of dangerous criminals. This was an open field in which medical men in the Midlands could make their reputations through original research. Each dissection venue thus became the site of speculation about the anatomy of the brain because it had ignited so much curiosity in nearby towns and cities. The criminal body became a benchmark for competing research agendas but whether this was what the legislation had intended was open to dispute.

In order to do brain-research penal surgeons at a criminal dissection had to make a basic decision. If they needed a skeleton for teaching purposes then there was little point in dissecting the head because it would destroy the skull. At Bury St, Edmunds it is worth recalling that at William

Corder's dissection in 1827 a decision was taken to preserve his skeleton and so his brain was left in-tact.[45] A cast was taken of his head and a phrenology examination made of its anatomical profile; likewise his corpse was skinned, and minute study was made of the muscles; but brain research was ruled out. This contrasts with Leicester Infirmary where medical men were excited by the sort of original brain dissections that were being done at Derby.[46] Today neurology is seen as a discipline of modernity[47] but it is possible to date its recognisably scientific research-stirrings to a selection of criminal brain dissections between 1807 and 1815 across the Midlands. This was a time when many provincial places started to specialize in vitality which in turn gave rise to a lot of disquiet about access to, and privilege knowledge of, original research on criminal heads. This debate was of course nothing new because from 1752 medical men had been expressing anxiety about the prurient interest of the crowds that accompanied the post-execution body. Yet, as brain death replaced heart-lung failure as an explanation for the boundaries of life/death there was a groundswell of opinion that only scientific eyes should be permitted to partake. At Leicester Infirmary on 23 March 1822 it was thus resolved that: 'the practice of exhibiting bodies of unfortunate persons given to the Infirmary for dissection appeared to us improper, and that in future no such public exhibition be permitted'.[48] This Midlands attitude was influential elsewhere. By 1827 the Royal Devon and Exeter Hospital followed suit declaring that the viewing of criminal corpses 'so much enjoyed by strangers... should be permitted only on the day of execution', and the 'doors be closed once dissection commenced'.[49] It would be a mistake however to conclude that generally crowds were excluded given what contemporary newspapers revealed in Section 1 of this chapter.

Popular Sunday journals like *The Age* explained that 'morbid curiosity' was a feature of many dissection venues and inside it was very difficult to control or predict. Often it was related to the criminal's visual appearance at a time when disease was said to make an appearance on the body. Returning briefly to the criminal dissection of murderers like Alexander Keand from Section 1 above, what dangerous criminals looked like physically when apprehended featured prominently in the reporting of post-mortem 'harm'. *The Age* elaborated on why Keand was such an intriguing anatomical subject: 'Alexander is 5 foot 7 inches, dark eyes, whiskers, broad set...generally wears gloves on his hands being scrofulous and scaly, and his nails bitten black at the bottom'.[50] The Sunday edition of *The Observer* newspaper meanwhile decided to publish some letters from medical correspondents elaborating on Alexander's brain dissection. Peering inside, his anatomy was normal but he had committed an abnormal crime; a disturbing finding. The Lancaster dissectors therefore agreed to two unnamed but respected local phrenologists preparing a separate report which was widely publicized for its 'moral' as well as 'scientific' value:

> The phrenologists have found, in the construction of Keand's head [Alexander] further evidence to support their system. His devotion in gaol [to his brother Michael] is explained by the singular prominence of the organ of veneration. His desperate resistance when taken into custody and the fight between the brothers on the first night after their apprehension are referred to the extraordinary development of the organ of combativeness, and his murderous propensity rendered evident by the fullness of the organ of destructiveness.[51]

Artistic sketches were also taken of Alexander's physiology but the *Manchester Guardian* reporter warned readers that 'the portraits that have been published do not resemble [him] at all'.[52] The prison surgeon instead noted that both Keand brothers had 'dark complexions and dark eyes' and spoke in 'a strong Scottish accent' (they came from Dumfries but moved for business reasons to the Manchester area). These characteristics were said to have added to their sense of menace. Evidence like this pieced together from a variety of reporting sources enables us to stand to one side of the crowd as they paraded past Alexander Keand in the yard of the Lancaster infirmary where he was laid out on a portable dissection table to view. He was not tall, slightly above the average height of a woman, muscular, with a stocky build. We learn he was hirsute with moustache whiskers but only his dark stubble would have visible after being shaved all over. He probably had some form of eczema, psoriasis or ring-worm, given how scaly his hands looked. His nails were bitten to stumps and blackened. Alexander had been a hawker. His sallow skin may therefore have been an occupational hazard, tanned from travelling the dusty roads in warmer weather-beaten climes. His facial expression was described as 'not prepossessing'. It was his ordinariness that was so disturbing, despite the various newspaper reports that took dramatic license with Alexander's determination to resist arrest, indifferent courtroom demeanour, and unrepentant final hours. Cases such as this and many like it in the archives explain why dissectors wanted to get inside the criminal brain. A logistical issue that had to be resolved was how to satisfy the 'natural curiosity' of the crowd to create a legal deterrent yet without fanning either 'morbid' or 'public curiosity' to such an extent that the agency of the crowd could overwhelm medico-legal officialdom. The question of the theatrical features of the criminal dissection showcase had therefore to be chosen carefully *in situ*.

Penal surgeons soon concluded that to do their public duty they had to enact a performance of criminal justice by staging retribution as an immersive form of theatre. This involved permitting the crowd to be in close proximity and looking carefully at the corpse often at arms-length. Yet, equally they had to grapple with the practical problem of what they could visibly actually show their audiences walking around the condemned body. When heart-lung failure was the standard explanation for medical death it was a physical shut-down open to everyone. Brain death happened deep inside the head and lacked visibility. It was much less accessible to laypeople, and yet potentially a very exciting area of new medical research. In satisfying the crowd that the criminal was a corpse about to be punished, penal surgeons lacked a graphic method of showing how the brain had died. In a craniotomy the top of the skull was sawn off and skilfully so, but what everyone was looking at in terms of grey matter remained mysterious. So increasingly many penal surgeons welcomed phrenologists and engaged in experimental work to try to ring-fence their medico-legal standing. A well-known example from Cambridge is illustrative of these brain research trends and the continuing role of the crowd's curiosity in shaping the choreography of criminal dissections.

The execution of Thomas Weems for murder on 6 August 1819 has become very famous in criminal histories.[53] His condemned body was subjected to a number of quasi-scientific experiments to explore the nature of electricity, resuscitation, and brain death, all associated with Mary Shelley's *Frankenstein* (1818). The homicide circumstances were akin to those of the John Holloway copycat killing of his rejected wife in Chapter 3. Weems had been compelled by parish officers to marry a young woman called Mary Ann whom he courted and allegedly got pregnant. After they were married she said that she had miscarried, but he did not believe her thinking he had been tricked. On his business travels (he was a miller) he met Maria Woodward a pretty girl from Edmonton in Bedfordshire. They started walking out together, fell in love, and he wanted to marry her, but was not free to do so. Before long, in despair he decided to murder his wife. Weems strangled Mary Ann, left her for dead in a shallow field-ditch, but was seen running away, arrested and apprehended by a local magistrate near Godmanchester in Huntingdonshire. General opinion held that his crime must fit the punishment, for he had 'with a garter...tied a slipping noose' around the victim's neck 'round which was a black mark' when it was removed. Murdering her by duplicating an execution method was for many local people unforgiveable. This context explains why there was so little sympathy expressed in the local press for the extensive nature of the post-mortem punishment rites at Cambridge where he was tried and found guilty of murder.

Weems was executed at 'a few minutes past noon' over the 'gateway of the county Gaol'.[54] An hour later he was cut down and paraded through the streets of Cambridge to the Botanic Gardens. These were at the time located next to a purpose-built anatomy theatre in the town-centre along Downing Street. By 1.45pm a reporter from the *Cambridge Chronicle* was on hand to record what happened next:

Friday 6 August 1819 -

The body, after being suspended an hour...was immediately conveyed in a cart to the Chemical Lecture-room of the Botanical Garden, where Professor Cumming had prepared a powerful galvanic battery (which formerly belonged to Professor Tennet) with the intention of repeating some of the experiments lately described by Dr Ure of Glasgow in the Journal of the Royal Institution.[55]

He then went on to describe how the usual anatomization procedure fused with a number of experimental research methods. They involved applying to various parts of the body some '220 pairs of double six inch plates charged with dilute sulphuric and fuming nitrous acid' to give the galvanic batteries 'intense action'. These were as follows:

Experiment 1 – One wire was applied to a small incision in the skin of the neck over the par vagum [pair of vegas nerves], and the other one made between the 6th and 7th rib; when at each discharge of the battery, the chest was disturbed in a manner similar to a slight shuddering from the cold; the period of shuddering corresponding to the number of plates struck by the operator in the trough.

Experiment 2 – The par vagum [pair of vegas nerves] was laid bare, and one of the wires passed under it; the other was placed in contact with the diaphragm, through an incision made deeper between the 6th and 7th rib. The contractions were evidently stronger than in the last experiment, and to all appearances confined to the same set of muscles. Not the smallest action of the diaphragm was perceptible.[56]

Experiment 4 was done on the 'supra-orbitary nerve' to see if there were any 'mental afflictions', but instead what happened physically was described as 'convulsive twitching'. Experiment 6 involved connecting up the legs and arms to see if they could be revived; whereas experiment 7 explored electricity and the 'spinal marrow'. All the experiments lasted together 'about an hour' until 2.45pm. Throughout the body temperature was recorded at '93 degrees' because the corpse was so fresh and it was summer. It was commented on that 'there was no dislocation of the neck' and 'no distortion of the countenance' either post-execution on arrival, or after the experiments. In other words, the body was in a good condition, could potentially have cheated medical death, and so was ideal for original research of this nature. Only once the experiments had been carried out, were 'the necessary dissections…executed by Mr. Okes in the presence of nearly all the medical men in Cambridge, many members of the University, and several of the most respectable inhabitants of the town and country'.[57] In all likelihood this was akin to *splanchnology* because the initial anatomization cuts were rather superficial, being subsumed with the experimental work to see if resuscitation was feasible. The exclusion of the crowd however from the spectacle was being talked about across town and did cause considerable disquiet. Questions were being asked in the local press about just how extensive was the dissection and what would happen next. The crowd were determined to have their curiosity satisfied.

The following day, Saturday, 'the body was exposed to public view in the same room in which the experiments had been performed'. To try to control the situation, 'Constables were placed at the gate of Downing-street, to prevent the room from being too much crowded, and the doors were opened from 12 noon' to the crowd. Reporters said:

The benches were instantly filled with spectators, whose countenances bespoke a *strange combination of curiosity*, disgust, and awe. Crowds of every description of persons continually succeeded each other until one o'clock and amongst the many hundreds who came to view the body, no one seemed moved by a feeling of pity for the fate of the criminal, so strong were the grounds for his condemnation. The doors were then shut, and at a

request of a large party of gentlemen, Mr Okes commenced a more extensive dissection of the body.[58]

By 1819 the medico-legal choreography that this book has retraced was
well-known to a gaol surgeon like Mr. Okes working for Cambridge Castle.
It was connected in the popular imagination to resuscitation checks and
only then was dissection done by degrees, first *splanchnology*, and then
cutting to the extremities.[59] The medical fraternity had wanted to exclude
the crowd to gain control of their research interests but they could not risk
doing so exclusively because of the threat of public riot. Ordinary people
of 'every description' remained interested, curious, but also repelled by
the disintegration of the criminal corpse in Cambridge. In this case, they
did not protest at being asked to leave once they had seen the criminal cut
open. The body was still in a state to recognise its face from the execution
as they departed. The constables on duty guarding the body could not
turn away 'many hundreds' without extra logistical support, but they also
did not have to do so. The crowd exited and then the doors were shut.
All was orderly since what happened next would look less than human, a
bloody mess. Yet, there was one last piece of the jigsaw puzzle of human
identity left over. At Trinity College a square piece of white-coloured
skin was preserved and sent to the University bookbinder. Handling this
criminal corpse became about rebinding fragments of the dissolute, dangerous, dead into a skin-hardcover of a book shelved for posterity in Sir
Christopher Wren's library. Before though examining those material afterlives in more detail, it is essential to first think about how this type of
original research started to reshape medical education and where exactly
the newly qualified felt they had to relocate to get on the career ladder in
eighteenth-century England.

A Disintegrating Corpse: *The Science of Extremities*

"In Mr Brooke's dissection room it was mutilated by the dissection knife"
The Guardian, 8 April 1822

Around the 1790s Thomas Bishopp, a medical student, wrote a series
of letters to his near relations at Cold Overton in Leicestershire.[60] He
explained about how his dissection training had progressed during the winter sessions after the Murder Act. Bishopp attended St Thomas's Hospital
as a pupil apprentice in the day. At night he paid to learn about anatomy,

dissection, and midwifery at private classes. Like many medical students at the time he thus wrote that he was 'a walking pupil' on the hospital wards.[61] From November 1792 he also described in detail how he had 'entered to the dissecting room, engaged for a subject, purchased a case of knives & ordered the necessary dress'.[62] The penal surgeons of London, he reported, had 'met with [the] difficulty of procuring bodies' by the 1790s because the murder rate could not keep pace with anatomical demand.[63] Original research had become the *raison d'etre* of the most ambitious surgeons and like many Bishopp paid to attend very crowded extra dissection lessons in 'Leicester Square' and on 'Windmill Street' around Soho where the Hunters conducted their famous anatomical classes. Bishopp needed though to ask for family money to be sent by return post to pay for body parts supplied from the gallows or illegally from graveyards. In April 1794 he thus wrote 'concerning his dissection of part of an extremity' that had become available and that 'Mr [Richard] Tookey [House Surgeon St George's hospital] had also given him the opportunity to acquire 'an extremity of a child [for dissection] & would I believe have given me more if they had not been so much distressed for subjects at Windmill Street where the dissecting room is'.[64] Evidence like this provides a window into a secretive world filled with curiosity and sets in context that 'dissection of the extremities and to the extremities' was standard on criminal corpses and resurrected body parts. By the close of 1794 Bishopp explained to his family that their financial support had not been in vain, for:

> I shall be sufficiently acquainted with practical dissection & anatomy to undertake all the common operations in surgery with good confidence in my qualification and I am not at all doubtful but that I understand the theory and practical department of midwifery as well as the majority of those who settle in the country.[65]

Here was a typical medical student preparing for provincial practice and trying to build a reputation for being skilled with the lancet whilst there was the opportunity to do criminal dissections in London. It was also important to remain mindful of the best places to learn to specialise in more original research. In terms of maximising future business acumen, the choice of the Midlands made sense since, as we have seen at Derby, reputations were being made because of better access to criminal dissections. Two strategic decisions taken by Bishopp at a formative career-stage illuminate the specialised medical market taking shape by the 1790s.

In October 1792, Thomas Bishopp, on first qualifying in medicine, took the opportunity to become an assistant to 'Dr Robert Chessher of Hinckley' (1750–1831). His medical employer and mentor was busy building an unparalleled reputation as the first really skilled orthopaedic surgeon in eighteenth-century England. The anatomical study of criminal bones gave men like Chessher the opportunity for original research on the 'mechanics of the extremities'. In particular, he specialised in spinal conditions and limb injuries. It is noteworthy that Chessher was the grandson of Lewis Whalley, a respected physician in Lancaster whom we encountered in Chapter 5 (see, page 190) working on criminal dissections in the North West. Knowledge of the criminal skull, spine, and bones, was in circulation between the provinces through family, as well as, professional medical networks. Thomas Bishopp was consequently typical of the sort of medical student who sought to advance his career by association with established experts. He subsequently took a decision in 1795 to return for a year to London for further his anatomical studies. This however did not resolve the question of how best to get started in business because he had to start earning to pay off his student debts accumulated from his expensive education.

Thomas Bishopp was a financial realist, recognising that the London medical scene was very competitive and overcrowded. His career strategy was therefore to buy into a provincial practice once he had experienced enough dissections and acquired skills in midwifery. In 1797 he wrote to John Frewen of Cold Overton, his wealthy cousin, that he had the 'prospect of purchasing a third share in Dr Chessher's practice' but was unsure whether to proceed given how hard it would be to make it financially viable in a small town such as Hinckley.[66] In the meantime he took on a fee-paying apprentice called 'Mr Beale' but the relationship turned sour because he proved to be lazy and devious about money:

> I am now persuaded that he means merely to deceive his friends. I have perhaps seen him 3 or 4 hours in one day in three, but certainly not more, although I call on him more frequently & have repeatedly told him to meet me at the proper hour every morning, where he goes to I know not nor can I undertake the care of him under these circumstances.[67]

Meantime, Bishopp was exploring 'the possibility of his purchasing [Edward] Le Grand's surgeon and apothecary's business at Canterbury', but again he had no luck.[68] To tide himself over financially after the protracted business negotiations, he took up a Frewen family offer to become surgeon to the Sapcote Sick Club in September 1799 near Cold Overton. By all accounts this then helped him to secure his own practice at Friar Lane in Leicester by 1800. It was hard work at first, for in 1802 he wrote home that: 'I fear I have little ground to boast, looking only to the better class of the trading part of the town'.[69] That final move brought him into the ambit of medical men working on the dissection of criminal brains and the fascinating subject of medical death and resuscitation in the Midlands.

By 1806 Thomas Bishopp was apologising for not writing to his relatives as often as he should because he had been very busy attending his private patients and trying to do original research in and around Leicester. He was intrigued for instance by new resuscitation techniques on attempted suicides: 'when I should have written to you, I was engaged in attempting to restore to life a young woman, servant to [Samuel Cheeke] Morris the surgeon [of Market Street Leicester], who had hanged herself'.[70] Living as he did in close proximity to the infirmary used for criminal dissections and coroners' cases, there was ample opportunity to make his medical mark in Leicester. Evidently this was beneficial for his reputation since he was delighted to report to his family sponsors that his business was buoyant with referrals. His growing reputation for innovation meant that: 'I will need to take on another apprentice, in a year or so'. Yet the credit control of fee-paying patients was a constant financial headache, and so he often had to take calculated risks. One further career move was therefore considered advantageous. Bishop was determined to apply for the position of surgeon to 'one or both of the gaols in Leicester' where he would have direct access to executed bodies and be paid a fixed-salary for prisoner-consultations.[71] His ambitions culminated in a letter dated 13 February in 1807 in which he thanked John Frewen warmly for his local patronage and was pleased to confirm that he had established his medical-standing in the Midlands.

Evidence like this private correspondence indicates that doctoring could be a steady occupation, but there was the ongoing problem of managing fickle consumer demand. It was imperative to stay ahead of the competition. It is noteworthy that John Bishopp's gains at Leicester were still fragile in 1807, for instance, hence he had to fall back on his Poor Law work: 'my pressure of business is of course very irregular & unsteady, otherwise I should think soon of giving up the care of All Saints Parish poor' in the town-centre. He admitted moreover that: 'I have been engaged with the poor when I have wanted for better patients'.[72] Gaining access to criminal bodies by becoming a gaol surgeon was there-

Table 6.2   Location of criminal dissections and their respective research activities by 1800

Location	Dissection Activity
Birmingham	Surgical amputations and trepanning of the skull
Bristol	Dissection of heart/lung/brain and midwifery
Bury St Edmunds	Muscles, heart-lung capacity, phrenology of head
Cambridge	Nerves, Galvanism, Spina Bifida
Colchester	Medical death, phrenology of head
Derby	Brain vitality & sensorium (nerves)
Hinckley	Bones & Orthopaedics of limbs and back
Ipswich	Gonorrhoea, Heart resuscitation, breast tissue
Leeds	Structure of the eye, hernia, limb amputations
Leicester	Brain research and medical death
Newcastle	Dentistry & morbid diseases of the coal industry
Northampton	Blood flow, mechanical physiology & secretions
Norwich	Skeleton, phrenology of head
Nottingham	Skeleton, phrenology of head
Plymouth	Skeleton, phrenology of head
Salisbury/Winchester	Fever, nerves, kidney & military wounds
York	Treatment of madness and maladies of the mind

fore a smart business move, embodying accessibility, exclusivity, and status. Against this backdrop, medical men in the Midlands tended to try to align closely with the criminal justice system, even though working on executed bodies might tarnish their public image. The opportunities criminal dissection afforded to develop a growing reputation for innovation out-weighed sensitive public relations considerations with regard to the post-execution crowd.

After the Murder Act the type of original research being done on criminal dissections was to gradually influence where ambitious new surgeons sought to locate themselves provided local business opportunities were favourable: the Table 6.2 above indicates key locations and the sorts of specialisms gaining a foothold. There was no distinctive regional patterning to particular types of anatomical interests, unlike the dissection venues in Map 5.1 page 183. The reason for this trend was that it really depended on the availability of criminal body types (age, gender, height and so on), and what then was feasible for local surgeons to do in terms of using dedicated dissection space. This influenced how their work related to the wider medical marketplace on location. Many wanted to pursue individual research but they had to be opportunists as most were working

in environments where scientific endeavour had to be paid for from business profits. If the executed criminal was very muscular, for instance, it
made sense to dissect the musculature in minute detail. If the crime was a
particularly gruesome homicide then a phrenology cast was usually taken
because the brain was the focus of intense interest. All of this type of post-
mortem work nonetheless involved taking apart the corpse by dissection
and dismemberment unless the skeleton was the main object. An added
complication was that medical students sometimes argued about getting
access to what were regarded as trophy heads to dissect. The *Morning Post*
of 15 September 1819 reported one such compliant that came before the
Middlesex Sessions:

> Mr James Luke, a young Gentleman, and one of the students of the London
> hospital was indicted for committing an assault on Mr Charles Roberts a
> fellow student, on the 7th June last in the Dissecting Room. Mr Charles
> Roberts, the Prosecutor, said the Defendant and himself were house pupils
> in the London Hospital when a misunderstanding arose upon the dissec-
> tion of a body. He was desirous, with a Mr Parrot of dissecting the brain,
> and Mr Luke said he had been told by Mr Andrews the surgeon, that the
> Prosecutor's skill was equal to the task. He saw Mr Andrew's who disclaimed
> having said anything of the kind. The assault took place two hours after. The
> Defendant, in the dissecting room, clenched his fist, and said he should like
> to lay both, meaning the Prosecutor and Mr Parrot flat on the floor...The
> Defendant struck the Prosecutor with a violent blow to the stomach.[73]

Evidence was heard from a senior surgeon that Mr Charles Roberts
was not telling the truth. He should not have been dissecting the brain
because it had been left for a coroner and experienced surgeon to examine. Then he stated that in any case dissecting a criminal head was 'a task
too delicate for a young man like the Prosecutor who had been no more
than six months at the hospital'. The Judge urged Roberts to apologise to
Luke, and asked the latter to make a charitable donation of '10/' to the
Hospital. He reminded them that their behaviour spoke of 'indecorum in
a place which may be called a place of education'. They shook hands and
the case was dismissed. Yet, the vignette was telling. Questions of medical
competency, competition to dissect brains, and a general fascination with
the criminal mind, were all aspects of dissection work that was secretive
but excited public curiosity. Standard techniques were well-known by the
1810s from anatomical books and manuals, but there was a lot of practical
information withheld from the general public about what amounted to an

'insatiable curiosity' for the torso and head-piece. One example stands in for many from the archives.

On 8 April 1822 a civil case, *King versus Cundice*, was brought before the Surrey Assizes. It exposed in court what it meant to dissect and dis-integrate a dead body convicted of a capital offence. In the case's open-ing remarks the prosecution declared that: 'the science of anatomy was most beneficial to mankind; but, its advancement ought not to be attained by the violation of those feelings which were the most scrupulous in the human breast'.[74] The facts were that a highwayman named Edward Lee was tried and found guilty of a capital offence. He was hanged and then his body was to be buried by 'William Walter, the keeper of Horsemonger-lane gaol'. He took a decision to delegate the internment to 'Edward Cundice, the undertaker' to be done in a 'decent and proper manner'. Walter was 'paid out of the rates of the county of Surrey' to do the burial recouped from gaol funds. Cundice however pocketed the money and sold on the body for dissection. The Judge had ordered that although Lee had not committed murder his body must still be anatomized. This was to make sure the convict underwent medical death, but he also stressed that it must not be dissected. Cundice ignored this stipulation and made a business deal with 'Mr Brooke's the anatomist' because supply-lines were low and there was a valuable profit to be made from the sale. In the course of giving evidence in a subsequent civil case about what happened at a criminal dissection the prosecuting counsel revealed that:

> Edward Lee was tried in Croydon on 23rd August and executed on the 10th September. The defendant Cundice was undertaker to the gaol and did all the carpenter's work...Three guineas was paid to the defendant for the burial...The body was traced to Brooke's dissection rooms where the body had been mutilated by the dissection knife.[75]

Local gossip alerted the family of the deceased that a false burial was staged and they called in 'James Glenman a police officer' who:

> Proved that he had made search after the body of Edward Lee and at length found it at Mr Brooke's dissecting rooms in Blenheim Street, Marlborough-street. The body had been operated upon; the top of the skull had been cut off but replaced. He was quite satisfied that it was the body of Edward Lee. On one arm was the initial E.L. tattooed with gunpowder.[76]

The body had been transported from the Surrey Assizes to be sold to a private anatomy school run for fee-paying students by Mr Joshua Brookes near Oxford Street in central London. Cundice defended that he had been duped. He claimed that the body had been stolen at night from his undertaking premises when he was asleep. As the victim of a dreadful trade, he panicked covering up the burglary with a false funeral. The defendant's counsel criticized the fact that 'Mr Brookes the imminent anatomist' and 'Fellow of the Royal College of Surgeons' was not being called to give evidence. The judge had already directed to the jury that official enquiries had been made informally. He decreed that the Crown was satisfied that he 'was not trafficking in human flesh' but had taken advantage of a gallows body becoming available which was not illegal. The view taken in court was that burying a coffin that 'contained nothing but earth' in capital offence cases was a common undertaking swindle. Sentiments were running high in final speeches to the jury: 'Reverence for the ashes of the dead was so deeply implanted in the human heart, that it was impossible upon such an occasion not to feel something like prejudice where its sanctity had been violated'. Despite an appeal by the defending barrister to consider the objective facts, Cundice was found guilty. He was sentenced to 6 months in prison for his undertaking scam. His successful conviction revealed that in all types of capital cases corpses were being sold for 'three guineas profit' by the 1820s. The extremities, marked by tattoos were 'mutilated' on the dissection table; the head was decapitated, placed back on top of the torso so a skeleton could be made. This type of material afterlife was often controversial, not just in the capital, but rural areas too, where we see equivalent examples occurring throughout Eastern counties.

In 1771, the Norwich and Norfolk Voluntary hospital was established for the treatment of the sick poor by voluntary subscriptions. By the 1790s when 'new anatomy' was being promoted in East Anglia the board met to decide what to do about requests to stage criminal dissections on their premises to further original research. On 12 August 1797 the governors thus stated: It is 'Ordered that the Physicians and Surgeons be requested to take into consideration the propriety of dissecting bodies at the hospital removed thither for that purpose'.[77] It was evidently not a foregone conclusion that when a judge sentenced someone to be dissected it would automatically happen where the legal authorities intended. This particular ruling gave rise to considerable medico-legal debate because on 19 August, seven days later, another clarification was made by the hospital board:

August 19 1797 – Ordered that no Malefactors shall be brought to the hospital for dissection without the consent of the majority of the physicians and surgeons at this hospital, such written consent to be lain before the next weekly board and entered amount the orders of that board.[78]

Yet it would be a mistake to take this situation either at face value or to regard it is as stable and unchanging. By the 1810s there was a new legal problem confronting penal surgeons. A local sheriff was very reluctant to do his job. *The Norfolk Annals* for 1811 explained why:

29 September 1811

Mr Francis Morse and Mr Thomas Troughton were sworn into the office as Sheriffs of Norwich. Mr Morse appeared in his shooting dress, namely a short coat, leather breeches and so on; the steward preceding, as usual, to invest him with the gold chain, he refused to put on what he termed a 'bauble', nor would he wear the gown, he said, unless it was absolutely necessary. Mr Steward Alderson observed that his refusal seemed to convey disrespect to the court. Mr Morse disavowed any individual disrespect and said he would perform his office irrespective of outward forms. It was forced upon him in the expectation of obtaining a fine of £80, as he was convinced there was not a Gentleman on the Bench who believed when the precept was sent to him that he would serve the Office.[79]

Morse was a colourful and controversial local character. He had already fought a duel on 3 September 1797 in Norwich over an election skirmish. Of independent-thinking, few expected a radical to want to condemn murderers to death at the gallows or to apply the capital code for lesser property offences. Morse (like many of the surgeons that should have served as Master of Anatomy at London's Surgeon's Hall) had been expected to pay a fine to avoid the duty of leaving business ties to attend to his civic duty as Sheriff. He had though taken umbrage at the high financial penalty and turned up to be inaugurated at Norwich to make a public statement about the absurdity of the elaborate ceremonial trappings. It is informative that although his political objection was about expense and he declared a strong disliking for the Bloody Code nevertheless once a felon was condemned for murder he swore an oath to do his civic duty. He did not shirk from the unsavoury however objectionable the ceremonial trappings. For as *The Norfolk Annals* explained people in Norwich were very used to seeing dead things in country life and they had a hardened attitude when it came to the death penalty. Most residents were not squeamish about death, dying or criminal dissections.[80] In the winter of 1811 at Xmas, for example, the Norwich Market was 'glutted' with dead birds from the annual pheasant shoots to be sold to London. There was by 22 December '800 hampers… having 10 stags' and altogether '720 horses' were needed to 'draw poultry, sausages and game sent within three days' of being killed in the fields of Norfolk 'from this city to the Metropolis'.[81] Norwich folk employed in the game industry were not generally sentimental about the visceral nature of the dead in animal or human welfare. These local findings add provincial colour to a sea-change by the 1820s in local hospital concerns about bad publicity surrounding human dissection. Criticisms had abated and been overtaken by a great deal of public interest in the medical death of the body, enthusiasm for the quasi-science of phrenology, and furthering medical education around Norwich. It was thus recorded that:

14 August 1829

At the Norwich Assizes, before Mr Justice Parkes, John Stratford (42) was found guilty of the murder of John Burgess, an inmate of Norwich Workhouse, by poisoning him with Arsenic on March 2nd. The execution took place on the roof of the new gaol on August 17th. After hanging the hour the body was removed to the lower court and conveyed to the lower court at the Guildhall where it was publically exposed for 2 hours. Thence it was conveyed to Norfolk and Norwich Hospital where Mazzotti, the modeller, took a cast of the head; and on the 18th Mr Crosse commenced a series on anatomical lectures at the dissection of the body. A public subscription was started for the family of the culprit[82]

Later we will return to the striking collection of criminal plaster-cast heads that have survived as material afterlives in regional museums. They attest to how by the 1820s expectations had changed at criminal dissections which had now become a different sort of performance of the teaching skills and research ambitions of penal surgeons in the dead-houses of provincial voluntary hospitals. Rather than trying to avoid such a duty it became *de rigueur*.

Touring a number of eighteenth-century provincial dissection venues in the archives, it becomes evident that criminal corpses were complicated medical research commodities. They were objectified in terms of curiosity and curated artefacts. After the Murder Act a new pathology of medical death required an accumulated anatomical knowledge and this became a competitive endeavour. Once medicine expanded her research horizons into sub-disciplines anatomists resented being stigmatized as instruments of legal redress. Rising consumer demand afforded different types of scientific opportunities with considerable business potential, but to realize these necessitated working on resurrected, as well as criminal dissections. From what was available at the gallows, given that the murder rate could not meet anatomical demand, provincial anatomists started to set their own research agendas. Many skilled practitioners came to the realization that a body of history was not simply in their collective keeping but in their individual making too. There was though an inherent contradiction in this research output. Defining the humanity of some depended on discovering the less than humane in others. Murderers destabilized what it meant to be a decent human being. Yet, their anatomical profiles had some features that were common to everyone when exposed to public view. For these complex reasons, the actions of the crowd at post-execution rites became a reflection of accepted modes of curiosity on which original research depended. Another example stands in for many in counties to the East of England, representative of what happened also in provincial areas bordering Suffolk, Essex, and Kent.

Lloyds Evening Post on 26 July 1769 reported that Benjamin Bush had been apprehended for assisting his lover in poisoning her husband, John Lott of Hythe in Kent.[83] Standing trial Bush was found guilty and ordered to be executed at Penenden Heath in front of a crowd of five thousand people gathered at Maidstone. Before sentencing however Bush appealed to the judge not to dissect his body. In a petition to the judge the condemned stated that as he did not actually administer the poison he should not undergo post-mortem punishment. He claimed that he was not a vicious, violent murderer but an unfortunate foolish wretch in love. He knew that his body was prized as he was in his early twenties and would thus be dissected in full public view at Maidstone Shire Hall. Confronted however under cross-examination Bush conceded 'in an angry tone' that he was 'Guilty enough' of pre-meditated homicide. The judge was not inclined to show mercy. This backdrop sets in context why Mr John Chubbe, a penal surgeon from Ipswich made a special request to receive the criminal corpse.

John Chubbe (1741–1811) worked as both a physician and a surgeon at Ipswich.[84] He trained at Colchester in Essex, spent some time in London, and built a reputation for being a skilled dissector by the time of Benjamin Bush's criminal dissection in 1769. The *Suffolk Garland* highlighted that he was a remarkably forward-thinking doctor determined to do original anatomical research.[85] Access to criminal dissections was essential if he was to prove that his theories had some basis in material reality. Between the 1760s and 1790s Chubbe had pursued three research foci: *A Treatise on the Inflammation of the Breasts of Lying-in Women* (London, 1779); articles published in the medical press on 'the use of digitalis for heart complaints and reviving patients that seemed dead'[86]; and a well-received book on improving sexual health called *An Inquiry into the Nature of Venereal Disease and the remedy made use of to prevent its effects, principally with lotions, unguents, remedies and injections, particularly addressed to young men* (London, 1782) . He had also earned a reputation for skilled operations on the spleen with good survival rates.[87] Tobias Smollett was praiseworthy in *The Critical Review* writing in 1783 that: 'Chubbe supports his opinions by plausible and ingenuous arguments'[88] There was he commented a lot of theoretical opinions about how to 'treat violent gonorrhoea' but Chubbe's originality derived from his anatomical expertise whereby he 'enters into a physiological discussion on the nature of the venereal poison, the structure of the penis, the manner in which it is received, with its progress, and mode of action'. Of course to be convincing Chubbe needed to be well-networked with medico-legal officials from the 1760s when he started to practice surgery. This meant that when criminal corpses like that of Benjamin Bush became available he had a good chance of obtaining the body officially; provided that is medical death had already been established in the vicinity of an execution. The body could then be moved by fast coach across county boundaries. Behind Benjamin Bush's anti-dissection plea was a very real sense that his anatomy would be studied in-depth and dissected extensively including his sexual organs because he was a valuable opportunity cost in a busy medical market.

It was important in many provincial areas for penal surgeons to work together to acquire and redistribute bodies between each other to enhance their credibility through original research. Word of mouth was an important mode of communication in county life. In areas where fewer bodies tended to become available it was vital to get this right. This set of motivations is apparent for instance in the *Transactions Book of the Huntingdon Medical and Surgical Society, 1792–1801* which contains medical cases and dissections of its founding members. Listed in its first printed rules of 1792 are the surgeons 'Samuel Allvey, Francis Hopkinson, John Smith, Joseph Vise, James Smyth, Joseph Michael, Richard Steward, Henry Oliver, Joseph Westbrook' (see Illustration 6.1, page 257).[89] It is noteworthy that rule number VII states that they have all agreed to work together in 'Lincoln, Huntingdon and Rutland, or the City of Peterborough' to petition the local legal authorities in charge of executions to make sure their surgical society got first call on whatever criminal corpses became available at the gallows. They agreed to meet every six months at 'Peterborough, Bourne, Stamford, and Stilton' where each, with the exception of Allvery from St. Neots, had a private practice. Their close co-operation would eventually result in the formation of the Eastern Provincial Medical and Surgical Society 'uniting the practitioners of Cambridge, Essex, Huntingdon, Lincoln, Norfolk, Suffolk and other eastern counties' by 1836.[90] By then, 'some 70 surgeons' had turned up to attend the opening meeting at Ipswich, providing evidence of how the criminal corpse in this wide expanse of the country was a catalyst for future professional ambitions.

Issues of age, gender, and the general condition of the body also impacted on the supply networks that developed between medical men. Yet, those practising surgery in East Anglia, and elsewhere, were all too aware that they had to take what they could get. Ethnicity for instance often determined the opportunity costs of research. In December 1771 it was thus reported in the London newspapers that:

> The Curiosity and Impatience of the People to see the dead bodies of the Jews exposed at Surgeon's Hall in Tuesday was so Great that it was with utmost Difficulty that the Gentlemen of the Faculty [surgeons and their pupils] could gain Admittance...The Professor of Anatomy and Mr Bromfield were obliged to climb in at the Window, to the no small Diversion of the Crowd, which at last became so Great that it was impossible to open the gates to any of the Sheriffs...[91]

Meanwhile, 'Two Eminent Teeth-Drawers of this Town led a scramble for the teeth of the four Jews'.[92] One of the gentleman managed to extract them first but not before the bereaved wife of one of the condemned begged to bury his body before dissection. He had been a doctor and she hoped that the surgeons would have some sympathy for her grieving plight. The widow was rejected on the grounds that the skeletons were needed for original research (Jews seldom being dissected since it was so offensive to their religious rites). It was explained to the relatives that as the crowd were determined to witness justice being done, it would have been foolhardy for the medico-legal officials to stop proceedings. Yet despite the obvious commotion and popular thirst for the spectacle once the standard cuts were made and the teeth stolen, the mob left the building. That so many ordinary spectators turned away from dissection to its extremities was again telling. None of the medical men present asked them to leave, nor on this occasion did the sheriff's men want to trigger a riot by acting with force. The crowd were in control of their curios-

At a Meeting of a Medical and Surgical Society established in the Country in the year 1792, it was Resolved that as the Intention of this Society is to improve its Members in the Practice of Medicine and Surgery, by receiving and communicating Medical and Surgical Information, the following Laws and Regulations be adopted.

I.

THAT each Member of this Society shall pledge himself to assist every other Member of the same, upon all Occasions, and in all Cases of Surgical Operations, or any other Case of Surgery, in which the Safety of a Patient may be affected, or the Character of any individual Member concerned by the misrepresentation of circumstances, without Fee or Reward.

II.

THAT this Society meet twice in the Year for the express Purpose of communicating the different Medical and Surgical Cases which have occurred in their practice, and to arrange such as may be worthy of Publication.

III.

THAT any Gentleman wishing to become a Member of this Society, must be proposed by a Member at one of the Half-yearly Meetings, who is expected then to produce a Thesis, written by the Candidate, upon some Medical or Surgical Subject, which Thesis shall be immediately read and discussed, and the Candidate ballotted for at the next Half-yearly Meeting, when one dissenting voice shall exclude him.

IV.

THAT Peterborough, Bourn, Stamford and Stilton, be the places for holding Half-yearly meetings, that each Place be taken in Rotation, and the day fixed by the Members present at the preceding Meeting.

V.

THAT if any Accident happens to a Member of this Society, so as to prevent him from attending the Duties of his Profession, then the different Members shall alternately assist him as far as they are able.

VI.

THAT in case a Patient is in Circumstances to pay a consulting Surgeon, and requests one to be called in on his own account, then each Member pledges himself to call in one of this Society.

VII.

THAT, as the advantages to be derived from the examination of Bodies after Death must be acknowledged by all, so the difficulty of procuring Subjects in the Country must be equally confessed, therefore in the Case of the condemnation of any Criminal with orders for Dissection, either in the Counties of Lincoln, Huntingdon, Rutland or the City of Peterborough, a petition be presented to the Sheriff or Magistrates in the name of the Society for the Body by some one of its Members, giving a concise history of the plan of the establishment and the advantage which will certainly accrue from the being supplied with the Bodies of Criminals for Dissection.

VIII.

THAT a Book be procured at the joint expence of the Society for the Purpose of inserting Cases, Medical information, the Minutes of each meeting, &c.

IX.

THAT a Secretary be appointed annually at whose House the Book shall be left to insert such Medical Information, as he may have received through the Channel of his Medical Correspondents.

SAMUEL ALLVEY, M. D. President,
FRANCIS HOPKINSON,
JOSEPH VISE,
RICHARD STEWARD,
JOHN SMITH,
JOSEPH WESTBROOK,

Illustration 6.1 © Huntingdon Record Office, Accession, 4715, Image taken 2013 by author of the 'Huntingdon Medical and Surgical Society, Transaction Book, 1792–1801'; Creative Commons Attribution-NonCommercial-ShareAlike 4.0 International License (CC BY-NC-SA 4.0)

ity, and they exited when the 'perils of curiosity' became too much. An excess of accessibility, the unpalatable side of human nature, informed the actions of the crowd and their natural impulses. To be enlightened might have been a powerful inducement to participate in the deterrence value of post-execution 'harm' but it remained nauseating too. According to the *Busy Body* of 1759 this material fact explains why it was that the material afterlives of criminals were so engrossing for the post-execution crowd. Those that could not face dissection were still curious about the criminal destroyed. They thus relied on a material culture to satisfy their inquisitiveness (Illustration 6.1).

The *Busy Body* reporter also elaborated how: 'the ornaments in the Old Bailey and the School of Anatomy at Oxford are the only objects which enable us to form any idea about their [original criminal] appearance'.[93] For many present the corpse became something 'other' and creating a material artefact made it somehow 'alive' in the popular imagination. Regardless of the unspoken reasons behind the timing of the crowd's departure from the dissection of the four Jewish criminals, the balance of evidence suggests that medicine did not simply distance itself from the mob to enhance professional-standing. Something much more complex happened because curiosity carried with it a complex array of emotional reactions and those present were seldom passive. Acquiring knife dexterity; cutting cleanly, working to the body-clock; putting in a good performance; being timely; co-ordinating with crowds of potential patients; appreciating that the social composition of audiences changed during a full-scale criminal dissection: these were all part of the choreography of punishment rites. Most contemporaries knew that it was what happened before the body was cut to its extremities that preserved in skeleton bone, tanned-skin, and plaster-cast material afterlives still identifiable in death. Otherwise most human remains simply disappeared down the drain, as criminal flesh disintegrated into dusty sweepings. There was then throughout our entire chronological focus a curious sense of respect for the dead, even for those that had led dissolute lives, and this came to be refabricated in museum settings too.

REMAINING HUMAN: *FACING A MATERIAL REALITY*

At Worcester Royal Infirmary, ten plaster casts taken from criminal dissections were found in the 1930s (Illustration 6.2).[94] This rogue's gallery makes a powerful statement about belonging and identity for those subject to the medical gaze of the Murder Act. Recently art historians, anthropologists, and archaeologists working in tandem have sought to inves-

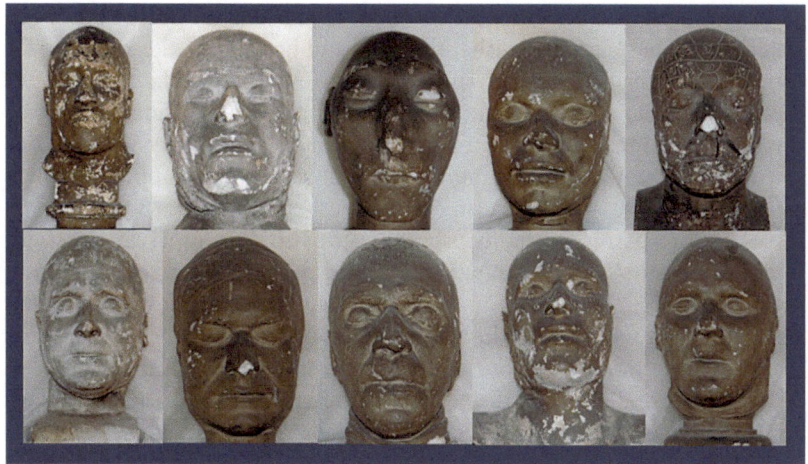

Illustration 6.2   © Charles Hastings Education Centre, Worcester, registered charity in England and Wales number 1074732 whose registered office is c/o John Yelland and Company, 22 Sansome Walk, Worcester, WR1 1LS. In conjunction with the George Marshall Medical Museum, Worcester, criminal death masks rediscovered at Worcester Royal Infirmary in the 1930s. Reproduced here by kind permission of the Trustees under Creative Commons Attribution-NonCommercial-ShareAlike 4.0 International License (CC BY-NC-SA 4.0). See, also http://www.medicalmuseum.org.uk/story/DeathMasksnew.htm

tigate a 'bodily hierarchy' that these heads of criminals and other social outcasts symbolise in a post-modern museum culture. Across Europe, there has been a renewed interest on the 'basis of beliefs, mythologies and traditions about 'a *cultural anatomy of the head*'.[95] A '*headless tradition*' still needs to be brought out of cold-storage to better inform criminal histories for it is apparent that 'headless dissection' was 'saying something different and distinctive from the skull' about the remaining humanity of the condemned. Being decapitated 'exploited the historicity of death' and its *momento mori* traditions by 'scrutinising mortality'. At the same time, at criminal dissections something very literal had to be faced as the head was separated from the torso. Plaster-casts embodied a re-assembling of the criminal-self to refabricate its human identity. There were three abstract things happening in the room when the head cast was taken of a condemned murderer.

Death masks of all descriptions generated from criminal dissections reflected a widespread belief that the head and heart battled each other for supremacy of the early modern mind.[96] At a time when explanations about medical death changed from heart-lung to brain-death taking a head cast was about the implied failure of personal moral responsibility and religious restraint. There was a very tangible medical sense that material afterlives came to represent a malfunction of a political economy that was losing its moral counterpart. Simultaneously there 'was a growing fascination with bodily fragments' and 'decapitation practices' within an 'increasingly strict judicial system'.[97] These cultural trends arose out of Enlightenment sensibilities about whether the head should, or indeed could, rule the heart. Plaster-casts tended therefore to be seen as tokens of punishment with an implicit moral message that when the heart ruled the head dangerous criminality was mindless. Aesthetically modellers had therefore to make a crucial decision about how much of the corpse to feature in a new plaster-cast. What became known as 'collar heads' tended to feature the cranium and shoulders of the condemned; any deep rope marks on the neck were also replicated. Prominent features like a large nose or Adam's apple were usually pressed into the plaster of paris and sometimes wax, since these were clear marks of a dissolute criminal life too. In condemned females, the modeller thumbed into the closed eye-lids, emphasising the visible art-form of lid-watching (see, Chapter 2). These reflected contemporary superstitions about sunken eye-lids being seen as a sign of moral ruin; the eyes being the window of the soul. They depicted in capital-death how the corrupt body-shell had failed to contain the human-spirit needing to return to God to seek redemption for awful misdeeds. Masks were thus seen as prototypes of a corrupt self-portraiture. These the legal process had exposed to moral redress in transformative material mediums. Above all, they exuded 'purity, danger and gender'.[98] And all of this complexity, containing implicit and explicit moral scruples, was reflective of the shared emotional codes and values of an early modern mind-set that came to dissections curious about each criminal's headless afterlife.

The *Liverpool Mercury* on 21 January 1825 was one of many local newspapers to report how the 'new science...of phrenology' was doing its rounds in provincial life and had been the cause of much debate given that dissection seemed to be necessary to know the true anatomy of the criminal brain. Those that practised phrenology were said to take care not to stray from 'the plain path of evidence'. It was a discipline that some claimed combined 'the feelings, dispositions and talents of our particular nature, in the endless and inconceivable diversity in which it appears in human society'. Its advantage was that it reconciled 'the Creator' (*old anatomy*) with the 'material nature of dissection' (*new anatomy*) by showing:

> That the mind is endowed with a plurality of innate faculties – Secondly, that each of these manifests itself through the medium of an appropriate organ, of which the brain is a congeries [conduit] – Thirdly, that the powers of manifesting each faculty bears a constant and uniform relation, *ceteris paribus*, to the size of the organ, or part of the brain, with which it is connected – Lastly, that it is possible to ascertain the relative size of the organ during life, by observing the different forms of the skull to which the brain gives shape.[99]

The argument ran that inside the headless was a mind-map of criminality if someone was trained to read it properly. These 'expert' views were naturally debated and disputed. The Royal Institute of Cornwall twelve months before the Liverpool meeting on 23 January 1824 invited Dr Percival Potts to speak. He travelled down to Truro by the post-chaise with his plaster casts taken from dissections of murderers done at St. Bartholomew's Hospital in London. Potts opened by saying that 'the novelty' of phrenology' was 'inviting, but its fundamental data was incapable of proof'; yet, it had captured the general public's imagination. He was sceptical about this but did express the view that based on his criminal dissection work he could not deny that:

> It does appear that those who have allowed the impulse of the animal propensities and desires, to the commission of crime, have been mostly found to have thick heads and small cerebrums…I have examined that of a murderer in many respects…and found grounds for the opinion I have formed.[100]

At Norwich meanwhile, where there had been a great deal of dispute as to whether criminal dissections should be permitted to take place in the hospital dead-house, by the 1820s the staff had started to accumulate an extensive collection of material afterlives. This included the collar-head of John Stratford executed in 1829 (refer Section 1) sent to the Norfolk and Norwich Hospital Museum where it remains carefully catalogued today.[101] It sits alongside:

> Entry 637 - Normal Anatomy Number 2: Skeleton of Mr [John] Pycraft who was executed for murder [August 1820], Mr Langstaff dissected

Entry 638 - Normal Anatomy Number 3: Skeleton of Mr Johnson who was executed for the murder of Mr Barker of Wells, Mr Langstaff dissected

Entry 1095 - Plaster Cast No. 135, Bust of Shalford executed for murder, Norfolk and Norwich Hospital dissection

Entry 1105 - Plaster Cast 145, Bust of Greenacre who was hanged for murder, Norfolk and Norwich Hospital dissection

Entry 1138 - Plaster Cast 178, Head of J B Rush – executed for the murder of Mr Jermsy [sic] and son of Stanfield Hall. Purchased.[102]

The longevity of these artefacts sets in context why when a renowned physician presented the Academy of Sciences in Paris with a 'piece of artificial anatomy, representing the body of a man according to its natural dimensions' and featuring 'the taking to pieces and putting together again all the various pieces of mechanism in their very fullest details', there was considerable interest in how 'the cranium may be opened and the brain taken out'.[103] The price was '3000 francs' and 'Mr M Ouroux' the inventor of a more interactive anatomical model to replace hand-held books generated considerable publicity in the London press.

In the English dispensary system meantime there was a great deal of interest in using new products other than wax to take casts of criminal heads. So much so that the Society of the Arts of London by 1841 had decided to award its gold medical to 'Mr Simpson...Surgeon to the Westminster General Dispensary' for the 'application of paper máché to the making of anatomical figures and models of morbid anatomy'.[104] In an eulogy at the medal ceremony the orator of the Society recalled that Simpson 'some years ago turned his attention to the difficulty and expense of constructing' reproductions 'in consequence of the difficulty and expense at the time on procuring subjects for dissection'. A praiseworthy newspaper article announcing the prize explained that: 'The materials in use for anatomical models were wax and plaster, of which the former was found to be too expensive to come within the means of lecturers and students in general and was too delicate to be handled in the lecture room without the chance of considerable damage'. The problem with Plaster of Paris casts was that they were 'objectionable too on account of their great weight and brittleness'. Simpson thus perfected using paper máché 'worked into moulds taken from dissections' and these had an 'extreme lightness' when handled and were not easily damaged. They could be 'painted in oil coloured' to represent dissected parts and these were removable revealing the internal structures of notably the brain, heart and lungs. The method soon proved to be a best-seller. Simpson traded on the publicity of his gold medal to promote the 'extreme cheapness' of his paper mâché design. He sold the collar-heads to the Royal Navy and East India Company keen to buy an alternative to wax that often softened in the hotter climate of the Far East. Material afterlives at home and far afield were about refabricating the criminal as public property. They were also about reproducing different price points—cheap paper mâché, more expensive Plaster of Paris and costly wax models—all from the opportunity costs presented to penal surgeons. We end this heady tour therefore with a personal history that brings the physical journey of those that became objects of curiosity in a material culture, full circle.

Dr Thomas Kirkland (1721–98) surgeon of Ashby-de-la-Zouche personifies many of the complexities, contradictions and speculations that criminal dissections afforded from 1760 once the medico-legal choreography of the Murder Act was established in provincial England. He has featured periodically in this book receiving condemned bodies for post-mortem punishment from Leicester gallows in Chapter 2 and then in Chapter 3 he gave critical medical evidence against Earl Ferrers whose method of criminal dissection set new standards on how to cut open the condemned body that became accepted practice in Chapter 4. Kirkland was at his most active between 1760 and 1790 at a critical time in the development of anatomy when its central purpose was reinvented. At first Kirkland did coroner's work before altering his methods from autopsy to anatomization and thence dissection after he became a penal surgeon. From this, he did original research on an eclectic and impressive range of conditions that reflected his patient caseload including: 'inter-cranial pressure arising from a head injury…industrial diseases…lumber abscess… scrofula…fractures…child-birth and puerperal fever'.[105] Thus the *Medical and Philosophical Commentaries* by 1792 announced that Kirkland was recommending 'the utility of Opium in certain species of Apoplexy and Palsy [stroke complaints from brain maladies].[106] If criticised for doing original dissection work on the brain, Kirkland always defended that: '*a peck of practice was worth a bushel of theory*'.[107] He likewise promoted the value of a private medical museum that became renowned in the Midlands.[108]

On Monday 30 August 1824 there was great excitement in Ashby-de-la-Zouche because the Kirkland family had decided the sell their famous medical collection accumulated under the Murder Act. Over four auction days 'upwards of 300 volumes of medical and surgical books in lots' were viewed, and then purchased for '£36 17s 6d' in 16 lots by the end of day two'. This left the contents of the museum to be sold on day four, 4th September. People came from far and wide across the Midlands to buy the eclectic items. Their so-called provenance was attached to some sensational eighteenth-century crimes, as well as the punishment rites exacted on condemned criminals:

Item Number in Sale Catalogue	Auction Prices		
5: Key of Newgate at time of the Rebellion		6s	6d
8: Sir Edward Verrey's Skull, Slain at Edgehill		9s	0d
9: Part of an Egyptian Mummy		1s	0d
29: Stuffed Birds, Beasts and Reptiles in glass cases	£25	13s	5d
30: Dresses and Shoes worn by several Dwarfs		7s	6d
37: Curious articles—includes Dick Turpin's rope	£1	0s	0d

Other curious items such as Earl Ferrers' rope, the Gunshot ball that shot his steward, skeletons from Criminal dissections and the entire Produce of the Museum

The auctioneer lists—Produce of Museum	£75	2s	7d
Expenses for the Sale Staff	(£10	8s	7d)
Total	**£64**	**14**	**0d**
First Day	£198	0s	6d
2nd Day	£36	17s	8d
3rd Day	£31	5s	7d
4th Day	£75	2s	3d
Total expenses	(£98	8s	5d)
Total Profit from entire Sale	**£242**	**17s**	**7d**[109]

In Chapter 2 we encountered the skeleton of Timothy Dunn sent for cleaning in January 1796. At the time of the medical museum sale it could not be located. It was to have featured as a star-item at the auction but it had taken on a material afterlife of its own. Kirkland had paid for the skeleton to 'be boiled down'. It was then sent back to be viewed by local people at the White Horse public house in the town where Kirkland had exposed the condemned body. Thence it was 'used for the furtherance of medical knowledge, passed from one doctor to another' in the Midlands. Having gone on its medical travels it eventually went missing in the 1820s when criminal dissections moved to purpose-built infirmaries in central England. Eventually it 'was found in a cupboard at 14 South Street Ashby-de-la-Zouche in 1930', but, by then, it was regarded with opprobrium and 'subsequently thrown down a well' in the town where by all accounts it disappeared from public view. As for the contents of the Kirkland medical museum that once occupied 63 Market street in Ashy-de-la-Zouche (the building was demolished in 1868), head-strung criminals were distributed far and wide, to be sold and resold, collected and dispersed over the next three generations. Even then, what remained became folklore. The 'public curiosity' that started with a crowded dramatic performance was refashioned into the tale of Thomas Kirkland that once did unsavoury criminal work and was Earl Ferrers' fatal undoing. The 'dead-end' of curiosity was seldom the end of the storyline.

CONCLUSION

The subject of the body divided is a familiar one in modern biomedicine.[110] Yet it was the afterlives of criminal dissections created by the capital code that were the start of this powerful narrative in a history of punishment. Those in the crowd absorbed material realities because they had ample opportunities to be confronted by the headless, gutless, and faceless creatures of a 'new anatomy'. By the end of dissection, when arteries were typically empty of blood (because blood pools in the veins after death) ordinary people could speak in some sense about the curious things happening. That discourse had a transitional language because cutting the corpse was about debasement of the condemned torso, defacement of the visage including the eyes once seen as the window of the soul, and defilement of the heart and mind. In other words, all the traditional seats of humanity and spirituality that 'old anatomy' had once served in a scheme of divine creation were being turned upside down by different sorts of emotional conversations in the name of crime, justice and science under the Murder Act. Predictably perhaps being involved with criminal dissections created the narrative subtext for disconcerting discussions about the familiar, but abnormal. In a strong oral culture these involved talking emotionally in some capacity about 'normal curiosity', 'public curiosity', and 'morbid curiosity'; so much so, that the 'perils of curiosity' expressed a heightened sensitivity. There was more body-awareness, and the great irony was that it took the morally deviant to trigger this level of *emotives*: a self-contradictory paradox perhaps during the Enlightenment project that promoted reason and rationality as its central *raison d'être*. There was the option of remaining silent but it is hardly credible to claim that all those that beat down the door to get inside a dissection venue staged their collective protest without emoting a single huff, puff, grunt, or word. It was likewise possible to be impervious but not likely given the synaesthesia stimulated in everyone. Accepting then that pushing and pulling to be on the inside gave voice to the archaeology of emotions considered normal at the time, makes it very necessary to revise an historical sense of the internal dynamics of the early modern crowd.[111] Post-execution the potential existed for emotional expressions to be self-exploring and self-altering, in ways that William Reddy has identified.[112] Those attending criminal dissections did not simply describe events but retold punishment dramas with human sentiments that were all about the essential quality of the immersive theatrical experience. Emotionally-speaking there was little point in going to a punishment rite if it was boring, banal, or baffling: the majority were 'out of body experiences', not in a modern quasi-spiritual sense, but in early modern terms of the staging of a public phantasm of criminal fleshiness to be visually consumed, materially composed, and emotionally recycled.

Recurrently there was a medical gaze at criminal dissections, but it was a very variable viewpoint. Audience members were repopulated over three to four days, peopled by those that came and went, and returned to see the body being punished post-execution. Not only was there a material reality to be faced as the body disintegrated but by preserving criminal heads for study a macabre sort of self-hood was recreated too.[113] This was not simply public engagement on the part of the anatomical fraternity but an act of co-creation in which the crowd accepted that what was being represented by anatomical modellers—whether in wax, Plaster of Paris or *paper maiche* in a collar head—was 'real', '*the dangerous dead*' and 'somehow alive though silenced forever'. This contemporary discourse was influenced by the quasi-sciences of galvanism and phrenology when electric impulses and nervous energy from brain dissection filled the punishment venue. Moving lips twitching seemed to make the headless speak and this set of experiences was enhanced by a sense of curiosity arising out of secrecy. Whichever sort of material afterlife was refabricated its artefacts were intimately related to research speculations that criminal dissections made feasible. Some of the work was perfunctory, some intriguing, some thought-provoking; each commodity was part of a mosaic of medical speculation and professional development. Anatomy under the Murder Act often resembled a macabre showcase or public drama of the unsavoury, as it had been for centuries. Yet, it is important to keep in mind that it was something far more intangible and curious too by the 1790s. Developing from a shared understanding that medical death was indefinite and had infinite varieties, the crowd and everyone present experienced something that every human being has always shared, and will arguably always do so. The journey from life to afterlife for many ordinary people generally leaves behind an emotionally-charged anatomical legacy of sorts.

NOTES

1. Anon. (1770) 'On the Dissection of a Body', *The Universal Magazine* (August issue), Vol. 46, 98–9.
2. See, Section 2, where this original phrase is elaborated later in the chapter.
3. I am grateful to Dr Heather Shore current Head of Anatomy at Leicester University Medical School for permitting me to interview her about best practice in dissection sessions with new medical students during the formative stages of writing this book. To maintain human dignity at all times, no more than one third of a donated body is dissected. There is then an annual service of commemoration and thanksgiving for the gift of each body to medical education.
4. Refer, E T Hurren (2011), *Dying for Victorian Medicine: English Anatomy and its Trade in the Dead Poor c. 1832–1929*, (Basingstoke: Palgrave Macmillan).
5. West Yorkshire Archive Service, Bradford Office, SpSt/11/5/1/2, William Hey Correspondence, Letter from Hey to William Stanhope, 21 May 1785.
6. 'Execution of Thomas for Murder', *The Times*, 31 August 1822, Issue 11339, p. 2.
7. All original quotes appear in Ernest Reginald Frizelle (1988), *The Life and Times of the Royal Infirmary at Leicester: The Making of a Teaching Hospital 1766–1800* (Leicester: Leicester Medical Society), chapter XXI, pp. 275–6. The book is based on primary sources located at Leicester Record Office [hereafter LRO], Leicester Infirmary Minute Books [hereafter LIMB], 17 December, & 29 December 1807, 8 June 1819, and 19 December 1822.

8. Frizelle, *Royal Infirmary*, p. 265, and discussed in LRO, LIMB, 20 August 1816, 8 April 1817, 3 August 1817, and 10 August 1817.

9. This sets in context why LRO, LIMB, 23 March 1822 states that: 'It being resolved unanimously that the practice of exhibiting the bodies of the unfortunate persons given to the Infirmary for dissection appeared to us improper and, that in future, no such exhibition is permitted'. The dissection room door was being closed in principle, though not necessarily in practice, as we shall see later.

10. On fatal medical accidents in the Darwin family history, see, 'On the Life and Writings of Erasmus Darwin' (1822), *The London Magazine*, (December issue), 520–9, medical death by dissection at p. 522.

11. Editorial feature (1788), 'The Life of Charles Darwin 1758–1778', *Medical and Philosophical Commentaries*, Vol. 5, 329–36.

12. Thomas Alcock (1827), *An Essay in the Use of Chlorurets of Oxide of Sodium and Lime as Powerful Disinfecting Agents and of the Chloruret of Oxide of Sodium more especially as a Remedy of Considerable Efficacy in the Treatment of Gangrene, Phagedenic, Syphilitic, and Ill-Conditioned Ulcers, Mortifications and Various Other Diseases dedicated to the Right Hon. Robert Peel* (Derby and London: Burgess and Hill), p. 14.

13. Ibid., p. 15.

14. See, Alcock, *Use of Chlorurets*, section 'On the Prevention of Putrefaction in conducted private Anatomical Studies', pp. 16–20.

15. See, Elizabeth Barry (2008), 'From epitaph to obituary: death and celebrity in eighteenth-century culture', *International Journal of Cultural Studies*, Vol. 11 (No. 3), 259–75, in which she argues that many of the attributes of modern fame can be identified as starting in the eighteenth-century when death became 'news' and capital punishment captured the headlines. Yet, the physical obituary of a criminal dissection happening before the crowd and its medical epitaph is seldom studied as an intrinsic aspect of the eighteenth-century fame culture.

16. See, Alcock, *Use of Chlorurets*, p. 21.

17. Ibid., pp. 21–2.

18. See, Alcock, *Use of Chlorurets*, p. 11.

19. Refer, 'Execution of A. and M. McKeand for the Murder at Winton', *The Observer*, 27 August 1826, pp. 4–5.

20. Refer, 'Execution of the Keands', *The Manchester Guardian*, 26 August 1826, pp. 1–2, quote at p. 2.

21. On his career, see, The National Archives [hereafter TNA], *BPP*, (1812–3), V, *Report of the Commissioners on the State Lancaster Prison and the Treatment of Prisoners Therein*, p. 34, and later printed as *State of Lancaster Gaol* (1812) (London: Lords and Commons Reports), entry 3 July 1812, pp. 897–8.

22. See, for instance, William M. Reddy (2001) *The Navigation of Feeling: A Framework for the History of Emotions*, (Cambridge: Cambridge University Press), p. 96.

23. 'Murder at Winton: Execution of the Keands', *The Age*, 26 August 1826, p. 542, Issue, 68.

24. Nottingham County Record Office, BB34.8, 'Gallows Rememberancer for 1779', pp. 47–8.

25. Neil Kenny (2004), *The Uses of Curiosity in Early Modern France and Germany*, (Oxford: Oxford University Press), especially pages 1–4, long quote at p. 1.

26. Ibid., p. 2.

27. Kenny, *Uses of Curiosity*, p. 4.

28. Ibid., p. 13.

29. Kenny, *Uses of Curiosity*, p. 159.

30. See, E. T. Hurren (27 July 2013), 'The Dangerous Dead: Dissecting the Criminal Corpse', *Lancet*, Vol. 382, Issue No. 9889, pp. pp. 302–3, on the dissections of William Hey at Leeds.

31. Reconstructed from, Leeds University Special Collections. MS 504/1/3, Manuscript of Mr William Hey, Senior Surgeon to the Leeds General Infirmary.

32. See also S. T. Anning and W. K. J. Walls (1982), *A History of Leeds Medical School* (Leeds: Leeds University Press).

33. Erasmus Darwin (1803), *The Temple of Nature or The Origins of Life* (London: Thomas Bensley Publishers), Cant. I. 1, 250, a poem published the year after his death.

34. Derbyshire General Infirmary was opened in October 1810 at a cost of £30, 000 raised by local subscriptions. It had two day rooms for convalescents, a fever house for infectious diseases, and could accommodate 80 patients on its main medical wards. They were attended by 'three physicians' (Bent, Fox junior and Baker), 'four surgeons' (Godwin, Wright, Douglas Fox, and another rotating surgeon) and a 'house apothecary' (Richard Dix), as well as Richard Forrester Forrester [sic] physician who acted as semi-retired 'consulting physician', see, Daniel and Samuel Lysons (1817), *Magna Britannia: Volume 5: Derbyshire*, (Derby: Walker and Co), pp. 94–129.

35. See, detailed letter on the dissection to the Editor of the *Medical and Physical Journal*, Volume 18, p. 157.

36. Darwin was enormously over-weight and probably suffered a stroke and collapsed lung caused by his obesity. At 24 stone, he usually sent ahead his assistant to check that when attending a private patient the wooden floor of their house could take his weight.

37. S. Glover and T Noble eds. (1829), *A History of the County of Derby Volume II*, (Derby: Henry Mozley and son) p. 593.

38. William Hutton (1817), *The History of Derby* (London: Nicols and son and Bentley), p. 202, written in 1791 but published in 1817.

39. Francis Fox senior had in fact been a prominent physician in Derby and bequeathed his business dealings to Francis junior who appears to have been related to Douglas Fox the surgeon by marriage. See, TNA, PROB/11/1212/32, Will of Francis Fox, gentleman of Derby, 2 December 1791.

40. Erasmus Darwin (1803), *The Temple of Nature or The Origin of Society*, "Additional Notes II, The Faculties of the Sensorium" to accompany the poem (London: Thomas Bensley Publishers).

41. Ibid., Additional Notes II, I-VI, explain how in the head there might remain 'a certain quantity of sensation' on the cusp of life-death in the brain's 'sensorium'.

42. Refer, C. U. M. Smith and Robert Arnott (2005), *The Genius of Erasmus Darwin*, (Farnham, Hampshire: Ashgate), on how Darwin's interest in resuscitation and galvanic experiments may have directly influenced Mary Shelley's *Frankenstein* (London, 1818).

43. See, endnote 34 above and Letters, 'To the Editors', *The Medical and Physical Journal*, Vol. 18, pp. 309–14.

44. Ibid.

45. Refer previous chapters and for a recent reappraisal, I am grateful to Shane McCorrsetine (2015) for sharing with me his new manuscript now published as, *William Corder and the Red Barn Murder: Journeys of the Criminal Body* (Basingstoke: Palgrave Pivot).

46. LRO, LIMB, LRO/13064/13, 23 June 1814–30 June 1819, entry dated 17 November 1815.

47. See, by way of example, Stanley Finger (2000), *Minds Behind the Brain: A History of the Pioneers and their Discoveries* (New York: Oxford University Press) and Carl Zimmer (2005), *Soul Made Flesh: The Discovery of the Brain and How it Changed the World* (New York: Simon and Schuster, Free Press).

48. LRO, LIMB, 23 March 1822.

49. As featured in, Frizelle, *Royal Infirmary Leicester*, p. 264.

50. 'Murder Near Manchester' *The Age*, 4 June 1826, p. 443, Issue 56.

51. Ibid.

52. 'Execution of the Keands', *The Manchester Guardian*, 26 August 1826, attached to which there was a sketch of the murderers being apprehended outside an Inn which warned readers not to believe that it was a good resemblance but was very defectively drawn.

53. See, Anon, (1819), *A Full Account of the Most Horrid Murder Committed by Thomas Weems or Weyms a Miller, Upon the body of his Wife by Strangling*

Her with a Garter near Puckeridge Hertfordshire (London: Charles Piggot publisher of Clerkenwell).

54. 'Execution of Weems and Galvanic Experiments upon the Body', Letter To the Editor, *Morning Chronicle*, Saturday 14 August, 1819, column 2, 15691.

55. 'Execution of Weems', *Cambridge Chronicle and Journal*, 13 August 1819 issue, carried a full account of the experimental stages.

56. Ibid.

57. Okes was the gaol surgeon at Cambridge Castle on a salary of £20 in 1819. By 1833, he worked at the Old Addenbrooke's Hospital on Trumpington Street near the Anatomy Theatre where he delivered medical lectures on surgery to a new school of medicine founded on the hospital site. He was described as a 'surgeon of high character' (1815) in the *Annals of Cambridge*, (Cambridge: Warwick and Co), p. 525.

58. 'Execution of Weems and Galvanic Experiments upon the Body', Letter To the Editor, *Morning Chronicle*, Saturday 14 August, 1819, column 2, 15691.

59. Members of the Okes doctoring family around Cambridge were also intrigued by original research on nerve and spinal complaints, branching out to work on spina bifida in 1812. One wrote that he was 'opposed to puncture or pressure' of the malformed spine based on what he had seen from bodies in fatal trauma, see, 'Intelligence Spina Bifida' in (1812) *The New England Journal of Medicine and Surgery and the Collateral Branches of Science conducted by a Number of Physicians, Volume 1*, (Boston: Wait & Co), pp. 98–102, q. at p. 101.

60. All the letters cited are part of a collection now retained at East Sussex Record Office [hereafter called ESRO], The Archive of the Frewen Collection, Family of Brickwall in Northiam [hereafter referred to as FC], more especially, ref. 2567–2658, Letters from Dr Thomas Bishopp of Leicester, c.1792–1811, and others cited per letter item.

61. ESRO, FC, FRE/2576, 7 October 1792, Thomas Bishopp to John Frewen.

62. ESRO, FC, FRE/2568, 29 November 1792, Thomas Bishopp to John Frewen.

63. ESRO, FC, FRE/2572 & FRE/2575, 10 April 1794 & 26 October 1794, Thomas Bishopp to John Frewen, reviews his dissection activities since 1792 and of late.

64. ESRO, FC, FRE/2572, 10 April 1794, Thomas Bishopp to John Frewen.

65. ESRO, FC, FRE2576, 7 November 1794, Thomas Bishopp to John Frewen.

66. ESRO, FC, FRE/2598, 24 October 1797, dated by post mark, Thomas Bishopp to John Frewen. Dr Chessner according to other letters was often jealous of his pupils, whether this was hearsay or based on personal experience is not clear, but see, FRE/2578, 14 Jan 1796.

67. ESRO, FC, FRE/2586, 6 March 1797, Thomas Bishopp to John Frewen.

68. Edward Le Grand died suddenly at Canterbury in 1797. He was described in *The Monthly Magazine or British Register Volume 3* for 1797 as 'a promising young man, whose endowments would have done honour to a riper age', but sadly no more, p. 169.

69. ESRO, FC, FRE/2658/97, no date, watermarked 1802, Thomas Bishopp to John Frewen.

70. ESRO, FC, FRE/2608, 2 June 1806, Thomas Bishopp to John Frewen.

71. ESRO, FC, FRE/2658/14, undated but likely to pre-date 1807, Thomas Bishopp to Frewen.

72. ESRO, FC, FRE/2603, 3 March 1804, Thomas Bishopp to John Frewen: the latter was taking the water cure at Bath staying at 7 South Parade and was very interested in medical fashions and the general business climate.

73. 'Middlesex Sessions', *Morning Post*, 15 September, 1819, Issue 15181, column 5.

74. 'Case of *King versus Cundice* (1822), Surrey Lent Assizes: Selling the body of a convict for dissection', *The Observer*, 8 April 1822.

75. Ibid.

76. See, endnote 73 above.

77. Norwich Record Office [hereafter NRO], NNH 1/6, Norwich and Norfolk Hospital Minutes 12 July 1794–13 July 1799, entry 12 August 1797.

78. NRO, NNH 1/6, Norwich and Norfolk Hospital Minute Books, 12 July 1794–13 July 1799, quoted entry 19 August 1797.

79. Charles Mackie (1901), *Norfolk Annals* (Norwich: Kessinger Publishing), entries 1811, p. 78.

80. Ibid., p. 74.

81. Mackie, *Norfolk Annals*, p. 79.

82. Ibid., p. 157.

83. Case reconstructed from *Lloyds Evening Post*, 28 July 1769 (a) and (b).

84. See, *Lloyds Evening Post*, 16 September 1769 (b).

85. James Ford (1801), *The Suffolk Garland* (Ipswich: John Row), pp. 180–1.

86. See for instance, *The London Medical and Physical Journal* (Volume 4), p. 137.

87. John Bensusan Butt edited by Shani D'Cruze (2009), *Essex in the Age of Enlightenment : Essays in Historical Biography* (Colchester: e-publisher lulu.com), p. 126.

88. Tobias George Smollett (1793), 'Book Review', *The Critical Review or Annals of Literature*, Vol. 54, p. 155.
89. Huntingdon Record Office, Accession 4715, Huntingdon Medical and Surgical Society, Transactions Book 1792–1801, acquired 2000, accessed 2013.
90. Announcement (1836), 'Eastern Provincial Medical and Surgical Society', *Medico-Chirurgical Review and Journal of Medical Science*, Vol. 28 (January), p. 166.
91. *Public Advertiser*, 14 December 1771.
92. Teeth were often extracted and sold on by barber-surgeons who did criminal dissections especially in London and Newcastle. In the capital John Hunter (1778) *The Natural History of Human Teeth I (with William Coombe)*, (London: J Johnson) stressed that good teeth were essential for healthy digestion. They soon became a valuable medical commodity, replacing bad teeth with a healthier set of criminal teeth. Newcastle barber surgeons concurred, selling criminal sets of teeth from their Surgeon's Hall and expanding dentistry in the vicinity by the 1770s. One Newcastle barber surgeon named Charles Edward Whitlock even ran a dentistry practice and a company of actors from the Hall to maximise business profits, see, John Boyes (1957), 'Medicine and Dentistry in Newcastle Upon Tyne in the Eighteenth-Century', *Proceedings of the Royal Society of Medicine*, Vol. 50 (April), 299–335.
93. *Busy Body*, 25 October, 1759.
94. Accessed at, http://www.medicalmuseum.org.uk/story/DeathMasksnew.htm.
95. Catrien Santing and Barbara Baert eds. (2013), *Disembodied Heads in Medieval and Early Modern Culture* (Oxfordshire: Brill Publications), p. 9.
96. A point well-made for instance in Ole M. Høystad (2007), *A History of the Heart* (London: Reaktion Books).
97. Santing and Baert, *Disembodied Heads*, p. 7.
98. Ibid., p. 12.
99. 'Phrenology lecture by Dr Cameron,' *Liverpool Mercury*, 21 January, 1825, Issue 713, column 3.
100. 'Royal Institution of Cornwall – Phrenology Speaker', *Royal Cornwall Gazette, Falmouth Packet and Plymouth Journal*, Truro, 3 January, 1824, Issue 1071, page 1, column 1.
101. Mackie, *Norfolk Annals*, entry 14 August 1829.
102. NRO, NNH 29/2, Catalogue of the Norfolk and Norwich Hospital Museum, where N indicates Normal Anatomy, P is a Plaster Cast.
103. 'Important Anatomical Invention', *The Morning Chronicle*, Saturday 2 April 1824, Issue 17460, page 1, column 1.

274 E.T. HURREN

104. 'Anatomical Models', *Manchester Guardian*, 26 May 1841, Issue 1841, p. 3, column 3.
105. R T Austin (1986), 'Dr Thomas Kirkland of Ashby-de-la-Zouche: 1721–98', *British Medical Journal*, Vol. 293 (25th October), 1075–6.
106. Announcement of Dr Kirkland's latest original work in (1792) *Medical and Philosophical Commentaries*, Vol. 16, p. 396.
107. Cited in a review of the career of Mr William Chessher, a fellow surgeon, in (1832) *The Annual Biography and Obituary for the Year of 1832*, Vol. 16, (London: Longman), p. 401.
108. Refer, TNA, PROB 11/1310/135, Will of Thomas Kirkland, Doctor of Medicine, Doctor of Physic, Ashby-de-la-Zouche, 21 July 1798.
109. LRO, P222, Photocopy of Catalogue Number 2222—*Copy of the Sale of Mr Thomas Kirkland's goods and Museum upon his death—sold on the premises by the family of Kirkland senior and junior- on the death of the latter in 1824.* The profit was a decent 3-year living for the Kirkham family.
110. See, Sarah Ferber and Sally Wilde eds., (2011), *The Body Divided: Human Beings and Human 'Material' in Modern Medical History* (Farnham, Surrey: Ashgate).
111. Refer, Sarah Tarlow (2012), 'The Archaeology of Emotion and Effect', *Annual Review of Anthropology*, Vol. 41, 169–85.
112. Again, see, Reddy, *The Navigation of Feeling* & his (2009),'Historical Research on the Self and Emotions', *Emotion Review*, I, 302–315.
113. A point made forcibly by Arjan R de Koomen 'The Self-Portrait 'En Décapité: Interpreting Artistic Self-Assertion' in Santing and Baert *Disembodied Heads*, chapter 7, pp. 191–222.

CHAPTER 7

'He that Hath an Ill-Name Is Half-Hanged': The Anatomical Legacy of the Criminal Corpse

Basic biology informed the legal limits of capital sentencing under the Murder Act in 1752.[1] Human beings sentenced to death shared physical traits when hanged. These had a biological timing that determined how to establish medical death. After which, the criminal corpse underwent post-mortem rites. By devoting so much scholarly attention since the 1960s to developing a theoretical apparatus to speculate about the crowd, embodiment and criminal justice in the eighteenth-century, ironically actual criminal bodies about to be dissected were left unattended. Once these were cut down from the gallows, they were neither organically stable nor individually invariant, but they did all have flesh, skin, blood, bones, brain matter, and viscera. Those corporeal features often appear in standard historical accounts but not in terms of actual medico-legal rituals that reconstruct punishment journeys from anti-mortem to post-mortem results. The 'historized body' still needs to be relocated in criminal histories and the medical humanities to reach a revisionist standpoint.[2] This book has therefore approached the condemned in distinctive ways by trying to envisage another side of the criminal justice process. One that was embodied upside down, outside inside, dehumanised to be refabricated,

© The Author(s) 2016
E.T. Hurren, *Dissecting the Criminal Corpse*, Palgrave
Historical Studies in the Criminal Corpse and its Afterlife,
DOI 10.1057/978-1-137-58249-2_7

275

in ways that resemble the image opposite (Illustration 7.1) of a flayed criminal
corpse drawn from a different angle on
a dissection table in 1815.[3]

After the Murder Act, historical
actors who attended executions and
criminal dissections both internalised
and externalised their 'natural curiosity'.
They had sensory agency to see and hear
a biological soundscape; to stare, gasp,
gossip, joke, and half-remember conversations amongst post-execution crowds
about how it made them feel to be part
of the spectacular. Lacking first-hand
evidence penned by ordinary people in
an oral culture, this book has explored
the physical qualities of a shared emotional history at criminal dissections
that was accessible and stimulating
in some respect to everyone. Seeing
something reachable did not necessarily make it a reductionist experience
in terms of medicine wielding power.
Instead when the criminal body was
about to be dissected, uppermost was
not the pathology of death. It was a

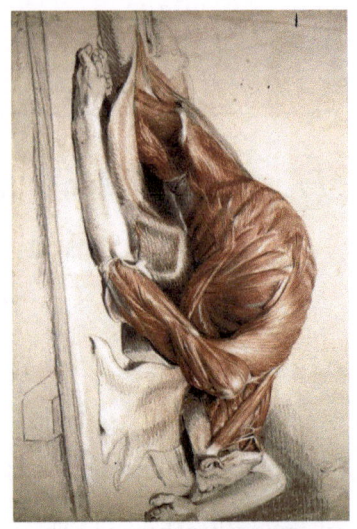

Illustration 7.1 © Wellcome Trust,
Image Collection, L0013340,
'*Lateral view of the trunk of a flayed
corpse*', by Charles Landseer, 1815;
Creative Commons Attribution-
NonCommercial-ShareAlike 4.0
International License (CC BY-NC-
SA 4.0)

head, hand, foot, and arm resembling fleshy normality that was danger-
ously deviant. The enticement to see a bloody mess made the picture of
punishment somehow 'alive'. This was the punitive price of being in close
proximity and those penal surgeons working in London knew it very well,
as the *Morning Chronicle* reported on 28 May, 1776: 'Yesterday being a
holiday for working men and apprentices great multitudes attended the
execution of the two men at Tyburn for murder, and not content with this
sight, great numbers returned to see them dissected at Surgeon's Hall'.[4]
The adjective used to describe what motivated the crowd is deliberate—it
was 'this sight'—a mental picture that animated an execution spectacle
and its equally spectacular post-mortem encore. Rethinking the material synaesthesia being staged, Adam Bencard has thought-provokingly
speculated:

There is a possible way in which the study of the body in history might be opened towards more direct understandings of embodiment...This notion of presence presents a possible way of refiguring the study of the body in history which embraces the inability of language to fully frame our interaction with reality and the importance of examining the unspoken, the unconscious, the sensual, the affective, the lived and the felt "stuff" of the body in history... presence is a result of our biographies, of history within the flesh. Presence does not transcend history, it is history; history as it works upon us, within and without language, both as individuals and as societies. Presence, then, offers a possible way of speaking about, examining and appropriating into the study of history the fleshy experience of history and the historical experience of fleshiness.[5]

Yet fresh flesh approaching organic death was kept animated in the popular imagination by the physical presence of the crowd and their human impulses that found expression in the criminal justice process. In this book liminal experiential spaces have been relocated in the archives to explore what William Reddy terms '*emotives*'—that is, a contemporary navigation of feelings, implicit and explicit, revealing emotional characteristics embedded in curiosity-driven embodied experiences at criminal dissections.[6] These the English state sought to harness with mixed-results, for there is little doubt that material stories were powerful tools in the hands of eighteenth-century medico-legal officials; nevertheless their cultural reception was difficult to control or predict for the forces of law and order. Arriving then at an empirical appreciation of punishment rites is about developing a more nuanced historical awareness of just how much post-mortem retribution was powerfully in transition once the Murder Act was enforced. Accepting that the anatomical sciences were modified—literally and symbolically—and went on being so—means taking a fleshy step-by-step approach; finding out precisely how medico-legal mechanisms were choreographed and where exactly, so that contemporaries could accumulate knowledge of post-mortem 'harm'. Potentially this was a discretionary form of popular justice at punishment events for everyone in Georgian life, a status quo that involves disembowelling historical clichés. Taken together six distinctive contributions encourage scholarship to look anew at 'the historized body' as we arrive at the end of a punitive enterprise and engage with its anatomical legacy.

This book, first and foremost, has stressed the vital importance of going back to the archives to better understand the biological basis of execution and post-mortem punishment in tandem. This needs to happen before historians can begin to re-assess the complex historical processes in a 'history of the body' deployed for socio-political and medico-legal ends after the Murder Act. In eighteenth-century criminal studies the intellectual baggage of academic life was loaded up onto a cartload of 'historized bodies' from the 1960s. Ironically researchers seldom thought anatomically about the criminal bodies they claimed to have rescued from the gallows.[7] As a theoretical corpus accumulated, its intellectual credentials seemed very persuasive even though few criminal studies examined the materialism of punishment taken to its logical conclusion. It is still rare to be able to engage with an organic instability that is inescapable in all human beings condemned to die, regardless of class, gender and ethnicity. This is what made Vic Gatrell's work exceptionally creative, and which this book has sought to encompass by going to the hanging tree and journeying with the condemned body until the punishment choreography was completed.[8] Admittedly, it is 'reductionist' in the history of medicine to start with a body to punish (its organic processes, confusing medical death, cut into pieces, subjected to original research, and refabricated for cultural consumption). Yet, in every chapter being 'essentialist' has been very necessary when the material basis of a criminal history is flesh, blood and bone. What then Michel Foucault felt so uneasy about in *Discipline and Punish* (1977)—an emphasis on 'the purely biological basis of existence' that had served 'historical processes'—turns out in this book to be a fundamental step in reappraising what it meant to take life, when, where and how to punish in death after the Murder Act.[9] Only then can a researcher attempt to do what Roger Cooter terms, 'interpret, problematize and destabilise...knowledge /power creation'.[10] Keeping that historical process 'alive' means starting at its 'dead-end'. To be 'anti-essentialist' (an intellectual position Foucault aspired to), entails first being 'essentialist'. A related contribution has been the ability to locate forgotten and overlooked sources like the Sherriff's Cravings at the National Archives and put them together with familiar sources such as newspaper accounts and court records in more creative ways. For the first time this study has handled the actual body material of a criminal history that has contained too many misleading medical oversimplifications. This sort of methodological advance moreover strongly suggests that the balance of the evidence presented supports an intellectual position that French thinkers perhaps should have taken literally in writing about the history of capital punishment, *'reculer pour mieux sauter'*.[11] In a history of the body it is sometimes indispensable to '*step backward in order to leap farther forward*'.

Having done so, a second major finding in this book is that the conundrum of medical death mattered a lot to the forces of law and order in early modern society. It is a crucial missing part of a medico-legal jigsaw puzzle that became the working choreography of the Murder Act. It is incongruous of criminal historians to have been so concerned with the wording of the new legislation—seeing it as substantially correct—that they neglected to question whether there was any material substance to the letter of the law after 1752. A related issue is that what has been misread as a catch-all medico-legal penalty—'*dissection and anatomization*'—was nothing of the sort. It has been mistakenly elided into a single, linear, post-mortem punishment rite, refuted from Chapter 2. The historical insights provided by Jonathan Sawday that the early modern body was 'a locus of all doubt' and Thomas Laqueur that 'becoming *really* dead... takes time' in the eighteenth-century, have been invaluable when researching the 'half-hanged', dying, and '*truly* dead'.[12] Again, the balance of the medical evidence presented indicates that although '*dissection and anatomization*' were interchangeable in the history of the body for centuries, under the Murder Act they came to mean something different, differential, and distinctive. By the time of Earl Ferrers' execution in the 1760s '*anatomization*' was '*splanchnology*'; it redefined the working definitions of the discretionary justice in the hands of penal surgeons. And that new anatomical procedure was one of seven types of post-mortem 'harm' that could be done for official purposes to the criminal body. A 'crucial incision' was a legal checking-mechanism for medical death once the condemned was moved from hangman's gallows to medical jurisdiction. This meant that '*anatomization*' did not simply mean '*dissection*', and it is erroneous to claim so in crime or cultural studies. They often took place in different medical venues, emphasising their separate punishment roles. The medical stipulations of the Murder Act are a classic case of what Joanna Innes calls an eighteenth-century statement of legal redress that was to be refined when the law was applied in provincial and metropolitan life.[13]

Throughout England, the history of early modern anatomy in relation to crime and justice became a very British enterprise. Penal surgeons adapted to fluid forms of retribution for murder because how to punish was in transition in the change-over from 'old' to 'new' anatomy by the 1790s: a time Andrew Cunningham has described as 'seismic'.[14] *Lex talionis* had considerable moral authority, especially if the method of homicide was brutal and violent, or indeed was a copy-cat killing that mirrored execution or dissection methods. *Kill and be killed* was a popular mantra that the medical profession carefully aligned with to avoid any further bad publicity on a troubled path to establishing a professional status quo. In all of this making and remaking of criminal justice from the margins or at the centre, the body was privileged because it had to be so. It was never inert or value-neutral. Yet, in archive searches for popular mentalities, medical discourses, and political representations of fleshiness, one finding remains constant. The timetabling of medical death by medico-legal officials was a scientific riddle in the Enlightenment. Recalibrating that finding in the history of crime and punishment involves questioning the timing of capital punishment *per se*. It is therefore worth keeping in mind that a distinguishing feature of scientific thinking is the search for falsifying, as well as confirming evidence. At times in the history of science, doctors and scientists have resisted new discoveries by selectively interpreting or ignoring unfavourable data. Historians of crime have seldom reflected that it is ahistorical to follow suit. If the material evidence for medical death was ambiguous in 1752 (and remained so by 1832) then relying on that ambiguity to support a statistical position that the criminal was a corpse when cut down from the gallows has created a 'confirmation bias' that has not served a history of punishment well. Penal surgeons knew they were sometimes in an ethical quandary of human vivisection—in one third of documented cases from Surgeon's Hall in London even by the 1810s—and therefore likely to be a much higher percentage of cases—perhaps as many as up to a third—before the introduction of the 'new drop'. That degree of medical complicity can only be investigated thoroughly by developing historical antennae more receptive to what penal surgeons said *and* did under the Murder Act.

A third contribution is that executed bodies that were left unattended at the gallows by crime histories have been repositioned in this book, transposed from anti-mortem to post-mortem settings. There were four official step-changes in the medico-legal status of a homicide culprit convicted under the Murder Act. Figure 7.1 (below) thus provides a more sophisticated paradigm of punishment for the world of eighteenth-century England. Procedurally, to this, it is essential to add another element when material afterlives were created.

Again, it is mistaken to see this as a set of linear punishment practices by penal surgeons: the choreography was triangulated by a quality of mercy that might be constrained by the body taking time to become a corpse.

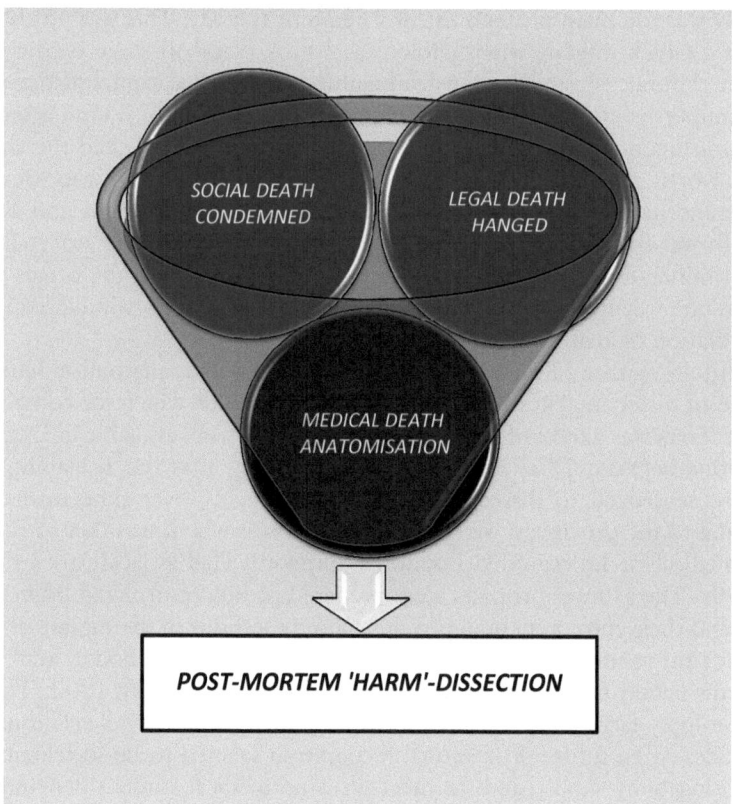

Figure 7.1   The Dangerous Dead under the Murder Act, c. 1752–1832.

It was funnelled through a set of physical procedures, until, that is, a criminal dissection could happen. In the process of which the body was cut a little, enough, or a lot. Those involved came and went, congregated and dispersed, got excited and recoiled, queued and jostled to position themselves. And it is credible that they did so, since human beings are motivated by curiosity, can act contradictory, and are often captivated by a deviant version of the 'wounded storyteller'.[15] The murderer condemned was not ill in a conventional medical sense, but they had corrupted the social well-being of a society that demanded physical restitution. Just as a patient with a fatal illness can transform their deadly diagnosis into a lived experience by becoming not a victim but a storyteller, so the murderer's body was the main protagonist in a penitent tragedy. This was not how-ever a Greek tragedy where violent acts took place off-stage because of refined public sensibilities. Instead eighteenth-century capital justice was an immersive form of theatre in which an eager audience was an actor in a punishment script co-ordinated by medico-legal officials and the state. It is worth recalling that 'pre-modern people had rich descriptors for dis-ease and its remedies' because there was so much uncertainty and since death was the ultimate 'uncertain certainty' the potential for storytelling at criminal dissections became extensive. Choreographing the rituals was then one way to try to stage-manage the knowledge flow stimulated by a circulation of anatomical storylines.

It follows that a fourth finding is that accumulating anatomical knowl-edge of dissection rites relied on body-supply trends which for convicted murderers was a total of 1, 150 criminal corpses made available in English regions between 1752 and 1832. In identifying that the availability of those sentenced to dissection shifted considerably over time from the capital to the provinces, we can begin to reassess why it was that so many in medical circles regarded London's Surgeon's Hall as lacklustre by the 1790s. The Hunter brothers were prepared to buy resurrected bodies to expand their entrepreneurial private anatomy sessions in the capital, at the same time as sheriffs in provincial life made more criminal bodies available for dissection compared to their London counterparts by 1800. These two supply factors led to a fundamental shift in central-local relations in decades when future professional recognition seemed to be so reliant on securing body-supply lines to meet growing medical student demand or to establish a business reputation for originality in a competitive medical marketplace. The symbolism of capital sentencing was thus influenced by four body-supply phases that delivered bodies for dissection.[16] Against this backdrop, crime historians need to revise the extent, nature and scale of post-mortem punishment rites, and their settings.

A related historical issue is that criminal historians have mistakenly elided a diminishment in the quantity of punishments with a qualitative dropping off in the 'harm' done to the criminal body, even though quantity and quality can go in different historical directions. Simon Devereaux has helpfully calculated that immediately after the Murder Act the execution rate increased.[17] Then it started to fall back until there was a crime wave in the 1780s when the number hanged rose sharply. By 1785 however the government had been criticised severely for the sheer number of people being executed under the Bloody Code. In response, in 1785–6 the pardoning rate rose and less people were hanged for capital offences but, as Devereaux points out, this did not lessen punishment on the condemned body especially for murder at a time of moral panic. If fewer criminals were being executed, but the legal state still wanted to be seen to be strong-armed in cases of homicide, then it was logical to sentence to death fewer but punish their bodies more. After all, those condemned represented the very worst type of criminality. The early modern sentiment—'*Hanging is not Enough*'—the subject of an upcoming book by Peter King—was all about making relevant punishment choices so that the Murder Act pained those it targeted.[18] In the history of pain the development of criminal dissections was a very divisive instrument of criminal justice that still needs to be resituated since it could differ a lot on location throughout eighteenth-century England. This observation leads us to the need for a fundamental reappraisal of post-mortem options and their public profile.

In crime history it has been commonplace to mistake gibbeting and dissection as punishment rites with different public and private functions respectively. Randy McGowan has for instance recently observed that at dissections: 'The seeing was limited to a select audience, whose viewing was not supposed to be voyeuristic but was intended rather to promote the progress of knowledge'.[19] This, he asserts, contrasted with the gib-beted body 'exposed to all eyes, the grim fact of the human mortality on display'. Yet, these generalisations are not based on substantial archive research of the sort presented in this book. The criminal corpse sent for dissection was punished over three-days in a high-profile set of public ritu-als. The audience composition contained different sorts of penal surgeons, many thousands of ordinary people, as well as local elites. It was a spectacle open to everyone and that was a deliberate policy. Medico-legal officials had autonomy which they deployed to stage procedures in medical spaces that were accessible to try to ensure that criminal dissection had sym-bolic meaning that was robust in the localities. This explains why in hang-ing-towns across the North of England like Lancaster, Leeds and York, post-mortem rites took place in small medical dispensaries until voluntary hospitals were built. The utility of Shire Halls for criminal dissections in the Midlands is likewise self-evident at Derby, Nottingham and Leicester. In major industrialising cities such as Manchester or mineral-rich econo-mies at Newcastle the pace of population growth justified the cost of pur-pose-built dissection spaces inside a major hospital site and at a dedicated Surgeon's Hall. Meantime, across areas of the West Country like Cornwall and Dorset, a gaol or business premises of a surgeon was utilised, unless the criminal dissection took place in Devon or Somerset where punishment rites could be hospitalised from the 1750s in premises already constructed. Gibbeting did of course display the body for a much longer period of physical time but at criminal dissections the body was nevertheless punished in a very public manner by a large community of interested parties. At many dissections ordinary people walked past the body on a table in a public space. This gave each person present an opportunity to compare their fleshiness to that of a deviant spectacle, with whom they shared an anatomical-design. And a crucial part of this face-to-face process of popular justice was taking the time to do so: clock-watching in a market-square before filing past the body at a rate of eighty-three people per minute for five thousand spectators in the first hour after hanging, up to a more manageable forty-two an hour for ten hours when crowds peaked at twenty five thousand in a day.[20] Small wonder perhaps that diarists of the period recorded curious spectators returning up to five times in one public session to get a proper look at the criminal corpse. The spatial dimensions of these dissection venues and their highly symbolic time-tabling are no less deserving of an accurate criminal history than the detailed geo-mapping of gibbet sites in the forthcoming work of Sarah Tarlow on *Hanging in Chains*.[21]

Turning then to a fifth contribution, when the criminal body became a narrative for punishment rites it had a life-story with a life-force capable of defying medical death. This then complicated the original research opportunities and material afterlives that were created by the remit of the Murder Act. The transition from a fatal diagnosis of heart-lung to brain death in many respects ran counter-intuitively to popularly held beliefs about latent 'sensibility' in the corpse. The mysterious nature of vitality was often associated with the Romantic notion of being 'heartfelt', whereas becoming 'mindful' of brain-death was akin to the Neo-Classical values of rationality and reason promoted by science and medicine in the Enlightenment. Just then as the quantity and quality of punishment rites could go in different directions at the same time, so too could confusion about the potency of heart/lung or head. These brought to the surface popular beliefs about the criminal corpse having medicinal powers: 'the science of extremities' had to somehow contain a 'cultural compost' of magic, medicine, and materialism.[22] Increasingly then medico-legal language was 'symptomatically and painfully status aware'[23]: moreover, 'material things had supposedly taken over from human relations' when a murderer was deprived of life, and, this, the anatomical sciences tried to control to enhance their professionalism.

In that process a lesson was learned that medicine could not instigate reform to improve its body-supply: a trend exemplified by the failure of Wilberforce's Anatomy Bill (1786) to expand dissection to all capital convictions. Instead the best way to proceed was to align with, and be the beneficiary of, future legislation associated with public health and welfare issues to which anatomy became attached.[24] Meantime brain dissection was a discipline-defining endeavour when quasi-sciences like phrenology produced plaster casts and brain matter was kneaded like dough at criminal dissections. In its broadest sense, neuroscience did take its first modern scientific steps and it is this under-researched perspective that collar heads displayed. Yet, a lot more still needs to be known about the career choices penal surgeons made from doing 'brain speculations' and whether any original research they carried out was robust scientifically to convince fee-paying patients. For on this punishment journey there was also a strong current of curiosity in which the crowd's actions in following the body throughout expressed almost every sort of feeling it was feasible to utter and this too energised retribution in death.

A sixth and final discovery is that this book ends up by giving us a completely different sense of how ordinary people experienced the drama of anatomy—it speaks to issues of power, professionalization and the relationship of the crowd—with a direct and appropriated arming of the state. The dramatic license of this rule of law was continually dependent on a relational sense of personal agency and its emotional representation expressed by physical actions. Emotions were continually exchanged and given free expression at capital punishment events in an English state that ironically sought to limit ordinary people's freedoms during a century in which revolutionary forces across Europe threatened the status quo. Ceremonies of capital punishment—whether *pre-mortem*, *peri-mortem* or *post-mortem*—facilitated different forms of cathartic release—giving voice to the social fabric of well-being in a world starting to be inverted by the secularism of the Enlightenment. Emoting feelings of anger and revenge, disgust and dread, as well as curiosity and prying over the criminal corpse had therefore to somehow be harnessed by the judicial system. Delving deeper into the archaeology of emotions expressed in this competitive atmosphere of conservative forces battling radical ideas involves the recognition that fear of the body had a large part to play in the Enlightenment endeavour. English penal surgeons were at pains to explain how viewing the dead was nothing to be frightened of since human anatomy was all about finding out more about the body's extraordinary capacity to surprise medicine and scientific thinking. Disease could be reduced to a site of scientific enquiry. And yet, this reason and rationality must begin with the fear-provoking prospect of seeing a vicious murderer pay for a heinous crime with post-mortem 'harm'. Logically an empirical approach to punishment meant that the person punished had forfeited their life, but their capital death espoused a darker medical truism too. Death has always had, and always will have, dominion over everyone. There was no escaping the fearful prospect that one day everyone in front of the criminal corpse would have to face an equivalent material fate. Engaging with the crowd's capacity for awe and horror, sympathy and pathos, as well as curiosity and fascination with the 'dangerous dead' hence meant a tacit acceptance that some sort of co-creation was inevitable in the exchange of medical and popular ideas. Spectators thus came to the spectacle of punishment with an inherent inquisitiveness, novelty, and wonder, and it was precisely these attributes that might in turn trouble the early modern state. So long as capital death had the capacity to be 'self-exploring and self-altering'—in ways that William Reddy has envisaged—agency by ordinary people—in their actions based on emotional encounters—could neither be controlled, nor predicted at criminal dissections.[25] This made the Murder Act melodramatic for audiences everywhere. In a sustained process of continuity and change, condemned body after condemned body imposed these deadly dramatics for criminal malady on everybody.

The criminal corpse dissected was an emblem of anatomy from Newcastle to Penzance, as much as London and Edinburgh, at times of moral panic and political upheaval. County-by-county the local attachment of medico-legal officials to the Murder Act was the culmination of a literal and figurative rendering of shared curiosity. These characterised 'the unforgiving assertion of a violent authority' on which rested a social order in transition from a moral to political economy.[26] Looking beneath the surface of the criminal copse ironically increased the 'vulnerability of what the forces of law and order sought to defend', their at times shaky grasp of a body politic. This was part of the 1790s paradox of being 'caught between radicalising political circumstances' and the religious requirement for good government of the self. As Adam Nicolson points out—'*Uneasy* [sic] is the eighteenth-century word' for it; and an apt medical one too.[27] So long as the criminal being cut open had a recognisable human face and body-shell, dramatic punishment for homicide remained deeply embedded in contemporary conversations about post-mortem 'harm' from

1752. Even a generation later, concerns about physical retribution being excessive did not diminish a public appetite for dissection spectacles. The irony was that as penal reform pressure gathered support from humanitarians, ordinary people continued to crowd out the criminal corpse whether in a 'good', 'bad' or 'contaminated' condition in provincial life. It was perfectly possible of course in the midst of all this ghoulish theatricality to adopt two frames of mind, one compelled to look, the other determined to look away.[28] To be Janus-like was 'sensible', 'polite', 'gentle-manly' and 'honourable'. In the political dance however between personal dignity and material fragmentation, the crowd stayed to play their part. Just as theatre-goers often paid extra to buy a ticket to sit on the actual stage with the actors in eighteenth-century playhouses, so the post-execution crowd leaned over the free-for-all or pay-as-you-go dissection drama that had a spectacular denouement.

Discretionary justice thus continued to be taken to its logical conclusion, especially in cases of ethnicity and gender, unless that is class intervened to pardon or lessen the cutting of the body in controversial cases like Earl Ferrers. Underlying moral qualities of good sense and decorum meantime was a harsh criminal reality that eighteenth-century life was competitive, diseased, and hazardous. The crowd 'contained such threats and were in turn the chief source of its greatest danger'.[29] So they had somehow to be encouraged to get involved, to shape the reception of discretionary justice alongside penal surgeons. The quandary was how to dramatize a material reality that was so culturally offensive yet the focus of curiosity, and at the same time required to show that physical retribution had substance that was trustworthy enough to maintain social equilibrium. Only by walking with the crowd from the gallows to the grave—capturing anew a synaesthesia at criminal dissections—are the shortcomings of the state's legal reach to enhance the deterrence value of punishment for murder fully exposed. Revealingly, public engagement by penal surgeons had to be an act of co-creation on a case-by-case basis, or it had little legal bite. It meant of course opening up a system of punishment to a disturbing conjunction of 'curiosity'—the deviant but normal—more alive than dead—compelling but repellent—disintegrated yet refabricated—faceless but not headless—shamed until forgotten.

We close this book at the 'dead-end' of all this punishment sightseeing with an old English proverb that still held sway for the early modern crowd but was freighted with anatomical meaning after the Murder Act. It ran: '*He that hath an ill-name, is half-hanged*'.[30] By 1754 when the

288 E.T. HURREN

skeleton of Ewen MacDonald (who opened this book) was dissected and displayed at Newcastle Surgeon's Hall, a sense of rank and place was distinguished by one's reputation in Georgian society. As a convicted murderer the nineteen year old had despoiled his moral reputation. His degraded body, displayed in a named niche, exuded ill-repute, and it did so for the foreseeable future. Yet, it was his disputed medical death as *Half-Hanged MacDonald* at a criminal dissection that truly earned him posthumous notoriety in the old-wives tales of the North-country. In *Dissecting the Criminal Corpse* however many scientific labels were attached to the condemned body in the name of good government, it was the emotional capacity of the crowd's 'curiosity' to appropriate the *'dangerous dead'* that went on testing the state's ability to trans-figure popular mentalities. For it was a compelling anatomical legacy that criminal justice in a punishment history of the body consigned to posterity from 1752 to 1832. Hence it happened that in the proverbial retelling—'*He that hath an ill-name, is half-hanged*'—the wordplay '*only half-hanged*' got added to a Georgian lexis. The criminal corpse that struck a note of infamy in popular culture was never punished cleanly by dissection under the Murder Act.[31]

NOTES

1. Refer to the last paragraph of this conclusion and *Oxford Book of Proverbs*, "Ho-so hath a wicked name Me semeth for sothe half hongid he is" [*a* 1400 in C. Brown *Religious Lyrics of XIVth Century* (1957) 193]; 'He that hath an yll name, is halfe hangd' [1546 J. Heywood *Dialogue of Proverbs* ii. vi. 12; 'It is a very ominous and suspitious thing to haue an ill name, The Prouerbe saith, he is halfe hanged', [1614 T. Adams *Devil's Banquet* iv. 156].
2. For an excellent summary of the theoretical debates, see, Roger Cooter, 'The Turn of the Body: History and the Politics of the Corporeal', *ARBOR Ciencia, Pensamiento y Cultural*, CLXXXVI 743 (May–June, 2010), 393–405, ISSN: 0210–1963, doi: 10.3989/arbor.2010.743n1204.
3. Illustration 7.1, ©Wellcome Trust, Image Collection, L0013340, '*Lateral view of the trunk of a flayed corpse*', by Charles Landseer, 1815; Creative Commons Attribution-NonCommercial-ShareAlike 4.0 International License (CC BY-NC-SA 4.0).
4. *Morning Chronicle*, 28 May 1776, column 1, page 1.

5. Adam Bencard, 'History in the Flesh', (unpublished PhD, Københavns Universitet. Institut for Folkesundhedsvidenskab. Medicinsk Museion, University of Copenhagen, 2008).

6. William Reddy, *The Navigation of Feeling: A Framework for the History of Emotions*, (CUP: Cambridge, 2001).

7. And the list was extensive: whether 'young', 'male', 'female', 'deviant', 'dangerous', 'bull-necked', strong-willed', 'Irish', 'Jewish', 'non-European', 'good bodies', 'bad bodies', 'contaminated bodies', 'extras', or 'social outcasts'.

8. V. A. C. Gatrell (1996 edition), *The Hanging Tree: Execution and the English People, 1770–1868*, (Oxford: Oxford University Press).

9. This he clarified in Michel Foucault, *The Order of Things* (Routledge: New York and London, 1970 edition, English translation), preface, p. xvii.

10. See, http://arbor.revistas.csic.es/index.php/arbor/article/viewFile/807/814, Roger Cooter, 'The Turn of the Body', p. 394–405, a central theme throughout his article.

11. It is also used figuratively to mean pick your battles or 'bide your time and wait for the opportune moment': both literal and figurative are relevant to the application of the Murder Act in England.

12. Jonathan Sawday (1996), *The Body Emblazoned: Dissection and the Human Body in Renaissance Culture* (New York and London: Routledge), p. 88; Thomas Laqueur (2011), 'The Deep Time of the Dead', *Social Research*, Vol. 78, Fall, III, 799–820, quote at p. 802.

13. Innes, J. (2009), *Inferior Politics: Social Problems and Social Policies in Eighteenth Century Britain*, (Oxford: Oxford University Press).

14. Andrew Cunningham, (2010), *Anatomy, Anatomiz'd' An Experimental Discipline in Enlightenment Europe* (Aldershot, Hampshire; Ashgate), cited in the preface.

15. Arthur W. Rank (1997), *The Wounded Storyteller: Body, Illness and Ethics*, (Chicago: University of Chicago Press), p 5.

16. These were: 17 bodies per year 1752 and 1800; a low of 4 in 1801 recovering to 15 by 1809; 18 a year by 1812, 25 by 1813, falling back to 15 in 1815, before doubling to a sharp peak of 30 in 1816; and finally 10 bodies per year on the eve of 1832, see chapter 5.

17. Simon Devereaux (2013), "England's 'Bloody Code' in Crisis and Transition: Executions at the Old Bailey, 1760–1837," *Journal of the Canadian Historical Association*, Vol. 2, Issue 34, 71–113.

18. I am grateful to Peter King for sharing with me his new research on the early modern sentiment '*Hanging is Not Enough*' (taken from the title of contemporary pamphlet literature), which he expands upon in his forthcoming book, King, P. J. R. (2016), *Criminal Justice, the Criminal Code*

and aggravated forms of Capital Punishment in England and Wales, 1700–1834 (Basingstoke: Palgrave Pivot).

19. McGowen, R. (2005), 'Making Examples' and the Crisis of Punishment in Mid-Eighteenth-Century England' in David Lemmings ed., *The British and Their Laws in the Eighteenth Century*, (Woodbridge: Boydell), pp. 182–205.

20. Refer, forthcoming article by this author, on 'Time, Spectatorship and the Criminal Corpse in Eighteenth and Nineteenth-Century England' on the time-management of criminal justice. It has not been discussed more fully here to avoid intellectual overlap.

21. I am grateful to Sarah Tarlow for sharing her new research on gibbeting for *Hanging in Chains: Gibbeting and the Murder Act* (forthcoming, Palgrave Pivot, 2016).

22. I am grateful to Owen Davies for sharing his latest research on the medicinal power of the criminal corpse in Davies, O. and Matteoni, F. (2016), *Executing Magic: The Power of Criminal Bodies* (Basingstoke: Palgrave).

23. These concluding reflections owe an intellectual debt to some of the broader eighteenth-century themes discussed throughout Adam Nicolson, (2011), *The Gentry: Stories of the English* (London: Harper Press), p. 318. The general intellectual framework of his and this book obviously differ quite a lot because of the medical context under discussion here; nevertheless, it should be acknowledged that Nicolson's framing and phrasing of ideas merits a scholarly credit.

24. Hurren, E. T. (2011), *Dying for Victorian Medicine: English Anatomy and its Trade in the Dead Poor c 1832–1929* (Basingstoke: Palgrave Macmillan), explains that anatomists never succeeded in enacting legislation to improve body supply in their own right.

25. Refer, endnote 6.

26. Nicolson, *The Gentry*, p. 12.

27. Ibid., p. 229.

28. Indeed as Nicolson, *The Gentry*, p. 209, points out the middling-sorts (the gentry class of magistrates and jurymen for instance) often did this as a matter of social discourse.

29. Again I am grateful here to Nicolson, *The Gentry*, who argues in the conclusion to his book that in England the gentry were intrinsic to a 'cloud-like economic constancy—rarely the same but always the same—... combined with a control of the political and judicial systems which placed them on little summits all over the English counties', p. 414. Yet they could never escape the agency of the crowd, since, as he also points out, behind all the rhetoric of husbandry and claims to landed status, it remained the case that—'Life is a struggle and community is political', whether in the eighteenth or twenty-first centuries, p. 418.

30. Under the Murder Act, the same was said of female criminals too.
31. Reflecting the central tension in *deterrent* or *retributive* theories of capital punishment, see, W R. P. Kaufman (2013), *Honour and Revenge: A Theory of Punishment* (London and New York: Springer Dordrecht).

BIBLIOGRAPHY

PRIMARY RESEARCH

Devon Record Office

59/7/4/10/1, Mortgage Bond, John Patch, Surgeon of Exeter, 13 March, 1740–41 and transfer of mortgage, 1142 B/T22/118-119, dated 1769.

Dorset Record Office

NG/PR1, Prison Admission and Discharge Registers, 1782–1901, 'Case of John Anderson Constable arrest of Elizabeth, otherwise known as Betty Marsh'.
D/SEN/3/7/8, 22 June 1792, bond, John Coombs and Philip Coombs, Dorchester surgeons.
D/SEN/3/7/10-11, 24–25 February 1794, property conveyance, John Coombs of Sturminster Newton and Philip Coombs residing at Shillingstone, after their respective marriages.

East Sussex Record Office

The Frewen Collection, Family of Brickwall in Northiam, FC/2567-2658, Letters from Dr Thomas Bishopp of Leicester, c.1792–1811, and others cited per letter item.

Huntingdon Record Office

Accession 4715, Huntingdon Medical and Surgical Society, Transactions Book 1792–1801.

© The Author(s) 2016   293
E.T. Hurren, *Dissecting the Criminal Corpse*, Palgrave
Historical Studies in the Criminal Corpse and its Afterlife,
DOI 10.1057/978-1-137-58249-2

Journals and Newspapers

Blackwood's Magazine
Brighton Herald
Cambridge Chronicle and Journal
Chester Annual Register
Drury Lane Journal
Edinburgh Advertiser
Examiner
Felix Farley's Bristol Journal
General Evening Post
Gentleman's Magazine and Historical Chronicle
Historical Magazine
Lancet
Law Times and Journal of Property
Leicester and Nottingham Journal
Leicester Journal
Leicester Mercury
Leicestershire Chronicle
Liverpool Mercury
Lloyds Evening Post
Local Record of Newcastle
London Evening Post
London Magazine
Manchester Guardian
Medical and Philosophical Commentaries
Medical and Physical Journal
Medico-Chirurgical Review and Journal of Medical Science
Midlands County Advertiser
Monthly Magazine or British Register
Morning Advertiser
Morning Chronicle
Morning Post
Newcastle Courant
Newcastle General Magazine
Notes and Queries
Public Advertiser
Public Ledger or The Daily Register of Commerce and Intelligence
Royal Cornwall Gazette, Falmouth Packet and Plymouth Journal
The Age
The Critical Review or Annals of Literature
The Observer
The Times

The Universal Magazine
Trewman's Flying Post
Westminster Journal
The York Herald and General Advertiser

Lambeth Palace Library

Medical Licences, Archbishop of Canterbury for Lincoln, 1535–1775, 4.19, pp. 36–7.

Lancaster Record Office

MSS, DDWh/4/99, Whittaker of Simonstone.

Leicestershire Record Office

DE3182/1, *The Diary of Thomas Kirkland Surgeon of Ashby and family, c. 1731–1931.*

P222 – *Copy of the Sale of Mr Thomas Kirkland's goods and Museum upon his death – sold on the premises by the family of Kirkland senior and junior – on the death of the latter 1824.*

DE107/261/1-10, 1734, Henry Halford MS and medical papers.

Leicester Infirmary Minute Books, LRO/13064/10, 17 December, & 29 December 1807, 20 August 1816, 8 April 1817, 3 August 1817, and 10 August 1817, 8 June 1819, 23 March 1822 and 19 December 1822.

Leicester Infirmary Minute Book, LRO/13064/13, 23 June 1814–30 June 1819, and entry dated 17 November 1815.

Leeds University Library Special Collections

MS 504/1/3, Manuscript of Mr William Hey, Senior Surgeon to the Leeds General Infirmary.

MS/1990/5, Hey Family of Leeds, Correspondence, 1828–42.

Lincoln Record Office

Summer Assizes, 4/8/1775, and reported in *Leicester and Nottingham Journal*, 12/8/1775 with no pardon given the brutal nature of the murder.

London Lives (and Online)

Fire Insurance Registers: 1777–86, policy number 380040, John Hopper the carpenter, near the George Drury Lane London.

Middlesex Coroners Court Records, CO/IC, 1747–1803, lists Thomas Pacey, John Mansell and John Hooper or Hopper acting as jury men on 'the View of a

Body of a Boy unknown then and there lying dead' who was discovered floating downstream in the River Thames.
Middlesex Sessions, Justices Working Papers, LMSMPS502480076, 23 November 1727 a signed deposition by Abraham Chovet the surgeon about a dangerous assault he witnessed.

Norfolk Record Office

NNH 1/6, Norwich and Norfolk Hospital Minute Books, 12 July 1794 –13 July 1799.
NNH 29/2, Catalogue of the Norfolk and Norwich Hospital Museum, where N indicates Normal Anatomy, P is a Plaster Cast.

Nottingham Record Office

BB34.8, 'Gallows Rememberancer for 1763', p. 25.
BB34.8, 'Gallows Rememberancer for 1779', pp. 25, 47–8.

Royal College of Surgeons

Annals of the Barber Surgeons of London, complied from their records and other sources by Sidney Young, one of the Court Assistants, the Worshipful Company of Barbers of London, illustrated by Austin T. Young (1890), published from original sources (London: Blades, East and Blades Publishers Ltd).
Messenger Monsey correspondence (1732–1788), MS0396, and obituary writer 'Mr Wadd'.
William Clift Collection (1775–1849), MS, Box 67.b.13, Elizabeth Ross case notes and sketch, 1832, anatomized and dissected.
William Clift Collection (1775–1849), MS0007/1/6/1/1, *Record of the Bodies of Murderers delivered to the College [of Surgeons] for Dissection*, written by hand inside the front cover of a dissection book starting 1800 and ending 1814.
William Clift Collection (1775–1849), MS0007/1/6/1/3, *Sketches of Murderers' Heads*, circa 1807–1832.
Royal College of Surgeons, Financial Records, COS/3/1, *Company of Surgeons, 1745–1778, Volume 1 (of six volumes)*, all original entries cited by date in chronological order.
Royal College of Surgeons, Museum Collection, MUS5/6, Museum letters, from William Clift to Sir Everard Home Clift (no date, but likely to be early 19th century), requesting several body parts of a man to be hanged the next day.
Royal College of Surgeons, Museum Collection, MUS5/6, Museum letters, from George William Clift, 22 December 1818, concerning his efforts to recover the 'hands' of an exhibition piece; and 29 February 1820, apologies for not making a meeting and on cleaning his macerating room following dissections.

The National Archives

British Parliamentary Papers, Manuscript Copy, Statute, 25 Geo II, c.37, 1752, "An Act for Regulating the Disposal after Execution of the Bodies of Criminals," HL/PO/JO/10/2/61, London, England.

British Parliamentary Papers (1812–3), V, Report of the Commissioners on the State Lancaster Prison and the Treatment of Prisoners Therein, p. 34, and later printed (1812) as State of Lancaster Gaol, (London: Lords and Commons Reports), entry 3 July 1812, pp. 897–8.

British Parliamentary Papers, Hansard, House of Commons Sitting (1819), HC debate, 6 July 1819, volume 40, cc 1518–36, testimony of Sir James Mackintosh that the murder rate was 67 murders and 57 executed 1755–1784, 54 murders and 44 executed 1874–1814, of which there was said to be on average 9 murders per annum in London.

British Parliamentary Papers (1819), The Committee for Investigating the Criminal Law.

Assizes Records, Ches[hire]/21, including murder cases ches/21/7/19, 21/7/53, 21/7/60 and 21/7/90.

Assizes Records Dur[ham] 15–16, including murder cases, durh16/1-4, 16/2, 16/3/29, 16/3/37, 16/5/2, 16/5/86, 16/5/348, 16/5/628.

Assizes Records, Lanc[aster], including murder cases PL 28/2/68, 28/2/103, 28/2/233, 28/2/240, 28/2/247, 28/2/272, 28/2/290, 28/2/316-317, 28/2/326, 28/2/333, 28/2/351, 28/2/351–352, 28/3/104, 28/3/261, 28/4/60, 28/4/93, and summary lists of Assizes, 1805–1811, PL 28/5/2–23.

Assizes Records Norf[olk], Gaol Books, ASSI–33.

Assizes Records Oxf[ord], Crown Minutes, ASSI–12.

PCC, Abstract of Wills: Wills Proved at the Prerogative Court of Canterbury, 1 April 1761, Edward Stanton, London's instrument Maker, buried Saint Mary Woolnoth, City of London.

PCC, Abstract of Wills, PROB11/1265, Will of John Bolding, Surgeon of Amphthill, Bedfordshire, 3 September 1795.

PCC, Abstract of Wills: Robert Patch, Surgeon of Exeter, PROB/11/1599, 17 December 1817.

PROB/11/1212/32, Will of Francis Fox, gentleman of Derby, 2 December 1791.

PROB 11/1310/135, Will of Thomas Kirkland, Doctor of Medicine, Doctor of Physic, Ashby–de–la–Zouche, 21 July 1798.

PROB 11/2066/422, Will of Dr Coryndon Rowe, Doctor in Medicine of Dockacre Launceston, Cornwall.

Sherriff Cravings, circa 1752–1832, E389–243 to E389–254; 90–147 to 90–169, primarily used to compile database of corpses sent for dissection (individual cases, cited in footnotes).

Selected case-studies cited, include: Sheriff Cravings for Devon, E389/242/227 (Assize Calendar), 20 March 1758; Sheriff Cravings for Devon, E389/252/217, (Assize Calendar), 6 Aug 1808; Sheriff's Cravings for Bedfordshire, 389/248/196, (Assize Calendar), dated March 1788; Sheriff's Cravings for Cornwall, T90/170, Burns, (Cornwall, 1814), ref. 7306891, expenses claimed back by Joseph Edwards, undersheriff; Sheriff's Cravings, E389/253/355; E389/253/359, 1814/03/31, Burns/Hamley dissection, is listed as doing dissections for 1820, 1821 and 1828.

Wellcome Trust

Wellcome Library Collection, Wellcome Image, V0042183, 'A Clock Dial on which a Skeleton holds an Oil Lamp, drawn by Caleb Elwin', reproduced in Cambridge, 1800.

Wellcome Library Collection, Wellcome Image, L0031335, the 'Dead-Alive' revive, and startle the living 18th/19th century.

Wellcome Trust, Image Collection, M0015855, 'Edward Stanton at the Saw and Crown' in Lombard Street, London.

Wellcome Trust, Image Collection, L0013340, Lateral view of the trunk of a flayed corpse, by Charles Landseer, 1815.

West Yorkshire Archive Service

Bradford Office, SpSt/11/5/1/2, William Hey Correspondence, Letter from Hey to William Stanhope, 21 May 1785.

Websites (Referencing Primary Research Material)

'Case of William Duell, Executed for Murder, Who Came to Life Again While Preparing for Dissection in Surgeon's Hall', e-transcript sourced online, see: http://www.bl.uk/learning/images/21cc/crime/transcript1595.html.

British History Online, http://www.british-history.ac.uk/report.aspx?compid= 65576, *Survey of London, volume 24, The parish of St Pancras, part 4, Kings Cross neighbourhood* (London, 1952), pp. 147–51.

Exeter Society, History of Doctors, http://www.exetercivicsociety.org.uk/ plaques/485/

Exhibition of 18th century plays at the Victorian and Albert Museum, London, see, http://www.vam.ac.uk/content/articles/0-9/18th-century-theatre/

http://www.18thconnect.org/news/?p=80yet, Murder Trial Documents, 'GEORGE CHENNEL AND J. CHALCRAFT, *Executed August, 1818 for the atrocious murder of Chennel's Father and his Housekeeper, at Godalming*' and e-transcript available on http://www.exclassics.com/newgate/ng577.htm

Nottingham contemporary court records and local crime histories can be accessed at: http://nottinghamhiddenhistoryteam.wordpress.com/page/11/

Old Bailey trials online, *The Proceedings of the Old Bailey, 1674–1913*, http://
www.oldbaileyonline.org/, cited by trial date reference.
On the medical development of therapeutic hypothermia in modern surgery, see:
http://www.sca-aware.org/sudden-cardiac-arrest-treatment
On the medical logistics of being hanged, see, Thomas Stuttaford, "Swift end rests
with skill of the hangman", *Times* newspaper online (1 January, 2007), http://
www.timesonline.co.uk/article/0,,3-2526006,00.html.

PRE-1900 PUBLISHED SOURCES

Alcock, T. (1827), *An Essay in the Use of Chlorurets of Oxide of Sodium and Lime
as Powerful Disinfecting Agents and of the Chloruret of Oxide of Sodium more
especially as a Remedy of Considerable Efficacy in the Treatment of Gangrene,
Phagedenic, Syphilitic, and Ill-Conditioned Ulcers, Mortifications and Various
Other Diseases dedicated to the Right Hon. Robert Peel* (Derby and London:
Burgess and Hill).

Anon., (1775, 1st edition), 'Hanging of William Farmery at Lincoln on 4 August
1775', *The Hibernian Magazine, Or, Compendium of Entertaining Knowledge,*
Vol. 5, (September), 562.

Anon., (1775, 2nd edition), 'Summer Assizes Report, Lincoln 1775', in *The
Annual Register or a View of the History, Politics, and Literature for the Year
1775* (London: J. Dodley of Pall Mall), pp. 154–5.

Anon., (1811), *The Royal Kalendar, or Complete and Correct Annual Register for
England, Scotland, Ireland, and America for the year 1811 [bound with] Rider's
British Merlin for 1811. Adorned with many delightful and useful Verities fitting
all Capacities in the Islands of Great Britain's Monarchy, With Notes of Husbandry,
Fairs, Marts, and Tables for many necessary uses* (London: J. Stockdale).

Anon., (1736), *A catalogue and particular description of the human anatomy in
wax-work, and several other preparations; to be seen at the Royal-Exchange*
(London: T White).

Anon., (1815), *Annals of Cambridge* (Cambridge: Warwick and Co).

Anon., (1819), *A Full Account of the Most Horrid Murder Committed by Thomas
Weems or Weyms a Miller, Upon the body of his Wife by Strangling Her with a
Garter near Puckeridge Hertfordshire* (London: Charles Piggot publisher of
Clerkenwell).

Anon., (1827), *The Strangers Guide through the town of Nottingham being a
description of the principle buildings and objects of curiosity in the ancient town*
(Nottingham: Sutton and Sons).

Bickersteth, E. (1822), *A Treatise on Prayer* (London: Private Publication).

Binns, J. (1807), 'Letter to Dr James Hamilton, Senior Physician to the Edinburgh
Royal Infirmary on the Cure of Scarlatina, from Physician Lancaster', *Edinburgh
Medical and Surgical Journal* Vol. 3, (April), 135–144.

Bissett, C. (1766), *Medical Essays and Observations* (Newcastle-upon-Tyne: Thompson Publishers), pp. 60–1.

Bourne, H. (1736), *The History of Newcastle upon Tyne or the Ancient and Present State of that Town by the Curate of All-Hallows in Newcastle* (Newcastle: John White Publishers).

Briscoe, J. P. (1895), *The Old Guild Hall and Prison of Nottingham* (Nottingham: Sutton and Sons).

Burke, E. (1775), *Dodsley's Annual Register* (London: Printed for R and J Dodsley in Pall Mall).

Campbell, E. H. (1831), 'Letter to Lancet: Occurrence of Menstruation during Hanging', *Lancet*, letter dated 20 August 1831 but published in the Sept. edition, p. 704.

Chalmers, A. (1814), *The General Biographical Dictionary Containing An Historical and Critical Account of Imminent Persons Volume XVII* (London: J. Nichols and Sons).

Chessher, William Mr [sic], (1832), *The Annual Biography and Obituary for the Year of 1832, Vol. 16* (London: Longman), p. 401.

Cooke, C. (1831), 'Letter to Lancet: Occurrence of Menstruation during Hanging', *Lancet*, letter dated 1 September, published in the Sept. edition, pp. 751–2.

Cooke, W. (1835), *A Brief Memoir of Sir William Blizard, read before the Hunterian Society October 7th 1835, with additional particulars of his Life and Writings* (London: Longman, Rees, Orme, Brown and Co).

Copland, J. (1833), *A Dictionary of Practical Medicine, Vols. 1-3* (London: Messrs Longman).

Crabtree, J. (1836) *A Concise History of the parish and vicarage of Halifax, in the county of York* (Halifax: Hartley and Walker Publishers).

Darwin, E. (1803), *The Temple of Nature or The Origin of Society*, "Additional Notes II, The Faculties of the Sensorium" [to accompany the poem] (London: Thomas Bensley Publishers).

Darwin, E. (1803), *The Temple of Nature or The Origins of Life* (London: Thomas Bensley Publishers), Cant. I. 1, 250, [being a poem published the year after his death].

Defoe, D. (1723), *Curious and Diverting Journeys through the Whole Island of Great Britain* (London: G. Parker).

Dover, T. (1762 edition), *The ancient physician's legacy to his country. Being what he collected himself in fifty-eight years of practice, and so on* (London: H. Kent, for C. Hitch, J. Brotherton and R. Minors).

Editorial (1762), *The Beauties of the Magazines selected including the several original comic pieces selected to be continued the Middle of every Month* (London: Waller publishers).

Editorial (1814), 'Murders in Cornwall', *Journal of The Universal Magazine of Knowledge and Pleasure, Provincial Occurrences*, December issue, 517–8.

Editorial (1819), 'Hiatus, a Play of Time's Magic Lantern: No. IX: *The Dissector, Blasquez, and Scholar*', *Blackwood's Magazine*, Vol., XXVI, May, Issue 5, 161–4.

Editorial Letters, (1807), *The London Medical and Physical Journal*, Vol. 108, XVII, 155–7.

Ford, J. (1801), *The Suffolk Garland* (Ipswich: John Row).

Glover, S and Noble, T. eds. (1829), *A History of the County of Derby Volume II* (Derby: Henry Mozley and Son).

Glover, S. (1829) *The history of the county of Derby, drawn up from actual observations and the best authorities* (Derby: Henry Morley and Son).

Hoare, R. [Baronet], Esq (1815), *Journal of the Shrivealty in the Years 1740–1* (Bath: Privately Printed).

Holloway, J. W. (1831), *An Authentic and Faithful History of the Atrocious Murder of Celia Holloway* (London: W Clowes and Son Ltd).

Hunter, J. (1778), *The Natural History of Human Teeth I (with William Coombe)* (London: J Johnson).

Hutton, W. (1817), *The History of Derby: From the Remote Ages of Antiquity* (London and Derby: Nicols & Son, and Bentley).

Knapp, A. and Baldwin, W. eds. (1828), *The Newgate Calendar, 1824–8* (London: J Robin and Co).

Lambert, W. (1831), 'Letter to Lancet: Effects of Hanging Upon the Organs of Secretions & so on', *Lancet* dated 14 Sept., published in the Sept. edition, p. 808.

Lysons, Rev. Daniel and Mr Samuel Lysons (1814), *Magna Britannica Volume 3, Cornwall* (London: Cadell and Davies in the Strand).

Lysons, Rev. Daniel and Mr Samuel Lysons (1817), *Magna Britannia: Volume 5: Derbyshire* (Derby: Walker and Co).

Mackensie, E. (1827), *A Descriptive and Historical Account of the Town of Newcastle-Upon-Tyne, including the Town of Gateshead, Volume I* (Newcastle: Mackensie and Dent Publishers).

Maginn, W. (1831 edition), *The Red Barn a Tale Founded on Fact* (London: John Bennett Publishers).

Medical Correspondent (1819), 'Consummate Depravity – Disinterestedness – Anecdotes', *The Imperial Magazine*, Vol. 3, March, Issue 1, 266.

Medical Correspondent (1829), 'Modern Medical Ethics; or State Maxims in Medicine by Philoethicus & C', *The Medico-Chirurgical Review and Journal of Practical Medicine*, October issue, 145–9.

Okes, Mr [sic],'Intelligence Spina Bifida' in (1812) *The New England Journal of Medicine and Surgery and the Collateral Branches of Science conducted by a Number of Physicians, Volume 1* (Boston: Wait & Co), pp. 98–102.

Orange, J. (1840), *History and Antiquities of Nottingham* (Nottingham: Nabu Press).

Pennant, T. (1790), *Account of London* (London: Robert Faulder printers).

Philip, A. P. W. (1831), 'On the Sources and Nature of Powers on which the Circulation of the Blood Depends', *Proceedings of the Royal Society of London*, Vol. 3, p. 64.

Philip, A. P. W. (1834), 'On the Nature of Death', *Philosophical Transactions of the Royal Society of London*, Vol. 123, 167–198.

Richardson, M. A. (1841–6), *The Local Historian's Table Book of Remarkable Occurrences Connected with the Counties of Newcastle-Upon-Tyne, Northumberland and Durham, Historical Division, Volume 1* (Newcastle Upon Tyne: Richardson Booksellers).

Richardson, M. A. (1843), *The Local Historian's Table Book of Remarkable Occurrences Volume II* (Newcastle-Upon-Tyne and London: Richardson & Smith Publishers).

Shelley, M. (1818, 1st edition; 2003, edition), *Two Works by Mary Shelley* (London: Peverell Press).

Smollett, T. B. (1793), 'Book Review', *The Critical Review or Annals of Literature*, Vol. 54, 155.

South, J. F. (1866), *Memorials of the Craft of Surgery in England* (London, Cassell & Co).

Stevenson, W. [of Hull] (1893), *Bygone Nottinghamshire* (Nottingham: Hard Press Publishing).

Thomson, H. (1827), "Le Revenant", *Blackwood's Magazine*, Vol. 27, 409–16.

Thornbury, W. (1878), 'St. Paul's Churchyard', *Old and New London: Volume 1*, (London, Cassell Ltd), chapter XXII, pp. 262–274.

Tulket, M. (1821), *A History of the Borough of Preston* (Preston and London: P. Whittle).

Wilson, Henry (1822), *Wonderful Characters, Comprising Memoirs and Anecdotes of the Most Eccentric Characters, Volume 3* (Boston, Lincolnshire: N. H. Whitaker).

Young, S. ed. (1890), *Annals of the Barber Surgeons of London compiled from their records and other sources* (London: Blades, East and Blades Publishers Ltd).

SECONDARY RESEARCH

Alberti, F. A. (2009), 'Bodies, Hearts and Minds: Why the History of Emotions Matters to Historians of Science and Medicine', *Isis (Chicago Journals, The History of Science Society)*, Vol. 100, Dec. IV, 798–810.

Amory, H. (1971), 'Henry Fielding and the Criminal Legislation of 1751–2,' *Philological Quarterly*, Vol. 50, 175–92.

Anning, S. T. and Walls, W. K. J. (1982), *A History of Leeds Medical School* (Leeds: Leeds University Press).

Auden, R. R. [FRCS] (1978), 'A Hunterian pupil: Sir William Blizard and the London Hospital', *Annals of the Royal College of Surgeons of England*, Vol. 60, 345–9.

Austin, R. T. (1986), 'Dr Thomas Kirkland of Ashby–de–la–Zouche: 1721–98', *British Medical Journal*, Vol. 293 (25th October), 1075–6.

Banner, S. (2003), *The Death Penalty: An American History* (New Haven: Harvard University Press), chapter 3, 'Degrees of Death' pp. 33–87.

Barber, P. (1998), *Vampires, Burial and Death: Folklore and Reality* (New Haven: Yale).

Barry, E. (2008), 'From Epitaph to Obituary: Death and Celebrity in Eighteenth–Century Culture', *International Journal of Cultural Studies*, Vol. 11 (No. 3), 259–275.

Barton, R. M. (1970), *Life in Cornwall in the Early 19th–Century: Being Extracts from the West Briton Newspaper in the Quarter Century from 1810–1835* (Cornwall: D. Bradford Barton Ltd).

Bates, A. W. (2008), '"Indecent and Demoralizing Representations": Public Anatomy Museums in mid-Victorian England', *Medical History*, Vol. 52, January, Issue 1, 1–22.

Beattie, J. M. (1986), *Crime and the Courts in England 1660–1800* (Princeton: Princeton University Press).

Beattie, J. M. (2001), *Policing and Punishment in London, 1660–1750: Urban Crime and the Limits of Terror Part II* (Oxford: Oxford University Press).

Beattie, J. M. (2002), *Policing and Punishment in London, 1660–1750* (Oxford: Oxford University Press).

Beattie, J., M. (2014 edition), *Urban Crime and the Limits of Terror, and The First English Detectives: The Bow Street Runners and the Policing of London, 1750–1840* (Oxford: Oxford University Press).

Bensuson Butt, J. edited by D'Cruze, S. (2009), *Essex in the Age of Enlightenment: Essays in Historical Biography* (Colchester: e-publisher lulu.com)

Blunt, W. S. (1909), '"John Baker's Horsham Diary [sic]" edited by Wilfred Scawen Blunt', *Sussex Archaeological Collections*, Vol. 52, pp. 38–83.

Bohstedt, J. (1993), *Riots and Community Politics in England and Wales, 1790–1810* (Harvard: Harvard University Press).

Bohstedt, J. (1994), 'The Dynamics of Riots: Escalation and Diffusion/Contagion' in M. Potegal and J.F. Knutson eds., *The Dynamics of Aggression: Biological and Social Processes in Dyads and Groups* (New Jersey: Psychology Press, Lawrence Erlbaum Associate Publishers), pp. 257–306.

Bohstedt, J. (2010), *The Politics of Provisions: Food Riots, Moral Economy, and Market Transition in England, c. 1550–1850* (London: Ashgate).

Bonner, N. (1995), *Becoming a Physician: Medical Education in Britain, France, Germany and the United States, 1750–1945* (Oxford: Oxford University Press).

Boyes, J. (1957), 'Medicine and Dentistry in Newcastle-Upon-Tyne in the 18th–Century', *Proceedings of the Royal Society of Medicine*, Vol. 50 (April), 299–335.

Brewer, J. (1979–80), 'Theatre and Counter-Theatre in Georgian Politics', *Radical History Review*, Vol. 32, 7–40.

Burwick, F. (2011), *Playing to the Crowd: London Popular Theatre, 1780–1830* (Basingstoke: Palgrave).

Castle, T. (1986), *Masquerade and Civilization: The Carnivalesque in Eighteenth-Century English Culture and Fiction* (London and Stanford: Stanford University Press).

Chaplin, S. (2012), 'The Divine Touch, or Touching Divines: John Hunter, David Hume and the Bishop of Durham's Rectum' in Mary Terrall and Helen Deutsch eds. (2012), *Vital Matters: Eighteenth–Century Views of Conception, Life and Death* (Toronto, Canada: University of Toronto Press), chapter 10, p. 222–46.

Clark, A., transcribed and edited by Trewin, I. (2002), *The Last Diaries: In and Out of the Wilderness: Alan Clark Diaries, Volume 3* (London: Phoenix Press).

Cohen, E. (1989), 'Symbols of Culpability and the Universal Language of Justice: The Rituals of Public Execution in Late–Medieval Europe', *History of European Ideas*, Vol. 11, I-VI, 407–16.

Collier, R. (1998), *Masculinities, Crime, and Criminology: Corporeality and Criminal(ised)* (London: Sage Publications).

Connors, R. (2002), 'Parliament and Poverty in Mid–Eighteenth Century England,' *Parliamentary History*, Vol. 21, 207–31.

Cooter, R. (2010), 'The Turn of the Body: History and the Politics of the Corporeal', *ARBOR Ciencia, Pensamiento y Cultural*, CLXXXVI, 743, May-June issue, 393–405.

Cozens-Hardy B., ed. (1908), *The Diary of Silas Neville, 1767–1788* (Oxford: Oxford University Press).

Craske, M. (2011), '"Unwholesome" and "Pornographic": A Reassessment of the Place of Rackstrow's Museum in the Story of Eighteenth–Century Anatomical Collection and Exhibition', *Journal of the History of Collections*, Vol. 23, I, 75–99.

Cunningham, A. (1997), *The Anatomical Renaissance: The Resurrection of the Anatomical Projects of the Ancients* (Aldershot, Hants: Scolar Press).

Cunningham, A. (2010), *Anatomy, Anatomiz'd' An Experimental Discipline in Enlightenment Europe* (Aldershot, Hampshire; Ashgate).

Davies, O. and Matteoni, F. (2015), '"A Virtue Beyond All Medicine:" The Hanged Man's Hand, Gallows Tradition and Healing in Eighteenth and Nineteenth-Century England', *Social History of Medicine*, (May, 2015), Issue 2, 1–37.

Davies, O. and Matteoni, F. (2016), *Executing Magic: The Power of Criminal Bodies* (Basingstoke: Palgrave).

Davies, O., Hurren, E. T., and Tarlow, S. eds., (2016) *The Power of the Criminal Corpse: English Capital Punishment, Medical Cures and Post-Mortem Harm in Perspective, 1700–1900* (Basingstoke: Palgrave).

de Koomen, A. R. 'The Self-Portrait 'En Décapité: Interpreting Artistic Self-Assertion' in Santing, C. and Baert, B. eds. (2013), *Disembodied Heads in Medieval and Early Modern Culture* (Oxfordshire: Brill Publications), chapter 7, pp. 191–222.

Desmond King-Hele ed. (2007), *The Collected Letters of Erasmus Darwin* (Cambridge: Cambridge University Press).

Devereaux, S. (2009), 'Recasting the Theatre of Execution: The Abolition of the Tyburn Ritual', *Past and Present*, Vol. 202, 127–74.

Devereaux, S. (2013), 'England's "Bloody Code" in Crisis and Transition: Executions at the Old Bailey, 1760-1837', *Journal of the Canadian Historical Association* Vol. 2, 24, 71–113.

Dobson, J. (1951), 'Cardiac Action after "Death" by Hanging', *Lancet*, 29 Dec., Vol. 258, No. 6696, pp. 1222–5.

Dowbiggin, I. (2007), *A Concise History of Euthanasia: Death, God and Medicine* (New York: Rowman & Littlefield).

Elias, N. (1978), *The Civilising Process: The History of Manners* (Basel: Urizen Books).

Ellwod, W.J. and Tuxford, A. F. eds (1984), *Some Manchester Doctors: A Biographical Collection to Mark the 150th Anniversary of the Manchester Medical School 1834–1984* (Surrey and Manchester: Unwin Brothers).

Farrer, W and Brownbill, J. (1911), 'The City and Parish of Manchester: Introduction', in *A History of the County of Lancaster: Volume 4* (London: Victoria County History Series), pp. 174–187.

Ferber, S. and Wilde, S. eds. (2011), *The Body Divided: Human Beings and Human 'Material' in Modern Medical History* (Farnham, Surrey: Ashgate).

Finger, S. (2000), *Minds Behind the Brain: A History of the Pioneers and their Discoveries* (New York: Oxford University Press).

Forsyth, M. (2013), *The Elements of Eloquence: How to Turn the Perfect English Phrase* (London: Icon Books).

Foucault, M. (1970 edition, English translation), *The Order of Things* (New York and London: Routledge).

Foucault, M. (1979, French edition) translated to English by Sheridan, A. (1995), *Discipline and Punish: The Birth of the Prison* (New York: Vintage Books).

French, R. (1999), *Dissection and Vivisection in the European Renaissance* (Aldershot, Hampshire: Ashgate).

Frizelle, E. A. (1988), *The Life and Times of the Royal Infirmary at Leicester: The Making of a Teaching Hospital, 1766–1800* (Leicester: Leicester Medical Society).

Garland, D. (2011), 'Modes of Capital Punishment: The Death Penalty in Historical Perspective', in D. Garland, R. McGowen, and M. Meranze eds., *America's Death Penalty: Past and Present* (New York: New York University Press), chapter 2, pp. 30–71.

Gatrell, V. A. C. (1996 edition), *The Hanging Tree: Execution and the English People, 1770-1868* (Oxford: Oxford University Press).

Gittings, R. (1973), 'John Keats, Physician and Poet', *Journal of the American Medical Association*, Vol. 223, 51–55.

Gorsky, M. and Sheard, S. (2006), *Financing Medicine: The British Experience since 1750* (London: Routledge).

Granger, J. (1907), 'The Old Streets of Nottingham', *Transactions of the Thornton Society, Volumes III and IV*, 3rd and 7th February issues, respectively.

Grell, Ole P., Cunningham, A. and Arrizabalaga, J. eds. (2010), *Centres of Medical Excellence? Medical Travel and Education in Europe, 1500-1789* (Surrey: Ashgate).

Guerriini, A. (2004), 'Anatomists and Entrepreneurs in Early Eighteenth–Century London', *Journal of the History of Medicine and Allied Sciences*, Vol. 59, April, II, 219–39.

Hanska, J. (2001), 'The Hanging of William Cragh: Anatomy of a Miracle', *Journal of Medieval History*, Vol. 27, 121–38.

Harrington, R. (2013), *Stress, Health and Well-Being: Thriving in the 21st century* (Belmont, USA: Wadsworth Publishers).

Hay, D., et al. (2011 edition), *Albion's Fatal Tree: Crime and Society in Eighteenth-Century England* (London: Verso).

Heller, B. (2010), 'The "Mene Peuple" and the Polite Spectator: The Individual in the Crowd at Eighteenth–Century Fairs', *Past and Present*, Vol. 208, I, 131–157.

Heron, C., Hunter, J., Knupfer, G., Martin, A. and Roberts, C. (1995), *Studies in Crime: An Introduction to Forensic Archaeology* (London and New York: Routledge).

Hillier, K. (1984), *The Book of Ashby–de–la–Zouche* (Buckingham: Barracuda Books Ltd).

Hills, T. E. (2010), 'Determining Brain Death: A Review of Evidence-Based Guidelines', *Nursing*, Vol. 40, (12th Dec. Issue), 34–40.

Horder, J. (2012), *Homicide and the Politics of Law Reform* (Oxford: Oxford University Press).

Howson, G. (2002), 'The Lancaster Doctors: Three Case Studies' in Sue Wilson and Jenny Loveridge eds., *Aspects of Lancaster: Discovering Local History* (Barnsley: Wharncliffe Books), chapter 5, pp. 53–63.

Høystad, O. M. (2007), *A History of the Heart* (London: Reaktion Books).

Hurren, E. T. (2011), *Dying for Victorian Medicine: English Anatomy and its Trade in the Dead Poor c. 1832 to 1929* (Basingstoke: Palgrave Macmillan).

Hurren, E.T. (27 July 2013), 'The Dangerous Dead: Dissecting the Criminal Corpse', *Lancet*, Vol. 382, Issue 9889, pp. 302–3.

Hutchinson, R. (2007), *Thomas Cromwell: The Rise and Fall of Henry VIII's Most Notorious Minister* (London: Weidenfdeld and Nicholson).

Innes, J. (2009), *Inferior Politics: Social Problems and Social Policies in Eighteenth–Century Britain* (Oxford: Oxford University Press).

Jansen, A. P. S., Van Nguyen, X., Karpitskiy, V., Mettenleiter, T.C. and Loewy, A. D. (1995), 'Central Command Neurons of the Sympathetic Nervous System: Basis of the Fight-or-Flight Response', *Science*, Oct. Vol. 27, 644–6.

Johnson, A. ed. (2001), *The Diary of Thomas Giordani Wright Newcastle Doctor 1826–1829* (Woodbridge, Suffolk: Boydell).

Jones, C. (2014), *The Smile Revolution in 18th–Century Paris* (Oxford: Oxford University Press).

Jones, C. and Porter, R. eds. (1994), *Reassessing Foucault: Power, Medicine, and the Body* (London: Routledge).

Jones, P. M. (2009), *Industrial Enlightenment: Science, Technology and Culture in Birmingham and the West Midlands, 1750–1820* (Manchester: Manchester University Press).

Kaufman, W. R. P. (2013), *Honour and Revenge: A Theory of Punishment* (London and New York: Springer Dordrecht).

Keats, J. (1934, 1st edition), *Anatomical and Physiological Notebook* (New York: Haskell House Publishers).

Kenny, N. (1998), *Curiosity in Early Modern Europe: Word Histories* (Wiesbaden: Harrassowitz).

Kenny, N. (2004), *The Uses of Curiosity in Early Modern France and Germany* (Oxford: Oxford University Press).

King, P. J. R. (2010), *Crime and Law in England, 1750–1840: Remaking Justice from the Margins* (Oxford: Oxford University Press).

King, P. J. R. (2010), 'The Impact of Urbanization on Murder Rates and on the Geography of Homicide in England and Wales 1780-1850', *Historical Journal*, Vol. 53, 1–28.

King, P. J. R. (2016), *Criminal Justice, the Criminal Code and Aggravated Forms of Capital Punishment in England and Wales, 1700–1834* (Basingstoke: Palgrave Pivot).

King S. A. (2000), *Poverty and Welfare in England, 1750–1850* (Manchester: Manchester University Press).

Knight, R. (2009), *Murder in the Shambles* (Milton Keynes: Author House UK Ltd).

Koestler, A. and Rolph, C. H. (1961, 1st edition), *Hanged by the Neck* (London: Penguin).

Laqueur, T. (1989), 'Crowds, Carnival, and the State in English Executions, 1604-1868', in A. L. Beier, D. Cannadine and J. Rosenheim eds., *The First Modern Society: Essays in English History in Honour of Lawrence Stone* (Cambridge: Cambridge University Press), pp. 305–56.

Laqueur, T. (2011), 'The Deep Time of the Dead', *Social Research*, Vol. 78, Fall, III, 799–820.

Lawrence, J. (1983), *The History of Capital Punishment* (London: Citadel Press).

Lawrence, S. C. (1996), *Charitable Knowledge: Hospital Pupils and Practitioners in 18th–Century London* (Cambridge: Cambridge University Press).

Linebaugh, P. (1975 edition), 'The Tyburn Riot Against the Surgeons' in Douglas Hay, Peter Linebaugh, John G. Rule, E. P. Thompson and Cal Winslow eds., *Albion's Fatal Tree: Crime and Society in Eighteenth–Century England* (London: Verso), pp. 65–117.

Linebaugh, P. (1996 edition), *The London Hanged: Crime and Civil Society in the Eighteenth-Century* (London: Verso).

Lizza, J. P. (1993), 'Persons and Death: What's Metaphysically Wrong with Our Statutory Definition of Death?' *Journal of Medical Philosophy*, Vol. 18, 351–74.

MacDonald, H. (2005), *Human Remains: Dissection and its Histories* (New Haven: Yale University Press).

Mackie, C. (1901), *Norfolk Annals* (Norwich: Kessinger Publishing),

Macleod, A.D. (2009), 'Eyelid Closure at Death', *Indian Journal of Palliative Care*, Vol. 15, July-December, II, 108–110.

Manns, A. (2012), 'Shorn Scalps and Perceptions of Male Dominance', *Social Psychology and Personality Science*, July, VII, 1–8.

Marland, H. (1987), *Medicine and Society in Wakefield and Huddersfield, 1780–1870* (Cambridge: Cambridge University Press).

Marsden, S. (2007), *Memento Mori: Churches and Churchyards of England* (London: English Heritage).

McCorrestine, S. (2014), *William Corder and the Red Barn Murder: Journeys of the Criminal Body* (Basingstoke: Palgrave Pivot).

McGowen, R. (2003), 'The Problem of Punishment in Eighteenth–Century England' in Simon Devereux and Paul Griffiths eds., *Penal Practice and Culture, 1500–1900: Punishing the English* (Basingstoke: Palgrave), pp. 210–31.

McGowen, R. (2005), '"Making Examples" and the Crisis of Punishment in Mid–Eighteenth–Century England' in David Lemmings ed., *The British and Their Laws in the Eighteenth Century* (Woodbridge: Boydell), pp. 182–205.

McKenna, A. (2006), 'God's Tribunal: Guilt, Innocence, and Execution in England, 1675–1775', *Cultural and Social History*, III, 121–144.

Munro, I. (2005), *The Figure of the Crowd in Early Modern London: The City and its Double* (Basingstoke: Palgrave).

Nicholson, A. (2011), *The Gentry: Stories of the English* (London: Harper Press).

O'Shaughnessy, T-L. (1987-8), 'A Single Capacity in the Beggar's Opera', *Eighteenth–Century Studies*, Vol. 21, II, Winter, 212–227.

Paley, M.D. (1996), *Coleridge's Later Poetry* (Oxford: Clarendon Press).

Patterson, A. F. (1954), *Radical Leicester: A History of Leicester, 1780–1850* (Michigan: University of Michigan Press).

Paulson, R. (1992), *Hogarth* (London: James Clarke & Co).

Pickstone, J. V. (1985), *Medicine and Industrial Society: A History of Hospital Development in Manchester and its region, 1752–1946* (Manchester: Manchester University Press).

Poole, S. (2015), '"For the Benefit of Example": Crime Scene Executions in England, 1720–1830' in R. Ward ed., *A Global History of Execution and the Criminal Corpse* (Basingstoke: Palgrave), pp. 71–101 [please note: this original conference paper was entitled 'Hanging at the Scene of the Crime' and was given at a keynote conference on this theme at Leicester University in 2013, and is thus referenced too in this book's endnotes – it is available on open access at: http://eprints.uwe.ac.uk/27986/].

Pooley, W. (2014), 'The History of the Body in 19th–Century Rural France', *Past and Future, Institute of Historical Research Magazine*, Vol. 16, Autumn/Winter Issue, 17–19.

Porter, R. (1988), 'Seeing the Past', *Past and Present*, Vol. 118, 186–203.

Porter, R. (1992), 'Medical Journalism in Britain to 1800' in W. F. Bynum, Stephen Lock and Roy Porter eds. *Medical Journals and Medical Knowledge: Historical Essays* (London and New York: Routledge), pp. 6–29.

Porter, R. (2001), 'History of the Body Reconsidered' in Peter Burke ed., *New Perspectives on Historical Writing* (London: Polity Press), pp. 232–260.

Porter, R. (2003), *Flesh in the Age of Reason: How the Enlightenment Transformed the Way We See Our Souls and Bodies* (London: W. W. Norton and Co).

Power, D. J. (1987), 'The Diagnosis of Brain Death in Adult Patients', *Journal of Intensive Care Medicine*, Vol. 2, 519–25.

Powner, D.J., Ackerman, B.M., and Grenvik, A. (1996), 'Medical Diagnosis of Death: Historical Contributions to Recent Controversies', *Lancet*, Nov. issue, panel 2, p. 1220.

Rank, A.W. (1997), *The Wounded Storyteller: Body, Illness and Ethics* (Chicago: University of Chicago Press).

Reddy, W. (1997), 'Against Constructionism: The Historical Ethnography of Emotions', *Current Anthropology*, Vol. 38, 327–51.

Reddy, W. (1998), 'Emotional Liberty: History and Politics in the Anthropology of Emotions', *Cultural Anthropology*, Vol. 14, 256–88.

Reddy, W. (2000) 'Sentimentalism and Its Erasure: The Role of Emotions in the Era of the French Revolution', *Journal of Modern History*, Vol. 72, 109–152.

Reddy, W. (2001), *The Navigation of Feeling: A Framework for the History of Emotions* (New York: Cambridge University Press).

Reddy, W. (2009), 'Historical Research on the Self and Emotions', *Emotion Review*, Vol. 1, 302–15.

Reinarz, J. (2009), *Healthcare in Birmingham: The Birmingham Teaching Hospitals, 1779–1939* (Woodbridge: Boydell).

Richardson, R. (2001 edition), *Death, Dissection and the Destitute* (London: Phoenix Press).

Roe, N. (2012), *John Keats* (New Haven: Yale University Press).

Rogers, N. (1990), 'Crowd and People in the Gordon Riots', in Eckhart Helimuth ed., *The Transformation of Political Culture: England and Germany in the late–Eighteenth Century* (Oxford: Studies of the German Historical Institute for Oxford University Press), pp. 503–33.

Rubenstein, R. L. (1983), *The Age of Triage: Fear and Hope in an Over-Crowded World* (Michigan, USA: Beacon Press, University of Michigan).

Rudé, G. F. (2005 edition), *The Crowd in History: A Study of Popular Disturbances in France and England, 1730–1848* (London: Serif Books).

Sachs, J. S. (2002), *Time of Death: The True Story of the Search for Death's Stopwatch* (New York: QPD for William Heinemann at the Random House Group Ltd).

Santing, C. and Baert, B. eds. (2013), *Disembodied Heads in Medieval and Early Modern Culture* (Oxfordshire: Brill Publications).

Sawday, J. (1996), *The Body Emblazoned: Dissection and the Human Body in Renaissance Culture* (New York and London: Routledge).

Sharp, J. A. (1990), *Judicial Punishment in England* (London: Faber and Faber).

Sharp, J. A. (1998 2nd edition) *Crime in Early Modern England, 1550–1750* (London: Longman).

Shelley, H. C. (2010), *Inns and Taverns of Old London, Part II: Coffee-Houses of London* (Bremen, Germany: Europaeischer Hochschulverlag GmbH & co).

Shoemaker, R. B. (2004), 'Streets of Shame? The Crowd and Public Punishments in London, 1700–1820' in Simon Devereaux and Paul Griffiths eds., *Penal Practice and Culture, 1500–1900: Punishing the English* (Basingstoke: Palgrave), pp. 232–57.

Shoemaker, R. B. (2004), *The London Mob: Violence and Disorder in Eighteenth-Century London* (London: Hambledon).

Shoia, M. M., Benninger, B., Aqutter, P., Loukas, M. and Tubbs, R. S. (2013), 'A Historical Perspective: Infection from Cadaveric Dissection from the 18th to 20th Centuries', *Clinical Anatomy*, Vol. 26, II, 154–60.

Smith, C. U. M. and Arnott, R. (2005), *The Genius of Erasmus Darwin* (Farnham, Hampshire: Ashgate).

Spierenburg, P. (2008), *The Spectacle of Suffering: Executions and the Evolution of Repression: From a Pre-Industrial Metropolis to the European Experience* (Cambridge: Cambridge University Press).

Spierenburg. P. (2001), 'Violence and the Civilising Process: Does it Work?' *Crime, Histories and Societies*, Vol. 5, II, 87–105.

Stevenson, J. (1985), 'The "Moral" Economy of the English Crowd: Myth and Reality', in Antony Fletcher and John Stevenson eds., *Order and Disorder in Early Modern England* (Cambridge: Cambridge University Press), pp. 218–38.

Stillinger, J. ed. (1978), *John Keats: The Complete Poems* (Cambridge, MA, USA: Harvard University Press).

Tarlow, S. (2012), 'The Archaeology of Emotion and Effect', *Annual Review of Anthropology*, Vol. 41, 169–85.

Tarlow, S. (2013), *Ritual, Belief and the Dead in Early Modern Britain and Ireland* (Cambridge: Cambridge University Press).

Tarlow, S. (2016), *The Golden Age of the Gibbet in Britain* (Basingstoke: Palgrave Pivot) [please note: this book previously had a working-title, *Hanging in Chain: Gibbeting and the Murder Act* – sometimes referred to in endnotes].

Teresi, D. (2012), *The Undead: Organ Harvesting, the Ice-Water Test, Beating Heart Cadavers – How Medicine is Blurring the Line between Life and Death* (New York and London: Vintage).

Truog, R.D. and Fackler, J.C. (1992), 'Rethinking Brain Death', *Critical Care Medicine*, Vol. 20, 1705–13.

Uglow, J. (2002), *The Lunar Men: Five Friends whose Curiosity Changed the World* (London: Faber and Faber).

Waddington, K. (2003), *Medical Education at St. Bartholomew's Hospital, 1123–1995* (Woodbridge: Boydell).

Wall, C. (1937), *The History of the Surgeon's Company, 1745–1800* (London: Hutchinson).

Walter, J. (2006), 'Crown and Crowd: Popular Culture and Popular Protest in Early Modern England', in John Walter ed., *Crowds and Popular Protest in Early Modern England* (Manchester: Manchester University Press), pp. 14–26.

Ward, R. (2014), *Print Culture, Crime and Justice in 18th–Century London* (London: Bloomsbury).

Watson, F. G. B. (1939–40), 'Thomas Patch (1725–1782): Notes on His Life, Together with a Catalogue of His Known Works', *The Volume of the Walpole Society*, Vol. 28, 15–50.

Whetstine, L. M. (2008), 'The History of the Definition(s) of Death: From the 18th–Century to the 20th–Century', in David W. Crippen ed., *End-of-Life Communication in the ICU* (New York: Springer-Verlag), pp. 65–78.

White, J. (2013), *A Great and Monstrous Thing: London in the Eighteenth–Century* (London: Random).

White, M. (2008), '"Rogues of the Meaner Sort?" Old Bailey Executions and the Crowd in the Early Nineteenth–Century', *London Journal*, Vol. 33, July, II, 135–153.

Wilf, S. (1993), 'Imagining Justice: Aesthetics and Public Executions in late–18th Century England', *Yale Journal of Law and the Humanities*, Vol. 5, I, 51–78.

Wilson, S. (2002), *Aspects of Lancaster: Discovering Local History* (Barnsley, South Yorkshire: Wharncliffe Books).

Wood, F. and Wood, K. eds. (1992), *A Lancashire Gentleman: The Letter and Journals of Richard Hodgkinson 1763–1847* (Stroud: Allan Sutton Press).

Young, L. (2002), *The Book of the Heart* (New York: Doubleday, Random Books).

Zimmer, C. (2005), *Soul Made Flesh: The Discovery of the Brain and How it Changed the World* (New York: Simon and Schuster, Free Press).

UNPUBLISHED ARTICLES

Ward. R. (2015), 'Wilberforce, Anatomists, and the Criminal Corpse: Parliamentary Attempts to Extend the Dissection of Offenders in Late Eighteenth–Century England', to be published in the *Journal of British Studies*, copy available at Leicester University, pp.1–28 [see, Vol. 54, I, 63-87 and on open access in accordance with Wellcome Trust funding at: https://www.researchgate.net/publication/273304230_The_Criminal_Corpse_Anatomists_and_the_Criminal_Law_Parliamentary_Attempts_to_Extend_the_Dissection_of_Offenders_in_Late_Eighteenth-Century_England].

UNPUBLISHED MA THESIS

Markless, R. E. (2012), 'Gender, Crime and Discretion in Yorkshire, 1735–1775', (unpublished MA dissertation, Department of the Humanities, Roehampton University).

UNPUBLISHED PHD THESES

Bencard, A. (2008), 'History in the Flesh', (unpublished PhD, Københavns Universitet. Institut for Folkesundhedsvidenskab. Medicinsk Museion, University of Copenhagen).

Chaplin, S. (2009), 'John Hunter and the Museum Œconomy [sic], 1750–1800' (unpublished PhD, King's College London).

INDEX[1]

---

[1] *Various versions of names were often used interchangeably in the original material, and these are rendered here as they appear in the source material, with the most frequently-used name first and alternatives in parenthesis. Aliases (indicated with 'aka') and nicknames (with inverted commas) appear in parenthesis. Finally, numbers in italics indicate images and the suffix 'n' indicates an endnote.*

© The Author(s) 2016   313
E.T. Hurren, *Dissecting the Criminal Corpse*, Palgrave
Historical Studies in the Criminal Corpse and its Afterlife,
DOI 10.1057/978-1-137-58249-2